Games and Information

GAMES AND INFORMATION

An Introduction to Game Theory

Eric Rasmusen

Basil Blackwell

First published 1989
Reprinted with corrections 1990

Basil Blackwell Ltd
108 Cowley Road, Oxford, OX4 1JF, UK

Basil Blackwell, Inc.
3 Cambridge Center
Cambridge, Massachusetts 02142, USA

British Library Cataloguing in Publication Data

A CIP catalogue record of this book is available from the British Library.

Library of Congress Cataloging-in-Publication Data

Rasmusen, Eric.
Games and information: an introduction to game theory / Eric Rasmusen.
p.   cm.
Bibliography: p.
Includes indexes.
ISBN 0-631-15709-3
1. Game theory. I. Title.
QA269.R37 1989
519.3–dc 19

Typeset in 10 on 12pt Computer Modern Roman
by Cambridge University Press
Printed in Great Britain by the University Press, Cambridge.

Contents

Preface   9
    Contents and Purpose   9
    Using the Book   10
    Acknowledgements   11
Introduction   13
    History   13
    Game Theory's Method   14
    No-Fat Modelling   14
    This Book's Style   16
    Notes   17

PART I GAME THEORY

1 The Rules of the Game   21
   1.1 Basic Definitions   21
   1.2 Dominant Strategies: the Prisoner's Dilemma   27
   1.3 Iterated Dominance: the Battle of the Bismarck Sea   30
   1.4 Nash Equilibrium: Boxed Pigs, the Battle of the Sexes,
       and Pure Coordination   32
   1.5 Focal Points   36
   Recommended Reading   37
   Problem 1   37
   Notes   38

2 Information   43
   2.1 Introduction   43
   2.2 The Normal and Extensive Forms of a Game   43
   2.3 Information Sets   48
   2.4 Perfect, Certain, Symmetric, and Complete Information   51
   2.5 Bayesian Games and the Harsanyi Transformation   54
   2.6 An Example: The Png Settlement Game   60
   Recommended Reading   65
   Problem 2   65

	Notes	65
3	Mixed and Continuous Strategies	69
	3.1 Introduction	69
	3.2 Mixed Strategies: the Welfare Game	69
	3.3 Chicken, the War of Attrition, and Correlated Strategies	73
	3.4 Continuous Strategies: the Cournot Game	76
	Recommended Reading	80
	Problem 3	80
	Notes	80
4	Dynamic Games with Symmetric Information	83
	4.1 Introduction	83
	4.2 Subgame Perfectness	83
	4.3 An Example of Perfectness: Entry Deterrence I	85
	4.4 Finitely Repeated Games and the Chainstore Paradox	88
	4.5 Discounting	89
	4.6 Infinitely Repeated Games and the Folk Theorem	91
	4.7 Reputation: the One-Sided Prisoner's Dilemma	94
	4.8 Product Quality in an Infinitely Repeated Game	96
	Recommended Reading	99
	Problem 4	100
	Notes	100
5	Dynamic Games with Asymmetric Information	107
	5.1 Perfect Bayesian Equilibrium: Entry Deterrence II and III	107
	5.2 Refining Perfect Bayesian Equilibrium: PhD Admissions	112
	5.3 The Importance of Common Knowledge: Entry Deterrence IV and V	116
	5.4 Incomplete Information in the Repeated Prisoner's Dilemma: the Gang of Four Model	118
	5.5 The Axelrod Tournament	119
	5.6 Evolutionary Equilibrium: the Hawk–Dove Game	121
	5.7 Existence of Equilibrium	123
	Recommended Reading	127
	Problem 5	127
	Notes	128

PART II   ASYMMETRIC INFORMATION

6	Moral Hazard: Hidden Actions	133
	6.1 Categories of Asymmetric Information Models	133
	6.2 A Principal–Agent Model: the Production Game	136
	6.3 The Self-Selection, Participation, and Competition Constraints	140
	6.4 State-Space Diagrams: Insurance Games I and II	142
	6.5 Optimal Contracts: the Broadway Game	147
	6.6 Institutions	151
	Recommended Reading	153
	Problem 6	154
	Notes	154

7 Moral Hazard: Hidden Information   159
   7.1   Pooling vs. Separating Equilibrium, and the Revelation
         Principle   159
   7.2   An Example: the Salesman Game   163
   7.3   Efficiency Wages   166
   7.4   Tournaments   167
   7.5   Monitoring   168
   7.6   Alleviating the Agency Problem   169
   7.7   Teams, and the Groves Mechanism   170
   Recommended Reading   175
   Problem 7   175
   Notes   176

8 Adverse Selection   181
   8.1   Introduction: Production Game V   181
   8.2   Adverse Selection under Certainty: Lemons I and II   182
   8.3   Heterogeneous Tastes: Lemons III and IV   186
   8.4   Adverse Selection under Uncertainty: Insurance Game III   189
   8.5   Other Equilibrium Concepts: Wilson and Reactive Equilibrium   193
   8.6   Applications   195
   Recommended Reading   199
   Problem 8   199
   Notes   199

9 Signalling   205
   9.1   Introduction   205
   9.2   The Informed Player Moves First: Signalling   205
   9.3   General Comments on Signalling in Education   210
   9.4   The Informed Player Moves Second: Screening   212
   9.5   Two Signals: Underpricing of Stock   218
   Recommended Reading   219
   Problem 9   221
   Notes   221

PART III   APPLICATIONS

10 Bargaining   227
    10.1 The Basic Bargaining Problem: Splitting a Pie   227
    10.2 The Nash Bargaining Solution   229
    10.3 Alternating Offers over Finite Time   231
    10.4 Alternating Offers over Infinite Time   234
    10.5 Incomplete Information   236
    Recommended Reading   241
    Problem 10   241
    Notes   242

11 Auctions   245
    11.1 Introduction   245
    11.2 Auction Classification and Private-Value Strategies   245
    11.3 Comparing Auction Rules   250

11.4 Common-Value Auctions and the Winner's Curse   251
11.5 Information in Common-Value Auctions   255
Recommended Reading   256
Problem 11   256
Notes   256

12 Pricing and Product Differentiation   259
12.1 Quantities as Strategies: the Cournot Equilibrium Revisited   259
12.2 Prices as Strategies: the Bertrand Equilibrium   263
12.3 Location Models   269
12.4 An Overlapping Generations Model of Consumer Switching Costs   274
12.5 Durable Monopoly   276
Recommended Reading   280
Problem 12   280
Notes   280

13 Industrial Organization   285
13.1 Predatory Pricing: the Kreps–Wilson Model   285
13.2 Credit and the Age of the Firm: the Diamond Model   288
13.3 Entry for Buyout   290
13.4 Innovation and Patent Races   294
13.5 Takeovers and Greenmail   301
Recommended Reading   306
Problem 13   306
Notes   306

Mathematical Appendix   309
Notation   309
Glossary   311
Fixed Point Theorems   315

Answers to Problems   317

References and Citation Index   321

Subject Index   345

Preface

Contents and Purpose

This book is about noncooperative game theory and asymmetric information. In the Introduction, I will say why I think these subjects are important, but here in the Preface I will try to help you decide whether this is the appropriate book to read if they do interest you.

Most of the topics covered here are absent from existing books on either game theory or information economics. There are many books on game theory, including Davis (1970), J. Friedman (1986), Harsanyi (1977), Luce & Raiffa (1957), Ordeshook (1986), Owen (1982), Rapoport (1970), Shubik (1982), Szep & Forgo (1985), and Williams (1966). But these books emphasize zero-sum and cooperative game theory, which are relatively unimportant in economics. Van Damme (1983) provides a worthwhile treatment of equilibrium concepts in noncooperative games, but readers must supply their own applications. The forthcoming *Handbook of Game Theory*, a collection of review articles on game theory edited by Aumann & S. Hart, and the Tirole (1988) text on theoretical industrial organization are close to the present book in style and content, but at a more advanced level. Fudenberg & Tirole (1986a) on oligopoly and McMillan (1986) on international economics are excellent little books of applications.

Books on information in economics have mainly been concerned with decision-making under uncertainty rather than asymmetric information. Exceptions include Bamberg & Spremann (1987) on agency theory, Hess (1983) on organization, and J. Green & Laffont (1979) on mechanism design. Binmore & Dasgupta (1986) and Diamond & Rothschild (1978) are useful collections of papers on game theory and the economics of information respectively, but they collect papers rather than explain them. Two books being written that will help remedy the lack are the information text by J. Hirshleifer & Riley and the *Handbook of Industrial Organization* edited by Schmalensee and Willig, but for now most topics are nowhere covered systematically.

I write as an applied theoretical economist, not as a game theorist, and readers in anthropology, law, physics, accounting, and management science have helped me to be aware of the provincialisms of economics and game theory. My aim is to present the game theory and information economics that currently exists in journal articles and oral tradition in a way that shows how to build simple models using a standard format. Journal articles are more complicated and less clear than seems necessary in retrospect; precisely because it is original, even the discoverer rarely understands a truly novel idea. After a few dozen successor articles have appeared, we all understand it and marvel at its simplicity. But journal editors are unreceptive to new articles that admit to containing exactly the same idea as old articles, just presented more clearly. At best, the clarification is hidden in some new article's introduction or condensed to a paragraph in a survey. Students, who find every idea as mystifying as everyone in the field did when they were new, must learn either from the confused original articles or the oral tradition of a top economics department. This book tries to present an alternative.

Using the Book

The book is divided into three parts: Part I on game theory; Part II on information economics; and Part III on bargaining, auctions, and industrial organization. Parts I and II are ordered sets of chapters, but neither the chapters nor all the sections within chapters of Part III are ordered. Chapters 12 and 13, in particular, may be studied before Part II.

Part I by itself would be appropriate for a course on game theory, and sections from Part III could be added for illustration. If students are already familiar with basic game theory, Part II can be used for a course on information economics, perhaps with the addition of Chapter 2 (information in games), Sections 4.1–4.3 and 5.1–5.2 (for perfectness), and Chapter 11 (auctions). Various sections would be useful for a course on industrial organization. These recommendations may be summarized as follows:

Game Theory:	1, 2, 3, 4, 5, 10
Game Theory (alternative):	1, 2, 3, 12, 4, 5
Information:	2, 4.1, 4.2, 4.3, 5.1, 5.2, 5.3, 6, 7, 8, 9, 11
Industrial Organization:	1, 2, 3, 12, 7.3., 7.4, 4, 5.1, 5.2, 5.3, 13

Recommended readings, an exercise, and notes follow each chapter. It is hard to thoroughly understand the material without reading articles, and I have made a few recommendations. My choices are not necessarily the most important or original articles: I looked for clear writing in widely available publications. I also recommend that readers try attending a seminar presentation of current research on some topic from the book; while most of the seminar may be incomprehensible, there is a real thrill in hearing someone attack the speaker with "Are you sure that equilibrium is perfect?" after just learning the previous week what "perfect" means.

The homework problems at the end of each chapter are relatively straightforward, putting slight twists on some of the concepts. Brief answers are given near the end of the book. The problems are especially recommended to readers learning this material by themselves.

The endnotes to each chapter include substantive material as well as references. Unlike the notes in many books, they are not meant to be skipped, except on first reading, since many of them are important but tangential, and some qualify statements in the main text. Less important notes supply additional examples or list technical results for reference. A mathematical appendix at the end of the book supplies technical references, defines certain mathematical terms, and lists some items—fixed point theorems, for example—for reference, even though they are not used in the main text.

The Level of Mathematics

In surveying the prefaces of previous books on game theory, I see that advising readers how much mathematical background they need exposes an author to charges of being out of touch with reality. The mathematical level here is about the same as in Luce & Raiffa (1957), and I can do no better than to quote the advice on p. 8 of their book:

> probably the most important prerequisite is that ill-defined quality: mathematical sophistication. We hope that this is an ingredient not required in large measure, but that it is needed to some degree there can be no doubt. The reader must be able to accept conditional statements, even though he feels the suppositions to be false; he must be willing to make concessions to mathematical simplicity; he must be patient enough to follow along with the peculiar kind of construction that mathematics is; and, above all, he must have sympathy with the method—a sympathy based upon his knowledge of its past sucesses in various of the empirical sciences and upon his realization of the necessity for rigorous deduction in science as we know it.

If you do not know the terms "risk averse," "first order condition," "utility function," "probability density," and "discount rate," you will not fully understand this book. Flipping through it, however, you will see that the equation density is much lower than in Varian (1984), the standard graduate microeconomics text. In a sense, game theory is less abstract than mainstream economics, since it deals with individual agents rather than anonymous points on a curve, and it is oriented towards explaining stylized facts rather than supplying econometric specifications.

Acknowledgements

I would like to thank, for advice taken and untaken, Dean Amel, Dan Asquith, Sushil Bikhchandani, Patricia Hughes Brennan, Paul Cheng, Luis Fernandez, David Hirshleifer, Jack Hirshleifer, Steven Lippman, Ivan Png, Benjamin Rasmusen, Marilyn Rasmusen, Ray Renken, Lloyd Shapley, Richard Silver, Yoon Suh, Brett Trueman, Barry Weingast, and students in Management 200a from Spring 1985 to Spring 1989. M. Dao, D. Koh, Jeanne Lamotte, In-Ho Lee, Loi Lu, Patricia Martin,

Timothy Opler, Sang Tran, Jeff Vincent, Tao Yang, Roy Zerner, and especially Emmanuel Petrakis helped me with research assistance at one stage or another.

I would also like to thank Robert Boyd, Mark Ramseyer, Ken Taymor, and John Wiley for their regular comments as each chapter appeared.

Eric Rasmusen

Introduction

History

Not long ago, the scoffer could say that econometrics and game theory were like Japan and Argentina. In the late 1940s both disciplines and both economies were full of promise, poised for rapid growth and ready to make a profound impact on the world. We all know what happened to the economies of Japan and Argentina. Of the disciplines, econometrics became an inseparable part of economics, while game theory languished as a subdiscipline, interesting to its specialists but ignored by the profession as a whole. The specialists in game theory were generally mathematicians, who cared about definitions and proofs rather than applying the methods to economic problems. Game theorists took pride in the diversity of disciplines to which their theory could be applied, but in none had it become indispensable.

In the 1970s, the analogy with Argentina broke down. At the same time that Argentina was inviting back Juan Peron, economists were beginning to discover what they could achieve by combining game theory with the structure of complex economic situations. Innovation in theory and application was especially useful for situations with asymmetric information and a temporal sequence of actions, the two major themes of this book. During the 1980s, game theory has become dramatically more important to mainstream economics. Indeed, it seems to be swallowing up microeconomics, just as econometrics has swallowed up empirical economics.

Game theory is generally considered to have begun with the publication of von Neumann & Morgenstern's *The Theory of Games and Economic Behaviour* in 1944. Although very little of the game theory in that thick volume is relevant to the present book, it introduced the idea that conflict could be mathematically analyzed and provided the terminology with which to do it. The development of the "Prisoner's Dilemma" (Tucker [unpub]) and Nash's papers on the definition and existence of equilibrium (Nash [1950b, 1951]) laid the foundations for modern noncooperative game theory. At the same time, cooperative game theory reached important results in papers by Nash (1950a) and Shapley (1953b) on bargaining games and Gillies (1953) and Shapley (1953a) on the core.

By 1953 virtually all the game theory that was to be used by economists for the next 20 years had been developed. Until the mid-1970s, game theory remained an autonomous field with little relevance to mainstream economics, important exceptions being Schelling's 1960 book, *The Strategy of Conflict*, which introduced the focal point, and a series of papers (of which Debreu & Scarf [1963] is typical) that showed the relationship of the core of a game to the general equilibrium of an economy.

In the 1970s, information became the focus of many models as economists started to put emphasis on individuals who act rationally but with limited information. When attention was given to individual agents, the time ordering in which they carried out actions began to be explicitly incorporated. With this addition, games had enough structure to reach interesting and non-obvious results. Important "toolbox" references include the earlier but long unapplied articles of Selten (1965) (on perfectness) and Harsanyi (1967) (on incomplete information), the papers by Selten (1975) and Kreps & Wilson (1982b) extending perfectness, and the article by Kreps, Milgrom, Roberts, & Wilson (1982) on incomplete information in repeated games. Most of the applications in the present book were developed after 1975, and the flow of research shows no sign of diminishing.

Game Theory's Method

Game theory has been successful in recent years because it fits so well into the new methodology of economics. In the past, macroeconomists started with broad behavioral relationships like the consumption function, and microeconomists often started with precise but irrational behavioral assumptions such as sales maximization. Now all economists start with primitive assumptions about the utility functions, production functions, and endowments of the actors in the models (to which must often be added the available information). The reason is that it is usually easier to judge whether primitive assumptions are sensible than to evaluate high-level assumptions about behavior. Having accepted the primitive assumptions, the modeller figures out what happens when the actors maximize their utility subject to the constraints imposed by their information, endowments, and production functions. This is exactly the paradigm of game theory: the modeller assigns payoff functions and strategy sets to his players, and sees what happens when they pick strategies to maximize their payoffs. The approach is a combination of the "Maximization Subject to Constraints" of MIT and the "No Free Lunch" of Chicago. We shall see, however, that game theory relies only on the spirit of these two approaches: it has moved away from maximization by calculus, and inefficient allocations are common. The players act rationally, but the consequences are often bizarre, which makes application to a world of intelligent men and ludicrous outcomes appropriate.

No-Fat Modelling

Along with the trend towards primitive assumptions and maximizing behavior has been a trend toward simplicity, which I call "no-fat modelling." No-fat modelling is an extension of Occam's razor and the *ceteris paribus* assumption so fundamental to economics or, indeed, to any kind of analysis. The heart of the approach is to discover the simplest assumptions needed to generate an interesting conclusion; the starkest, barest, model that has the desired result. This desired result is the answer to some relatively narrow question: Could education be just a signal of ability? Why do bid–ask spreads exist? Is predatory pricing ever rational?

The modeller starts with a vague idea such as "People go to college to show they're smart." He then models the idea formally in a simple way. The idea might survive intact, it might be found formally meaningless, it might survive with qualifications, or its opposite might turn out to be true. The modeller then uses the model to come up with precise theorems, whose proofs may tell him still more about the idea. After the proofs, he goes back to thinking in words, trying to understand more than whether the proofs are mathematically correct.

A critic of the use of the mathematical approach in biology has compared it to an hourglass (Slatkin [1980]). First, a broad and important problem is introduced. Second, it is reduced to a very special but tractable model that hopes to capture its essence. Finally, in the most perilous part of the process, the results are expanded to apply to the original problem. No-fat modelling also uses this approach.

The process is one of setting up "If–Then" statements, whether in words or symbols. To apply such statements, their premises and conclusions need to be verified, either by casual or careful empiricism. If the required assumptions seem contrived or the assumptions and implications contradict reality, the idea should be discarded. If "reality" is not immediately obvious and data is available, econometric tests may help show whether the model is valid. Predictions can be made about future events, but that is not usually the primary motivation: most of us are more interested in explaining and understanding than predicting.

The method just described is close to how, according to Lakatos (1976), mathematical theorems are developed. It contrasts sharply with the common view that the researcher starts with a hypothesis and proves or disproves it. Instead, the process of proof helps show how the hypothesis should be formulated.

An important part of no-fat modelling is what Kreps & Spence (1984) have called "blackboxing": treating unimportant subcomponents of a model in a cursory way. The game Entry for Buyout of Section 13.3, for example, asks whether a new entrant would be bought out by a market's incumbent, something that depends on duopoly pricing and bargaining. Both pricing and bargaining are complicated games in themselves, but if the modeller does not wish to deflect attention to those topics, he can use the simple Nash and Cournot solutions to those games and go on to analyze buyout. If the entire focus of the model were duopoly pricing, then using the Cournot solution would be open to attack, but as a simplifying assumption, rather than one that "drives" the model, it is acceptable.

Intelligent laymen have objected to the amount of mathematics in economics since at least the 1880s, when George Bernard Shaw said that as a boy he (1) let someone assume that $a = b$, (2) permitted several steps of algebra, and (3) found he had accepted a proof that $1 = 2$. Forever after, Shaw distrusted assumptions and algebra. Despite the effort to achieve simplicity (or perhaps because of it), mathematics is essential to low-fat modelling. The conclusions can be retranslated into words, but rarely can they be found by verbal reasoning. The economist Wicksteed put this nicely in his reply to Shaw's criticism:

Mr Shaw arrived at the sapient conclusion that there "was a screw loose somewhere"—not in his own reasoning powers, but—"in the algebraic art"; and thenceforth renounced mathematical reasoning in favour of the literary method which enables a clever man to follow equally fallacious arguments to equally absurd conclusions *without seeing that they are absurd.* This is the exact difference between the mathematical and literary treatment of the pure theory of political economy. (Wicksteed [1885] p. 732)

With no-fat modelling, one can still rig a model to achieve a wide range of results, but it must be rigged by making strange primitive assumptions. Everyone knows that the place to look for the source of suspicious results is the description at the start of the model. If that description is not clear, the reader deduces that the model's counterintuitive results arise from bad assumptions concealed in poor writing. Clarity is therefore important, and the somewhat inelegant Players–Actions–Payoffs presentation used in this book is useful not only for helping the writer, but for persuading the reader.

This Book's Style

Substance and style are closely related. The difference between a good and a bad model is not just whether the essence of the situation is captured, but in how much froth covers the essence. In this book, I have tried to make the games as simple as possible. They often, for example, allow each player a choice of only two actions. Our intuition works best with such models, and continuous actions are technically more troublesome. Other assumptions, such as zero production costs, rely on trained intuition. To the layman, the assumption that output is costless seems very strong, but a little experience with these models teaches that it is the constancy of the marginal cost that usually matters, not its level.

What matters more than what a model says is what we understand it to say. Just as an article written in Tibetan is useless to me, so is one that is excessively mathematical or poorly written, no matter how rigorous it seems to the author. Such an article leaves me with some new belief about its subject, but that belief is not sharp, or precisely correct. Overprecision in sending a message creates imprecision when it is received. A simpler model would leave the same impression with less time wasted. I have tried to simplify the structure and notation of models while giving credit to their original authors, but I must ask forgiveness of anyone whose model has been oversimplified or distorted. In trying to be understandable, I have taken risks with respect to accuracy. My hope is that the impression left in the readers' minds will be more accurate than if a style more cautious and obscure had left them to devise their own errors.

Readers may be surprised to find occasional references to newspaper and magazine articles in this book. I hope these references will be reminders that models ought eventually to be applied to specific facts, and that a great many interesting situations are waiting for our analysis. The principal–agent problem is not found only in back issues of *Econometrica*: it can be found on the front page of today's *Wall Street Journal* if one knows what to look for.

I have been forced to exercise more discretion over definitions than I had hoped. Many concepts have been defined on an article-by-article basis in the literature, with no consistency and little attention to euphony or usefulness. Other concepts, such as "asymmetric information" and "incomplete information," have been considered so basic as to not need definition, and hence have been used in contradictory ways. I use existing terms whenever possible, and synonyms are listed.

I have named the players Smith and Brown rather than 1 and 2 so that the reader's memory will be less taxed in remembering which is a player and which is a time period. I hope also to reinforce the idea that a model is a story made precise; we begin with Smith and Brown, even if we quickly descend to s and b. Keeping this in mind, the modeller is less likely to build mathematically correct models with absurd action sets, and his descriptions are more pleasant to read. In the same vein, labelling a curve "$U = 83$," sacrifices no generality: the phrase "$U = 83$ and $U = 66$" has virtually the same content as "$U = \alpha$ and $U = \beta$, where $\alpha > \beta$," but uses less short-term memory.

A danger of this approach is that readers may not appreciate the complexity of some of the material. While journal articles make the material seem harder than it is, this approach makes it seem easier (a statement that can be true even if readers find this book difficult). The better the author does his job, the worse this problem becomes. Keynes (1933) says of Alfred Marshall's *Principles*,

> The lack of emphasis and of strong light and shade, the sedulous rubbing away of rough edges and salients and projections, until what is most novel can appear as trite, allows the reader to pass too easily through. Like a duck leaving water, he can escape from this douche of ideas with scarce a wetting. The difficulties are concealed; the most ticklish problems are solved in footnotes; a pregnant and original judgement is dressed up as a platitude.

That is why the problems and recommended readings are important. The problems force readers to think carefully and slowly, and the readings show what thorough presentations look like. The efficient way to learn how to do research is to start doing it, not to read about it, and after reading this book, if not before, many readers will want to build their own models. But my purpose here is to show them the big picture, to help them understand the models intuitively, and to give them a feel for the modelling process.

Notes

- Perhaps the most important contribution of von Neumann & Morgenstern (1944) is the theory of expected utility (see Section 2.4). Although they developed the theory because they needed it to find the equilibria of games, it is today heavily used in all branches of economics. In game theory proper, they contributed the framework to describe games, and the concept of mixed strategies (see Section 3.2).
- On method, see the dialogue by Lakatos (1976), or Davis & Hersch (1981), Chapter 6 of which is a shorter dialogue in the same style. M. Friedman (1953) is the classic essay on a different methodology: evaluating a model by testing its predictions. Kreps & Spence (1984) is a discussion of what I call no-fat modelling by two of the scholars most important to its development.
- Because style and substance are so closely linked, writing style is important. For advice

on writing, see McCloskey (1985, 1987) (on economics), Basil Blackwell (1985) (on books), Bowersock (1985) (on footnotes), Fowler (1965), Fowler & Fowler (1949), Halmos (1970) (on mathematical writing), Strunk & White (1959), Weiner (1984), and Wydick (1978).

- **A Fallacious Proof that 1=2.** Suppose that $a = b$. Then $ab = b^2$ and $ab - b^2 = a^2 - b^2$. Factoring the last equation gives us $b(a - b) = (a + b)(a - b)$, which can be simplified to $b = a + b$. But then, using our initial assumption, $b = 2b$ and $1 = 2$. (The fallacy is division by zero.)

Part I

Game Theory

Game Theory

1 The Rules of the Game

1.1 Basic Definitions

Game theory is concerned with the actions of individuals who are conscious that their actions affect each other. When the only two publishers in a city choose prices for their newspapers, aware that their sales are determined jointly, they are players in a game with each other. They are not in a game with the readers who buy the newspapers, because each reader ignores his effect on the publisher. Game theory is not useful when decisions are made that ignore the reactions of others or treat them as impersonal market forces. The best way to understand which situations can be modelled as games is to think about examples. Consider the following:

(1) OPEC members choosing their annual output.
(2) General Motors purchasing from US Steel.
(3) Two manufacturers, one of nuts and one of bolts, deciding whether to use metric or American standards.
(4) A board of directors setting up a stock option plan for the Chief Executive Officer.
(5) United Fruit Company hiring workers in Honduras in the 1930s.
(6) An electric company deciding whether to order a new power plant given its estimate of demand for electricity in ten years.

The first four examples are games. (1) OPEC members are playing a game because Saudi Arabia knows that Kuwait's oil output is based on Kuwait's forecast of Saudi output, and the output from both countries matters to the world price. (2) A significant portion of American trade in steel is between General Motors and US Steel, companies which realize that the quantities traded by each of them affect the price. One wants the price low, the other, high. (3) The nut and bolt manufacturers are not in conflict, but the actions of one affects the desired actions of the other. (4) The board of directors chooses a stock option plan anticipating the effect on the actions of the CEO.

Game theory is inappropriate for modelling the final two examples. (5) Each individual worker affects United Fruit insignificantly, and each worker makes his employment decision without regard for the impact on United Fruit's behavior. (6) The electric company faces a complicated decision, but it does not face another rational agent. Changes in the important economic variables could turn examples (5) and (6) into games. The appropriate model changes if United Fruit faces a plantation workers' union or if the public utility commission pressures the utility to change its generating capacity.

Game theory as it will be presented in this book is a modelling tool, not an axiomatic system. The presentation in this chapter is unconventional. Rather than starting with mathematical definitions, or simple little games of the kind used later in the chapter, we will start with a situation to be modelled, and build a game from it step by step.

Describing a Game

The essential elements of a game are **players**, **actions**, **information**, **strategies**, **payoffs**, **outcomes**, and **equilibria**. At a minimum, the game's description must include the players, strategies, and payoffs, for which the actions and information are building blocks. The players, actions, and outcomes are collectively referred to as the **rules of the game**, and the modeller's objective is to use the rules of the game to determine the equilibrium.

We will define the terms using a game we will call OPEC Model I as an example.

*The **players** are the individuals who make decisions. Each player's goal is to maximize his utility by choice of actions.*

In OPEC Model I, we specify the players to be Saudi Arabia (S) and Others (O) (referring to the other members of OPEC). Let us assume that each player's utility is the sum of his oil revenues in 1988 and 1989. Passive individuals like the American consumer, who react predictably to oil price changes without any thought of trying to change anyone's behavior, are not players, but environmental parameters. Sometimes it is useful to explicitly include individuals in the model called **non-players** whose actions are taken in a purely mechanical way.

Nature *is a non-player who takes random actions at specified points in the game with specified probabilities.*

In OPEC Model I, we will assume that the strength of world demand for oil, denoted D, can take one of two permanent values. At the beginning of the game, Nature randomly decides whether oil demand will be *Weak* or *Strong*, assigning, let us assume, probabilities of 70 and 30 percent. Even if the players always took the same actions, this random move means that the model would yield more than just one prediction. We say that there are different **realizations** of a game depending on the results of random moves.

*An **action** or **move** by player i, denoted a_i, is a choice he can make.*

*Player i's **action set** $A_i = \{a_i\}$ is the entire set of actions available to him.*

An **action combination** *is an ordered set* a = $\{a_i\}$, $(i = 1, \ldots, n)$ *of one action for each of the* n *players in the game.*

For our model we specify the same action sets for both Saudi Arabia and Others: an oil output that is either *High* or *Low* in each year. We will use the notation $Q_{country,year} = level$: If Saudi output in 1988 is *High*, we say that $Q_{S,8} = H$.

Besides specifying the actions available to a player, we must specify when they are available: the **order of play**. We will say that a country chooses its outputs afresh each year rather than choosing both years' outputs at the start of the game. The order of play is therefore

(0) Nature picks demand, D, to be *Weak* or *Strong*.
(1) Saudi Arabia and Others simultaneously choose their individual 1988 outputs from the action sets
$$\{Q_{S,8} = L, Q_{S,8} = H\} \text{ and } \{Q_{O,8} = L, Q_{O,8} = H\}.$$
(2) Saudi Arabia and Others simultaneously choose their individual 1989 outputs from the action sets
$$\{Q_{S,9} = L, Q_{S,9} = H\} \text{ and } \{Q_{O,9} = L, Q_{O,9} = H\}.$$

An alternative specification, appropriate if the technology of oil production required advance planning, is that a country chooses its output for both years at the start of the game. Then the order of play would have just two elements:

(0) Nature picks demand, D, to be *Weak* or *Strong*.
(1) Saudi Arabia chooses its individual 1988 and 1989 outputs from the action set
$$\left\{ \begin{array}{l} (Q_{S,8} = L, Q_{S,9} = L), (Q_{S,8} = L, Q_{S,9} = H), \\ (Q_{S,8} = H, Q_{S,9} = L), (Q_{S,8} = H, Q_{S,9} = H) \end{array} \right\},$$

Others simultaneously chooses actions from its equivalent action set.

Information is modelled using the concept of the **information set**, which we will define precisely in Section 2.3. For now, think of a player's information set as his knowledge at a particular time of the values of different variables. The elements of the information set are the different values that the player thinks are possible. If the information set has many elements, there are many values the player cannot rule out; if it has one element, he knows the value precisely. Let us specify that after Nature moves, Saudi Arabia knows whether world oil demand is *Strong* or *Weak*, but Others cannot rule out either possibility. The information sets are

Others: $\{D = Strong, D = Weak\}$;
Saudi Arabia: $\{D = Strong\}$ or $\{D = Weak\}$, depending on demand.

A player's information set includes not only distinctions between the values of variables like the strength of oil demand, but also knowledge of what actions have previously been taken, so his information set changes over the course of the game.

*Player i's **strategy** s_i is a rule that tells him which action to choose at each instant of the game, given his information set.*

*Player i's **strategy set** or **strategy space** $S_i = \{s_i\}$ is the set of strategies available to him.*

*A **strategy combination** $s = (s_1, \ldots, s_n)$ is an ordered set consisting of one strategy for each of the n players in the game.*

Since the information set includes whatever the player knows about the previous actions of other players, the strategy tells him how to react to their actions. In the OPEC game the actions are to produce *High* or *Low* in 1988 and 1989. One strategy in Saudi Arabia's strategy set is .

$$
\begin{cases}
Q_{S,8}(D) = \begin{cases} L & \text{if } D = Weak \\ H & \text{if } D = Strong \end{cases} \\[2ex]
Q_{S,9}(D, Q_{S,8}, Q_{O,8}) = \begin{cases} L & \text{if } D = Weak, Q_{S,8} = L, \text{ and } Q_{O,8} = L \\ H & \text{otherwise} \end{cases}
\end{cases}
$$

This strategy gives Saudi Arabia's 1988 action as a function of the strength of demand alone (since Others's action is not yet known), and its 1989 action as a function of demand, its own 1988 action, and Others's 1988 action. A strategy is a function only of observed history, not of current actions or of another player's strategy. Saudi Arabia's strategy cannot be specified to give its 1989 action as a function of Others's 1989 action or Others's strategy. Such misspecification is a common source of confusion. In the simple games of the next few sections the distinction between actions and strategies is not important, but in later chapters it will be quite helpful. The concept of the strategy is useful because the action a player wishes to pick depends on the past actions of Nature and the other players. Only rarely can we predict a player's actions unconditionally. More often we can predict how he will respond to the outside world.

Another source of confusion is that a player's strategy is a complete set of instructions for him, which tells him what actions to pick in every conceivable situation, even if he does not expect to reach that situation. Strictly speaking, even if a player's strategy instructs him to commit suicide in 1989, it ought to also specify what actions he takes if he is still alive in 1990. Besides being necessary to fit the definition, this kind of carefulness will be important when we look at subgame perfect equilibrium in Chapter 4. The completeness of the description also means that strategies, unlike actions, are unobservable. A strategy is mental; an action is physical.

*By player i's **payoff** $\pi_i(s_1, \ldots, s_n)$, we mean either:*

(1) The utility he receives after all players and Nature have picked their strategies and the game has been played out; or

(2) The expected utility he receives as a function of the strategies chosen by himself and the other players.

Definitions (1) and (2) are distinct and different, but in the literature and this book the term "payoff" is used for both the actual payoff and the expected payoff. The context will make clear which is meant.

We assumed above that the payoffs of Others and Saudi Arabia were the sums of their oil revenues over the two years of production. If we were not merely sketching the model, we would specify π_S and π_O as functions relating the sum of revenues to the strength of demand and the outputs in the two years.

The **outcome** *of the game is a set of interesting elements that the modeller picks from the values of actions, payoffs, and other variables after the game is played out.*

The definition of the outcome for any particular model depends on what variables are interesting to the modeller. One outcome of OPEC Model I is

$$Q_{S,8} = L, \ Q_{S,9} = H, \ Q_{O,8} = H, \ Q_{O,9} = L, \ D = L, \ \pi_S = 100, \ \pi_O = 80, \quad (1.1)$$

where 100 and 80 are the values specified by the payoff functions. The outcome could be more narrowly defined as just the set of payoffs or the levels of output. Which definition is chosen depends on what you think is interesting about OPEC. This entire model, in fact, is only one of the many possible models of OPEC. For contrast, another game representing OPEC is OPEC Model II.

OPEC Model II

Players ·

Saudi Arabia, Libya, Venezuela, Kuwait, Nigeria.

Information

All players know the value of demand, but they choose their actions simultaneously.

Actions and Events

(1)   The players simultaneously choose supply schedules consisting of the quantity each will supply at each possible market price in 1988 (this model does not concern itself with 1989).
(2)   Nature picks world demand for oil to be either *Weak* or *Strong*, with equal probability.

Strategies

The strategies are the same as actions, since no information is revealed that might affect the action a player chooses.

Payoffs

For each country the payoff is +100 if its oil revenue is above a country-specific amount necessary to avoid a military coup, −100 if revenue is lower.

Outcomes

The outcome includes the quantities supplied, the state of demand, the resulting revenues and market price, and whether or not each country has a coup.

Choosing the suitable model is where much of the talent of a modeller is displayed, since he must trade off realism, ease of solution, and clarity of presentation. How would *you* decide between the two OPEC models?

Equilibrium

To predict the outcome of a game, the modeller focusses on the possible strategy combinations, since it is the interaction of the different players' strategies that determines what happens. The distinction between strategy combinations, which are sets of strategies, and outcomes, which are sets of values of whichever variables are considered interesting, is a common source of confusion. Often different strategy combinations lead to the same outcome. In OPEC Model I, the single outcome

$$(Q_{S,8} = L, \; Q_{O,8} = L, \; Q_{S,9} = L, \; Q_{O,9} = L, \; D = Strong, \; \pi_S = 100, \; \pi_O = 80)$$
$$(1.2)$$

is produced by either of the two strategy combinations, the Golden Rule or An Eye for an Eye.

The Golden Rule: (*Low output no matter what happens*)

$$\begin{cases} \text{Saudi Arabia}: & (Q_{S,8} = L; Q_{S,9} = L), \\ \text{Others}: & (Q_{O,8} = L; Q_{O,9} = L); \end{cases}$$

An Eye for an Eye: (*Retaliate*)

$$\begin{cases} \text{Saudi Arabia}: & (Q_{S,8} = L; Q_{S,9} = L \text{ if } Q_{O,8} = L, \; Q_{S,9} = H \text{ otherwise}), \\ \text{Others}: & (Q_{O,8} = L; Q_{O,9} = L \text{ if } Q_{S,8} = L, \; Q_{O,9} = H \text{ otherwise}). \end{cases}$$

Under the Golden Rule, both players always choose low output, so output is low in both years. Under An Eye for an Eye (a variant of the "tit-for-tat" strategy of Section 4.6), both players choose low output in 1988, and in 1989 each chooses the output the other had chosen the previous year, so output is also low.

> An **equilibrium** $s^* = (s_1^*, \ldots, s_n^*)$ *is a strategy combination consisting of a best strategy for each of the* n *players in the game.*

The **equilibrium strategies** are the strategies players pick in trying to maximize their individual payoffs, as distinct from the many possible strategy combinations obtainable by randomly choosing one strategy per player. Equilibrium is used differently in game theory than in other areas of economics. In a general equilibrium model, for example, an equilibrium is a set of prices resulting from optimal behavior by the individuals in the economy. In game theory, that set of prices would be the **equilibrium outcome**, but the equilibrium itself would be the strategy combination—individuals' rules for buying and selling—that generated the outcome.

To find the equilibrium it is not enough to specify the players, strategies, and payoffs, because the modeller must also decide what "best strategy" means. He does this by defining a solution concept.

An **equilibrium concept** *or* **solution concept** $F : \{S_1, \ldots, S_n, \pi_1, \ldots, \pi_n\}$ $\rightarrow s^*$ *is a rule that defines an equilibrium based on the possible strategy combinations and the payoff functions.*

Only a few equilibrium concepts are generally accepted, and the remaining sections of this chapter are devoted to finding the equilibrium using the two best-known concepts: dominant strategy and Nash equilibrium.

Uniqueness

Because the accepted solution concepts do not guarantee uniqueness, lack of a unique equilibrium is a major problem for game theory. Often the solution concept employed leads us to believe that the players will pick one of the two strategy combinations A or B, not C or D, but we cannot say whether A or B is more likely. Sometimes we have the opposite problem and the game has no equilibrium at all; by this is meant either that the modeller sees no good reason why one strategy combination is more likely than another, or that some player wants to pick an infinite value for one of his actions, something we will discuss further in Section 5.6.

A model with no equilibrium or multiple equilibria is underspecified. The modeller has failed to provide a full and precise prediction for what will happen. One option is to admit that his theory is incomplete: an admission of incompleteness like the Folk Theorem of Section 4.6 is a valuable negative result. Or perhaps the situation being modelled really is unpredictable. Another option is to renew the attack by changing the game's description or the solution concept. Preferably it is the description that is changed, since economists look to the rules of the game for the differences between models, and not to the solution concept, and the reader is likely to feel tricked if an important part of the game is concealed under the definition of equilibrium.

1.2   Dominant Strategies: the Prisoner's Dilemma

In discussing equilibrium concepts, it is useful to have shorthand notation for "all the other players' strategies."

For any vector $y = (y_1, \ldots, y_n)$, *denote by* y_{-i} *the vector* $(y_1, \ldots, y_{i-1}, y_{i+1}, \ldots, y_n)$, *which is the portion of* y *not associated with player* i.

Using this notation, s_{-i}, for instance, is the combination of strategies of every player except player i. That combination is of great interest to player i, because he uses it to help choose his own strategy, and the new notation helps define his best response.

> Player i's **best response** *or* **best reply** *to the strategies* s_{-i} *chosen by the other players is the strategy* s_i^* *that yields him the greatest payoff, that is,*

$$\pi_i(s_i^*, s_{-i}) \geq \pi_i(s_i', s_{-i}) \quad \forall s_i' \neq s_i^*. \tag{1.3}$$

The best response is strongly best if no other strategies are equally good, and weakly best otherwise.

The first important equilibrium concept is the dominant strategy equilibrium.

> *The strategy* s_i^* *is a* **dominant strategy** *if it is a player's strictly best response to any strategies the other players might pick, in the sense that whatever strategies they pick, his payoff is highest with* s_i^*. *Mathematically,*

$$\pi_i(s_i^*, s_{-i}) > \pi_i(s_i', s_{-i}) \quad \forall s_{-i}, \quad \forall s_i' \neq s_i^*. \tag{1.4}$$

> *His inferior strategies are* **dominated strategies**.

> *A* **dominant strategy equilibrium** *is a strategy combination consisting of each player's dominant strategy.*

A player's dominant strategy is his strictly best response even to very stupid actions by the other players. Most games do not have dominant strategies, and the players must try to figure out each others' actions to choose their own.

In developing OPEC Model I, we incorporated considerable complexity to illustrate such things as information sets and the time sequence of actions. To illustrate equilibrium concepts we will use simpler games such as the Prisoner's Dilemma. In the Prisoner's Dilemma, two prisoners, Messrs Row and Column, are being interrogated separately. If both confess (*Fink*) they are each sentenced to eight years in prison; if both hold out (*Cooperate*), each is sentenced to one year. If just one confesses, he is released, but the other prisoner is sentenced to ten years. The Prisoner's Dilemma is an example of a **2-by-2 game**, because each of the two players—Row and Column—has two possible actions in his action set—*Fink* and *Cooperate*. The payoffs are given by Table 1.1.

Each player has a dominant strategy. Consider Row. Row does not know which action Column is choosing, but if Column chooses *Cooperate*, Row faces a *Cooperate* payoff of -1 and a *Fink* payoff of 0, whereas if Column chooses *Fink*, Row faces a *Cooperate* payoff of -10 and a *Fink* payoff of -8. In either case Row does better with *Fink*, and since the game is symmetric, Column's incentives are the same. The dominant strategy equilibrium is (*Fink, Fink*), and the equilibrium payoffs are $(-8, -8)$, which is worse for both players than $(-1, -1)$. Sixteen, in fact, is the greatest possible combined total of years in prison.

The result is even stronger. Because the equilibrium is a dominant strategy

Table 1.1 The Prisoner's Dilemma

Column

		Cooperate	*Fink*
	Cooperate	−1, −1	−10, 0
Row			
	Fink	0, −10	**−8, −8**

Payoffs to: (Row, Column)

equilibrium, the information structure of the game does not matter. If Column is allowed to know Row's move before taking his own, the equilibrium is unchanged. Row still chooses *Fink*, knowing that Column will surely choose *Fink* afterwards.

If the outcome does not seem right to you, you should realize that very often the chief usefulness of a model is to induce discomfort. Discomfort is a sign that your model is not what you think it is—that you left out something essential to the result you expected and didn't get. Either your original thought or your model is mistaken; and finding such mistakes is a real, if painful, benefit of model-building.

The Prisoner's Dilemma crops up in many different situations, including oligopoly pricing, auction bidding, salesman effort, political bargaining, and arms races. Whenever you observe individuals in a conflict that hurts all of them, your first thought should be of the Prisoner's Dilemma.

Cooperative and Noncooperative Games

What difference would it make if the two prisoners could talk to each other before making their decisions? It depends on the strength of promises—if promises are not binding, then although the two prisoners might agree not to fink, they would fink anyway when the time came to choose actions.

A **cooperative game** *is a game in which the players can make binding commitments, as opposed to a* **noncooperative game**, *in which they cannot.*

This definition draws the usual distinction between the two theories of games, but the real difference lies in the modelling approach. Both theories start off with the rules of the game, but they differ in the kinds of solution concepts employed. Cooperative game theory is axiomatic, frequently appealing to Pareto-optimality (see note N1.3), fairness, and equity. Noncooperative game theory is economic in flavor, with solution concepts based on players maximizing their own utility functions subject to stated constraints. Except for Section 10.2 of the chapter on bargaining, this book is concerned exclusively with noncooperative games.

In applied economics, the most commonly encountered use of cooperative games is to model bargaining. The Prisoner's Dilemma is a noncooperative game, but it could be modelled as cooperative by allowing the two players not only to communicate, but to make binding commitments. Cooperative games often allow players to split the gains from cooperation by making **side-payments**—transfers between themselves that change the prescribed payoffs. Cooperative game theory generally incorporates commitments and side-payments via the solution concept, which can become very elaborate, while noncooperative game theory incorporates them by adding extra actions. The distinction between cooperative and noncooperative games does *not* lie in conflict or absence of conflict, as is shown by the following examples of situations commonly modelled one way or the other:

A cooperative game without conflict: Members of a work force choose which of equally arduous tasks to undertake to best coordinate with each other.

A cooperative game with conflict: Bargaining over price between a monopolist and a monopsonist.

A noncooperative game with conflict: The Prisoner's Dilemma.

A noncooperative game without conflict: Two companies set a product standard without communication.

1.3   Iterated Dominance: the Battle of the Bismarck Sea

Very few games have a dominant strategy equilibrium, but the Battle of the Bismarck Sea shows how dominance can be used even when it does not resolve things quite so neatly as in the Prisoner's Dilemma. The game is set in the South Pacific in 1943. Admiral Imamura has been ordered to transport Japanese troops across the Bismarck Sea to New Guinea, and Admiral Kenney wishes to bomb the troop transports. Imamura must choose between a shorter Northern route or a longer Southern route, and Kenney must decide where to send his planes to look for the Japanese. If Kenney sends his planes to the wrong route he can recall them, but the number of days of bombing is reduced.

The players are Kenney and Imamura, and they each have the same action set, {*North, South*}, but their payoffs, given by Table 1.2, are never the same. Imamura loses when Kenney gains. Because of this feature, the payoffs could be represented using just four numbers instead of eight, but listing all eight payoffs in Table 1.2 saves the reader a little thinking.

Table 1.2 The Battle of the Bismarck Sea

		Imamura	
		North	*South*
Kenney	*North*	**2, −2**	2, −2
	South	1, −1	3, −3

Payoffs to: (Kenney, Imamura)

Strictly speaking, neither player has a dominant strategy. Kenney would choose *North* if he thought Imamura would choose *North*, but *South* if he thought Imamura would choose *South*. Imamura would choose *North* if he thought Kenney would choose *South*, and he would be indifferent between actions if he thought Kenney would choose *North*. But we can still find a plausible equilibrium, using the concept of weak dominance, which differs from strong dominance only by replacing a strong inequality with a weak one.

Strategy s_i^* *is a* **weakly dominant strategy** *if it is a player's best response to any strategies the other players might pick, in the sense that whatever strategies they pick, his payoff is no smaller with s_i than with any other strategy, and is greater in some strategy combination. Mathematically,*

$$\pi_i(s_i^*, s_{-i}) \geq \pi_i(s_i', s_{-i}) \quad \forall s_{-i}, \forall s_i',$$

and $\pi_i(s_i^*, s_{-i}) > \pi_i(s_i', s_{-i}) \quad \forall s_i'$, *for some* s_{-i}. \hfill (1.5)

An **iterated dominant strategy equilibrium** *is a strategy combination found by deleting a weakly dominated strategy from the strategy set of one of the players, recalculating to find which remaining strategies are weakly dominated, deleting one of them, and continuing the process until only one strategy remains for each player.*

Admiral Imamura does have a weakly dominant strategy. *North* is weakly dominant, because Imamura's payoff from *North* is no lower than his payoff from *South*, and is greater if Kenney picks *South*. Suppose that Kenney realizes this and decides that Imamura will pick *North*, deleting "Imamura chooses *South*" from consideration. Having deleted one column of Table 1.2, Kenney now has a strongly dominant strategy (where by "strongly" we mean that this strategy achieves payoffs strictly greater than any others), and he chooses *North*. The strategy combination (*North*,*North*) is an iterated dominant strategy equilibrium, and despite all the qualifying adjectives it seems a good prediction. And indeed (*North*,*North*) was the outcome in 1943.

It is interesting to consider modifying the order of play or the information structure in the Battle of the Bismarck Sea. If Kenney moved first, rather than simultaneously with Imamura, (*North*,*North*) would remain an equilibrium, but (*North*,*South*) would also become one. The payoffs would be the same for both equilibria, but the outcomes would be different.

If Imamura moved first, then (*North*,*North*) would be the only equilibrium. Imamura moving first is also equivalent to both players moving simultaneously when both know that Kenney has cracked the Japanese code and knows Imamura's plan. In both situations, Kenney's information set becomes either {Imamura moved *North*} or {Imamura moved *South*}, so Kenney's equilibrium strategy is specified as (*North* if Imamura moved *North*, *South* if Imamura moved *South*).

Although the 2-by-2 games in this chapter may seem facetious, they are simple enough to adapt to economic situations. The Battle of the Bismarck Sea, for example, can be turned into a game of corporate strategy. Two firms, Kenney and Imamura, are trying to maximize their shares of a market of constant size by choosing between the two product designs *North* and *South*. Kenney has a marketing advantage, and would like to compete head-to-head, while Imamura would rather carve out its own niche. The equilibrium is (*North,North*).

The Battle of the Bismarck Sea is special because the payoffs of the players always sum to zero. This feature is important enough to deserve a name.

> A **zero-sum game** *is a game in which the sum of the payoffs of all the players is zero whatever strategies they choose. A game which is not zero-sum is* **non-zero-sum.**

In a zero-sum game, what one player gains, another player must lose. The Battle of the Bismarck Sea is a zero-sum game, but the Prisoner's Dilemma and OPEC Models I and II are not. Although zero-sum games have fascinated game theorists for many years, they are uncommon in economics. One of the few examples is the bargaining game between two players who divide a surplus, but even this is usually modelled nowadays as a non-zero-sum game in which the surplus shrinks as the players spend more time deciding how to divide it.

1.4   Nash Equilibrium: Boxed Pigs, the Battle of the Sexes, and Pure Coordination

For the vast majority of games, which lack even iterated dominant strategy equilibria, we use Nash equilibrium, the most important and widespread equilibrium concept. To introduce Nash equilibrium we will use the game Boxed Pigs (Baldwin & Meese [1979]). Two pigs are put into a Skinner box with a special panel at one end and a food dispenser at the other. When the panel is pressed, at a utility cost of 2 units, 10 units of food are dispensed. One pig is "dominant" (let us assume he is larger), and if he gets to the dispenser first, the other pig will only get his leavings, worth 1 unit. The small pig does somewhat better if he gets there first, eating 4 units of food, and even if they arrive at the same time he can eat 3 units. Table 1.3 summarizes the payoffs for the strategies *Press* the panel and *Wait* by the dispenser.

Table 1.3 Boxed Pigs

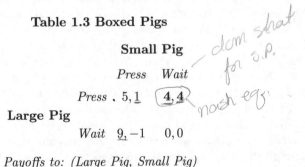

		Small Pig	
		Press	*Wait*
Large Pig	*Press*	. 5,<u>1</u>	(4,4)
	Wait	<u>9,</u>−1	0,0

Payoffs to: (Large Pig, Small Pig)

Boxed Pigs has no dominant strategy equilibrium, because what the large pig chooses depends on what he thinks the small pig will choose. If he believed that the small pig would press the panel, the large pig would wait by the dispenser, but if he believed that the small pig would wait, the large pig would press. There does exist an iterated dominant strategy equilibrium, $(Press, Wait)$, but we will employ a different line of reasoning to justify that outcome.

The equilibrium concept that is standardly employed is Nash equilibrium, which is less obviously correct than dominant strategy equilibrium, but more often applicable. Nash equilibrium is so widely accepted that the reader can assume that if a model does not specify which equilibrium concept is being used it is Nash.

*The strategy combination s^* is a **Nash equilibrium** if no player has incentive to deviate from his strategy given that the other players do not deviate. Formally,*

$$\forall i, \quad \pi_i(s_i^*, s_{-i}^*) \geq \pi_i(s_i', s_{-i}^*), \quad \forall s_i'. \tag{1.6}$$

The strategy combination $(Press, Wait)$ is a Nash equilibrium. The way to approach Nash equilibrium is to propose a strategy combination and test whether each player's strategy is a best response to the others' strategies. If the large pig picks $Press$, the small pig, who faces a choice between a payoff of 1 from pressing and 4 from waiting, is willing to wait. If the small pig picks $Wait$, the large pig, who has a choice between a payoff of 4 from pressing and 0 from waiting, is willing to press. This confirms that $(Press, Wait)$ is a Nash equilibrium, and in fact it is the unique Nash equilibrium.

The pigs in this game have more thinking to do than the players in the Prisoner's Dilemma. They have to realize that the only set of strategies which is supported by self-consistent beliefs is $(Press, Wait)$. The definition of Nash equilibrium lacks the extra "$\forall s_{-i}$" of dominant strategy equilibrium, so a Nash strategy need only be a best response to the other Nash strategies, not to all possible strategies. And although we talk of "best responses," the moves are actually simultaneous, so the players are predicting each others' moves. If the game were repeated, or the players communicated, Nash equilibrium would be especially attractive, because it is even more compelling that beliefs should be consistent.

Every dominant strategy equilibrium is a Nash equilibrium, but not every Nash equilibrium is a dominant strategy equilibrium. If a strategy is dominant it is a best response to *any* strategies the other players pick, including their equilibrium strategies. If a strategy is part of a Nash equilibrium, it need only be a best response to the other players' *equilibrium* strategies. This is shown by the Nash Puzzle in Table 1.4, which has $(Up, Left)$ and $(Down, Right)$ as Nash equilibria, and $(Down, Right)$ as an iterated dominant strategy equilibrium. Consider the Nash equilibrium $(Up, Left)$. Neither Smith nor Brown has incentive to unilaterally deviate from it; Smith's payoff would remain 0, and Brown's would drop from 1 to 0. Despite forming part of a Nash equilibrium, Up is a weakly dominated strategy for Smith: it yields him either 0 or -2, depending on what Brown does, while $Down$ yields him 0 or -1. But if Smith expects Brown to choose $Left$, as he does in equilibrium, there is no reason for Smith to prefer $Down$. In fact, the Nash equilibrium $(Up, Left)$ Pareto-dominates $(Down, Right)$; it is preferred by both players.

Like a dominant strategy equilibrium, a Nash equilibrium can be either weak

Table 1.4 The Nash Puzzle

Brown

Left   *Right*

		Left	*Right*
	Up	0, 1	−2, 0
Smith	*Down*	0, −1	−1, 0

N.E

Payoffs to: (Smith, Brown)

or strong. The definition above is for a weak Nash equilibrium. To define the strong Nash equilibrium, make the inequality strict; that is, say that no player is indifferent between his equilibrium strategy and some other strategy.

The Battle of the Sexes

The third game we will use to illustrate Nash equilibrium is the Battle of the Sexes, a conflict between a man who wants to go to a prize fight and a woman who wants to go to a ballet. While selfish, they are deeply in love, and would, if necessary, sacrifice their preferences to be with each other. Less romantically, their payoffs are given by Table 1.5.

Table 1.5 The Battle of the Sexes

Woman

		Prize Fight	*Ballet*
	Prize Fight	**2, 1**	−1, −1
Man	*Ballet*	−5, −5	**1, 2**

Payoffs to: (Man, Woman)

The Battle of the Sexes does not have an iterated dominant strategy equilibrium. It has two Nash equilibria, one of which is the strategy combination (*Prize Fight,Prize Fight*). Given that the man chooses *Prize Fight*, so does the woman; given that the woman chooses *Prize Fight*, so does the man. The strategy combination (*Ballet, Ballet*) is another Nash equilibrium, by the same line of reasoning.

How do the players know which Nash equilibrium to choose? Going to the fight and going to the ballet are both Nash strategies, but for different equilibria. Nash equilibrium assumes correct and consistent beliefs. If they do not talk beforehand, the man might go to the ballet and the woman to the fight, each mistaken about the other's beliefs. But even if the players do not communicate, Nash equilibrium is sometimes justified by repetition of the game. If the couple do not talk, but repeat the game night after night, one may suppose that eventually they settle on one of the Nash equilibria.

Each of the Nash equilibria in the Battle of the Sexes is Pareto-efficient; that is, no other strategy combination increases the payoff of one player without decreasing that of the other. In many games the Nash equilibrium is not Pareto-efficient: (*Fink, Fink*), for example, is the unique Nash equilibrium of the Prisoner's Dilemma, although its payoffs of $(-8, -8)$ are Pareto-inferior to the $(-1, -1)$ generated by (*Cooperate, Cooperate*). And we just saw in the Nash Puzzle that one Nash equilibrium can Pareto-dominate another.

Who moves first is important in these games. In the Nash Puzzle, if either player moved first we would expect (*Up, Left*) to be the equilibrium. In the Battle of the Sexes, if the man could buy the fight ticket in advance, his commitment would induce the woman to go to the fight. But if, instead, the woman could buy the ballet ticket in advance, her commitment would induce the man to go to the ballet. In many games, but not all, the player who moves first (which is equivalent to commitment) has a **first-mover advantage.**

The Battle of the Sexes has many economic applications. One is the choice of an industrywide standard when two firms have different preferences but both want a common standard to encourage consumers to buy the product. A second is to the choice of language used in a contract when two firms want to formalize a sales agreement but they prefer different terms.

Pure Coordination

Sometimes one can use the size of the payoffs to choose between Nash equilibria. In the following game, players Smith and Brown are trying to decide whether to design the computers they sell to use large or small floppy disks. Both players will sell more computers if their disk drives are compatible. The payoffs are given by Table 1.6.

Table 1.6 Pure Coordination

Brown

		Large	*Small*
	Large	**2, 2**	$-1, -1$
Smith			
	Small	$-1, -1$	**1, 1**

Payoffs to: (Smith, Brown)

The strategy combinations (*Small, Small*) and (*Large, Large*) are both Nash equilibria, but (*Large, Large*) Pareto-dominates (*Small, Small*). Unlike in the Nash Puzzle, no strategy is weakly dominant. Both players prefer (*Large, Large*), and most modellers would use the Pareto-efficient equilibrium to predict the actual outcome. We could imagine that it arises from pregame communication between Smith and Brown taking place outside of the specification of the model, but the interesting question is what happens if communication is impossible. Is the Pareto-efficient equilibrium still more plausible? The question is really one of psychology rather than economics.

1.5   Focal Points

Although Thomas Schelling's book *The Strategy of Conflict* was published almost 30 years ago, it is surprisingly modern in spirit. Schelling is not a mathematician but a strategist, and he examines such things as threats, commitments, hostages, and delegation, which we will examine in a more formal way in the remainder of this book. He is perhaps best known for his coordination games. Take a moment to decide on a strategy in each of the following games, adapted from Schelling, which you win by matching your response to those of as many of the other players as possible.

(1)   Name Heads or Tails.
(2)   Name Tails or Heads.
(3)   Circle one of the following numbers 7, 100, 13, 261, 99, 666.
(4)   You are to meet somebody in New York City. When? Where?
(5)   You are to split a pie, and get nothing if your proportions add to more than 100 percent.

Each of the games above has many Nash equilibria: if I think you will choose 666, and you think I will choose 666, we both choose it. But to a greater or lesser extent they also have Nash equilibria that seem more likely. Certain of the strategy combinations are **focal points:** Nash equilibria which for psychological reasons are particularly compelling. Formalizing what makes a strategy combination a focal point is hard and depends on the context. In Example (3), Schelling found 7 to be the most common strategy, but in a group of Satanists, 666 might be the focal point. In repeated games, focal points are often provided by past history. If we split a pie once, we are likely to agree on 50:50. But if last year we split a pie in the ratio 60:40, that provides a focal point for this year.

The **boundary** is a particular kind of focal point. If player Russia chooses the action of putting his troops anywhere from one inch to 100 miles away from the Chinese border, player China does not react. If he chooses to put troops from one inch to 100 miles *beyond* the border, China declares war. There is an arbitrary discontinuity in behavior at the boundary. Another example, quite vivid in its arbitrariness, is the rallying cry, "Fifty-Four Forty or Fight!," which refers to the geographic parallel claimed as the boundary by jingoist Americans in the Oregon dispute between Britain and the United States in the 1840s.[1]

Once the boundary is established, it takes on additional significance, because behavior with respect to the boundary conveys information. When Russia crosses an established boundary, that tells China that Russia intends to make a serious incursion further into China. Boundaries must be sharp and well-known if they are not to be violated, and a large part of both law and diplomacy is devoted to clarifying them. Boundaries can also arise in business: two companies producing an unhealthful product might agree not to mention relative healthfulness in their advertising, but a boundary rule like "Mention unhealthfulness if you like, but do not stress it," would not work.

---

[1]   The threat was not credible: that parallel is now deep in British Columbia.

Mediation and **communication** are both important in the absence of a clear focal point. If players can communicate, they can tell each other what actions they will take, and sometimes, as in Pure Coordination, this works, because they have no motive to lie. If the players cannot communicate, a mediator may be able to help by suggesting an equilibrium to all of them. They have no reason not to take the suggestion, and they would use the mediator even if his services were costly. Mediation in cases like this is as effective as arbitration, in which an outside party imposes a solution.

One disadvantage of focal points is that they lead to inflexibility. Suppose the Pareto-superior equilibrium (*Large, Large*) was chosen as a focal point in Pure Coordination, but the game was repeated over a long interval of time. The numbers in the payoff matrix might slowly change until (*Small,Small*) and (*Large,Large*) both had payoffs of 1.5, and (*Small,Small*) might start to dominate. When, if ever, would the equilibrium switch?

In Pure Coordination, we would expect that after some time one firm would switch and the other would follow. If there were communication, the switch point would be at the payoff of 1.5. But what if the first firm to switch is penalized more? Such is the problem in oligopoly pricing. If costs rise, so should the monopoly price, but whichever firm raises its price first suffers a loss of market share.

Recommended Reading

Bernheim, B. Douglas (1984a) "Rationalizable Strategic Behavior" *Econometrica.* July 1984. 52, 4: 1007–28.
Schelling, Thomas (1960) *The Strategy of Conflict.* Cambridge, Mass.: Harvard University Press, 1960.

Problem 1

A Discoordination Game

Suppose that a man and a woman each choose whether to go to a prize fight or a ballet. The man would rather go to the prize fight, and the woman to the ballet. What is more important to them, however, is that the man wants to show up to the same event as the woman, but the woman wants to avoid him.

(1) Construct a game matrix to illustrate this game, choosing numbers to fit the preferences described verbally.
(2) If the woman moves first, what will happen?
(3) Does the game have a first-mover advantage?
(4) Show that there is no Nash equilibrium if the players move simultaneously.

You will discover how to find a Nash equilibrium in random strategies for games like these in Chapter 3.

Notes

N1.1   Basic Definitions

- The standard descriptive names help both the modeller and his readers. For the modeller, the names are useful because they help ensure that the important details of the game have been fully specified. For his readers, they make the game easier to understand, especially if, as with most technical papers, the paper is first skimmed quickly to see if it is worth reading. The less clear a writer's style, the more closely he should adhere to the standard names, which means that most of us ought to adhere very closely indeed.

  Think of writing a paper as a game between author and reader, rather than as a single-player production process. The author, knowing that he has valuable information but imperfect means of communication, is trying to convey the information to the reader. The reader does not know whether the information is valuable, and he must choose whether to read the paper closely enough to find out. What are the possible equilibria?

- The equilibrium concepts that will be mentioned in this book, with the sections in which they are introduced, are dominant strategy (1.2), iterated dominant strategy (1.3), Nash (1.4, rationalizable (N1.4), Bayesian (2.5), correlated strategy (3.3), subgame perfect (4.2), coalition-proof Nash (N4.2), minimax (N4.6), maximin (N4.6), perfect Bayesian (5.1), trembling hand perfect (5.1), sequential (5.1), proper (N5.2), divine (N5.2), intuitive (N5.2), evolutionarily stable strategy (ESS) (5.6), reactive (8.5), and Wilson (8.5). A recent book on equilibrium concepts is Harsanyi & Selten (1988), which emphasizes bargaining games.

- In OPEC Model I, the notation "$Q_{S,8} = H$" was used to denote "Saudi Arabian oil output in 1988 is High." A logically equivalent model uses the notation "$X_{1,1} = 2$" to denote "Country 1's oil output in Period 1 is 2." Does it make any difference which notation is used?

N1.2   Dominant Strategies: the Prisoner's Dilemma

- The Prisoner's Dilemma was named by Albert Tucker in an unpublished paper, although the particular 2-by-2 matrix, discovered by Dresher and Flood, was already well-known. Tucker was asked to give a talk on game theory to the psychology department at Stanford, and invented a story to go with the matrix. Straffin (1980) tells this history.

- Herodotus describes an early example of the reasoning in the Prisoner's Dilemma in the conspiracy of Darius against the Persian emperor. A group of nobles met and decided to overthrow the emperor, and it was proposed to adjourn till another meeting. Darius then spoke up and said that if they adjourned, he knew that one of them would go straight to the emperor and fink on them, because if nobody else did, he would himself. Darius also suggested a solution—that they immediately go to the palace and kill the emperor.

  The conspiracy also illustrates a way out of coordination games. After killing the emperor, the nobles wished to select one of themselves as the new emperor. Rather than fight, they agreed to go to a certain hill at dawn, and whoever's horse neighed first would become emperor. Herodotus tells how Darius's groom manipulated this randomization scheme to make him the new emperor.

- Philosophers are intrigued by the Prisoner's Dilemma: see Campbell & Sowden (1985), a collection of articles on the Prisoner's Dilemma and the related Newcombe's paradox.

- Many economists are reluctant to use the concept of cardinal utility, and even more reluctant to compare utility across individuals (see Cooter & Rappoport [1984]). Non-cooperative game theory never requires interpersonal utility comparisons, and only ordinal utility is needed to find the equilibrium in the Prisoner's Dilemma. So long as each player's rank ordering of payoffs in different outcomes is preserved, the payoffs can be altered without changing the equilibrium. In general, the dominant strategy and pure strategy Nash equilibria of games depend only on the ordinal ranking of the payoffs, but the mixed strategy equilibria depend on the cardinal values (see Section 3.2, and compare Chicken [Section 3.3] with the Hawk-Dove Game [Section 5.6]).

- If we consider only the ordinal ranking of the payoffs in 2-by-2 games, there are 78 distinct games in which each player has a strict preference ordering over the four outcomes (listed and described in Rapoport & Guyer [1966]), and 726 distinct games (Guyer & Hamburger [1968]) if we allow ties in the payoffs .
- The Prisoner's Dilemma is not always defined the same way. If we consider just ordinal payoffs, then the game in Table 1.7 is a Prisoner's Dilemma if $c > a > d > b$. If the game is repeated, the cardinal values of the payoffs can be important. The requirement $2a > b + c > 2d$ should be added if the game is to be a standard Prisoner's Dilemma, in which $(Cooperate, Cooperate)$ and $(Fink, Fink)$ are the best and worst possible outcomes in terms of the sum of payoffs. Section 4.7 will show that an asymmetric game called the One-Sided Prisoner's Dilemma has properties similar to the standard Prisoner's Dilemma, but does not fit this definition.

  Sometimes the game in which $2a < b + c$ is also called a Prisoner's Dilemma, but then the sum of the players' payoffs is maximized when one finks and the other cooperates. If the game were repeated, or they could use the correlated equilibria defined in Section 3.3, they would prefer taking turns being finked against, which would make the game a coordination game similar to the Battle of the Sexes. David Shimko has suggested the name "Battle of the Prisoners" for this (which is more dignified than "Sex Prisoners' Dilemma").

Table 1.7 The Prisoner's Dilemma with general payoffs

Column

		Cooperate	*Fink*
	Cooperate	a, a	b, c
Row			
	Fink	c, b	d, d

Payoffs to: (Row, Column)

N1.3   Iterated Dominance: the Battle of the Bismarck Sea

- The Battle of the Bismarck Sea can be found in Haywood (1954).
- The 2-by-2 form with just four entries that could be used for the Battle of the Bismarck Sea and other zero-sum games is a **matrix game**, while the equivalent table with eight entries is a **bimatrix game**. Games can be represented as bimatrix games even if they have more than two moves, so long as the number is finite.
- The dominant strategy equilibrium of any game is unique, if it exists. Each player has only one strategy (or none) whose payoff in any strategy combination is strictly higher than the payoff from any other strategy, so only one strategy combination can be formed out of dominant strategies. A game may have several weak dominant strategy equilibria, however, because a single player can have two or more weakly dominant strategies.

  An iterated strong dominant strategy equilibrium is unique, if it exists. An iterated weak dominant strategy equilibrium is not always unique, because the order in which strategies are deleted can matter to the final solution. If, however, all the weakly dominated strategies are eliminated simultaneously at each round of elimination, the iterated equilibrium is unique.
- If a game is zero-sum the utilities of the players can be represented so as to sum to zero under any outcome. Since utility functions are to some extent arbitrary, the sum can also be represented to be non-zero even if the game is zero-sum. Often modellers will refer to a game as zero-sum even when the payoffs do not add up to zero, so long as the payoffs add up to some constant amount. The difference is a trivial normalization.

- If outcome X **strongly Pareto-dominates** outcome Y, then all players have higher utility under outcome X. If outcome X **weakly Pareto-dominates** outcome Y, some player has higher utility under X, and no player has lower utility. A zero-sum game does not have outcomes that even weakly Pareto-dominate other outcomes. All its equilibria are Pareto-efficient, because no player gains without another player losing.

N1.4   Nash Equilibrium: Boxed Pigs, the Battle of the Sexes, and Pure Coordination

- I invented the payoffs for Boxed Pigs from the description of one of the experiments in Baldwin & Meese (1979). They do *not* think of this as an experiment in game theory, and they describe the result in terms of "reinforcement." I invented the Nash Puzzle and Pure Coordination for this book. The Battle of the Sexes is taken from p. 90 of Luce & Raiffa (1957). I have changed their payoffs of $(-1, -1)$ to $(-5, -5)$ to fit the story.
- Some people prefer "equilibrium point" to "Nash equilibrium," but the latter is convenient, since the name is "Nash" and not "Mazurkiewicz."
- Bernheim (1984a) and Pearce (1984) use the idea of mutually consistent beliefs to arrive at a different equilibrium concept than Nash. They define a **rationalizable strategy** to be a strategy which is a best response for some set of rational beliefs in which a player believes that the other players choose their best responses. The difference from Nash is that not all players need have the same beliefs concerning which strategies will be chosen, nor need their beliefs be consistent.

   This idea is attractive in the context of Bertrand games (see Section 12.2). The Nash equilibrium in the Bertrand game is weakly dominated—by picking any other price above MC, which yields the same profit of zero as does the equilibrium. Rationalizability rules that out.
- One of the best-known coordination problems is that of the QWERTY typewriter keyboard, developed in the 1870s when typing had to proceed slowly to avoid jamming. QWERTY became the standard, although a US Navy study found in the 1940s that the faster speed possible with the Dvorak keyboard would amortize the cost of retraining full-time typists within ten days (David [1985]). Why large companies have not retrained their typists is a mystery.
- A well-known discoordination game without conflict is the **Highway Game**. 10,000 commuters decide each morning whether to drive on the San Diego Freeway or Sepulveda Boulevard. A commuter's payoff is higher if he chooses the road that has fewer drivers. What does he do? Discoordination games have no symmetric equilibria except in randomized strategies (see Section 3.2), but there are many asymmetric equilibria in which 5,000 commuters are on each road.
- Another well-known discoordination game is **Matching Pennies**. In this zero-sum, 2-by-2 game, players Smith and Brown each choose *Heads* or *Tails*. If they choose the same action, Smith pays 10 to Brown. If they choose different actions, Brown pays 10 to Smith.
- O. Henry's story, "The Gift of the Magi" is about a coordination game that is noteworthy for the reason why communication is ruled out. The husband sells his watch to buy his wife combs for Christmas, while she sells her hair to buy him a watch fob. Communication would spoil the surprise, a worse outcome than discoordination.
- Macroeconomics has more game theory in it than is readily apparent. The macroeconomic concept of *rational expectations* faces the same problems of multiple equilibria and consistency of expectations as Nash equilibrium.
- Standard setting is an active topic of research at the time I write this. Recent examples are Katz & Shapiro (1985) and Farrell & Saloner (1985).
- In Section 3.3 we return to problems of coordination to discuss the concepts of "correlated strategies" and "cheap talk."

N1.5   Focal Points

- Besides his 1960 book, Schelling has written books on diplomacy (1966) and the oddities of aggregation (1978).   Political scientists are now looking at the same issues more technically; see Brams & Kilgour (1988) and Ordeshook (1986).
- In Chapter 12 of *The General Theory*, Keynes suggests that the stock market is a game with multiple equilibria, like a contest in which a newspaper publishes the faces of 20 girls, and contestants submit the name of the one they think most people would submit as the prettiest. When the focal point changes, big swings in prices result.
- Not all of what we call boundaries have an arbitrary basis. If the Chinese cannot defend themselves as easily once the Russians cross the boundary at the Amur River, they have a clear reason to fight there.

2 Information

2.1 Introduction

This chapter is about games in which the order of moves is important and some actions are not instantly observed by every player. The game tree and extensive form are introduced in Section 2.2 as ways to depict such games. Using the game tree, we formally define which players know what in Section 2.3, and in Section 2.4 we construct various informational categories for games. Games of asymmetric information are especially important, so we analyze them in greater depth in Section 2.5 and show how to use Bayes's Rule to update players' beliefs after they observe other players' moves. Section 2.6 ends the chapter with an extended example, the Png model of out-of-court settlement.

2.2 The Normal and Extensive Forms of a Game

The Normal Form and the Outcome Matrix

Games with several moves in sequence require more care in presentation than single-move games. In Section 1.4 we used the 2-by-2 form, which for the game Pure Coordination was what is shown in Table 2.1.

Because strategies are the same as actions in Pure Coordination and the outcomes are simple, the 2-by-2 form in Table 2.1 accomplishes two things: it relates strategy combinations to payoffs, and action combinations to outcomes. These two mappings are called the normal form and the outcome matrix, and in more complicated games they are distinct from each other. The normal form shows what payoffs result from each possible strategy combination, while the outcome matrix shows what outcome results from each possible action combination. The definitions below use n to denote the number of players, k the number of variables in the outcome vector, p the number of strategy combinations, and q the number of action combinations.

Table 2.1 Pure Coordination

Brown

		Large	Small
	Large	**2, 2**	$-1, -1$
Smith			
	Small	$-1, -1$	**1, 1**

Payoffs to: (Smith, Brown)

The **normal form** *consists of*

(1) All possible strategy combinations $s^1, s^2, \ldots, s^p$.

(2) Payoff functions mapping s^i onto the payoff n-vector π^i, $(i = 1, 2, \ldots, p)$.

The **outcome matrix** *consists of*

(1) All possible action combinations $a^1, a^2, \ldots, a^q$.

(2) Outcome functions mapping a^i onto the outcome k-vector z^i, $(i = 1, 2, \ldots, q)$.

Consider the following game based on Pure Coordination, which we will call "Follow the Leader I" since we will create several variants of it. The difference from Pure Coordination is that Smith moves first, committing himself to a certain disk size no matter what size Brown chooses. The new game has an outcome matrix identical to Pure Coordination, but its normal form is different, because Brown's strategies are no longer single actions. Brown's strategy set has four elements,

$$
\left\{
\begin{array}{l}
\text{(If Smith chose \textit{Large}, choose \textit{Large}; if Smith chose \textit{Small}, choose \textit{Large}),} \\
\text{(If Smith chose \textit{Large}, choose \textit{Large}; if Smith chose \textit{Small}, choose \textit{Small}),} \\
\text{(If Smith chose \textit{Large}, choose \textit{Small}; if Smith chose \textit{Small}, choose \textit{Large}),} \\
\text{(If Smith chose \textit{Large}, choose \textit{Small}; if Smith chose \textit{Small}, choose \textit{Small}),}
\end{array}
\right\}
$$

which we will abbreviate as

$$
\left\{
\begin{array}{l}
(\textit{Large, Large}), \\
(\textit{Large, Small}), \\
(\textit{Small, Large}), \\
(\textit{Small, Small}).
\end{array}
\right\}
$$

Follow the Leader I illustrates how adding a little complexity can make the normal form too obscure to be very useful. The normal form is shown in Table 2.2, with equilibria boldfaced and labelled X, Y, and Z.

Table 2.2 Follow the Leader I

Brown

		(*Large*, *Large*)	(*Large*, *Small*)	(*Small*, *Large*)	(*Small*, *Small*)
	Large	**2, 2 (X)**	**2, 2 (Y)**	$-1, -1$	$-1, -1$
Smith					
	Small	$-1, -1$	$1, 1$	$-1, -1$	**1, 1 (Z)**

Payoffs to: (Smith, Brown)

Equilibrium	Strategies	Outcome
X	{*Large*, (*Large*, *Large*)}	Both pick *Large*.
Y	{*Large*, (*Large*, *Small*)}	Both pick *Large*.
Z	{*Small*, (*Small*, *Small*)}	Both pick *Small*.

Consider why X, Y, and Z are Nash equilibria. In Equilibrium X, Brown will respond with *Large* regardless of what Smith does, so Smith quite happily chooses *Large*. Brown would be stupid to choose *Large* if Smith chose *Small* first, but that event never happens in equilibrium. In Equilibrium Y, Brown will choose whatever Smith chose, so Smith chooses *Large* to make the payoff 2 instead of 1. In Equilibrium Z, Smith chooses *Small* because he knows that Brown will respond with *Small* whatever he does, and Brown is willing to respond with *Small* because Smith chooses *Small* in equilibrium. Equilibria X and Z are not completely sensible, but except for a little discussion in the context of the game tree we will defer to Chapter 4 the discussion of how equilibrium can be redefined to rule them out.

The Extensive Form and Game Tree

Two other ways to describe a game are the extensive form and the game tree. First we need to define their building blocks, but as you read the definitions you may wish to refer to Figure 2.1 as an example.

A **node** *is a point in the game at which some player or Nature takes an action, or the game ends.*

A **successor** *to node X is a node that may occur later in the game if X has been reached.*

A **predecessor** *to node X is a node that must be reached before X can be reached.*

A **starting node** *is a node with no predecessors.*

An **end node** *or* **end point** *is a node with no successors.*

A **branch** *is one action in a player's action set at a particular node.*

Figure 2.1   Follow the Leader I

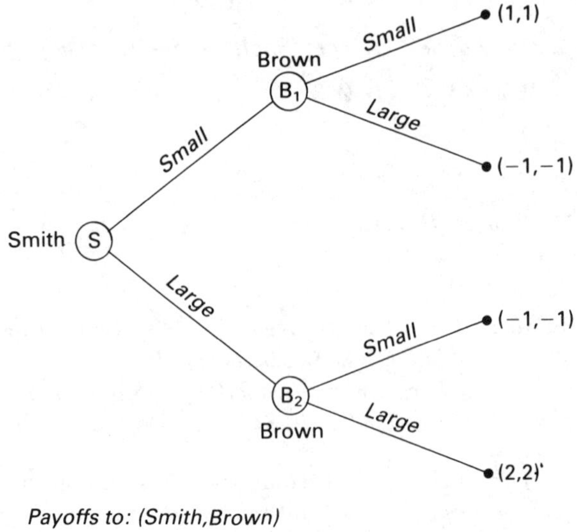

Payoffs to: (Smith,Brown)

A **path** *is a sequence of nodes and branches leading from the starting node to an end node.*

Using these concepts we can define the extensive form and the game tree.

The **extensive form** *is a description of a game consisting of*

(1) A configuration of nodes and branches running without any closed loops from a single starting node to its end nodes.

(2) An indication of which node belongs to which player.

(3) The probabilities that Nature uses to choose different branches at its nodes.

(4) The information sets into which each player's nodes are divided.

(5) The payoffs for each player at each end node.

The **game tree** *is the same as the extensive form except that (5) is replaced with (5') The outcomes at each end node.*

If the outcome is defined as the payoff combination, the extensive form is equivalent to the game tree.

The extensive form for Follow the Leader I is shown in Figure 2.1. We can see why Equilibria X and Z of Table 2.2 are unsatisfactory even though they are Nash equilibria. If the game actually reached Nodes B_1 or B_2, Brown would have dominant actions, *Large* at B_2 and *Small* at B_1, but X and Z specify other actions at those nodes. In Section 4.2 we will return to this game and show how the Nash concept can be refined to make Y the only equilibrium.

The extensive form for Pure Coordination, shown in Figure 2.2, adds dotted lines to the extensive form for Follow the Leader I. Each player makes a single decision between two actions. The moves are simultaneous, which we show by letting Smith move first, but not letting Brown know how he moved. The dotted line shows that Brown's knowledge stays the same after Smith moves. All Brown knows is that the game has reached some node within the information set defined by the dotted line; he does not know the exact node reached.

Figure 2.2   Pure Coordination

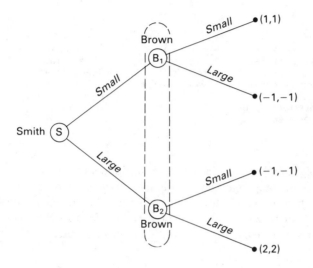

Payoffs to: (Smith,Brown)

The Time Line

The **time line**, a line showing the order of events, has become popular in recent years as a way to help describe games. Time lines are particularly useful for games with continuous strategies, exogenous arrival of information, and multiple periods, games that are frequently used in the accounting and finance literature. A typical time line is shown in Figure 2.3a, which represents a game that will be described in Section 9.5.

The time line illustrates the order of actions and events, not necessarily the passage of time. Certain events occur in an instant, others over an interval. In Figure 2.3a, Events 2 and 3 occur immediately after Event 1, but Events 4 and 5 might occur ten years later. We sometimes refer to the sequence in which decisions are made as **decision time** and the interval over which physical actions are taken as **real time.** A major difference is that players put higher value on payments received earlier in real time because of time preference, something we will explore further in Section 4.5.

A common and bad modelling habit is to use the dates of the time line only to separate events in real time. Events 1 and 2 in Figure 2.3a are not separated by real time: as soon as the entrepreneur learns the project's value, he offers to sell stock.

Figure 2.3   Underpricing: (a) a good time line; (b) a bad time line

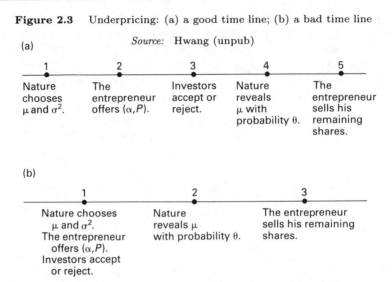

(a)   *Source:* Hwang (unpub)

1	2	3	4	5
Nature chooses μ and σ^2.	The entrepreneur offers (α, P).	Investors accept or reject.	Nature reveals μ with probability θ.	The entrepreneur sells his remaining shares.

(b)

1	2	3
Nature chooses μ and σ^2. The entrepreneur offers (α, P). Investors accept or reject.	Nature reveals μ with probability θ.	The entrepreneur sells his remaining shares.

The modeller might foolishly decide to depict his model by a picture like Figure 2.3b in which both events happen at Date 1. Figure 2.3b is badly drawn, because readers might wonder which event occurs first or whether they occur simultaneously. In more than one seminar, 20 minutes of heated and confusing debate could have been avoided by ten seconds' care to delineate the order of events.

2.3   Information Sets

A game's information structure, like the order of its moves, is often obscured in the normal form. To closely analyze information, we need technical definitions to say precisely who knows what. We will do this by using the "information set," the set of nodes a player thinks the game might have reached, as the basic unit of knowledge.

> *Player i's* **information set** ω_i *at any particular point of the game is the set of different nodes in the game tree that he knows might be the actual node, but between which he cannot distinguish by direct observation.*

As defined here, the information set for player i is a set of nodes belonging to one player (not necessarily i), but on different paths. Historically, i's information set has been defined to include only nodes at which player i moves, which is appropriate for single-person decision theory, but leaves a player's knowledge undefined for most of any game with two or more players. The broader definition above allows comparison of information across players, which under the older definition is a comparison of apples and oranges.

Figure 2.4 shows part of a game in which Smith moves at Node S_1 in 1984 and Brown moves at Nodes $B_1, B_2, B_3,$ and B_4 in 1985 or 1986. The dotted lines show that Smith knows his own move, but Brown can tell only whether Smith has chosen the moves which lead to B_1, B_2, or "other"; he cannot distinguish between B_3 and B_4. If Smith has chosen the move leading to B_3, his own information set is simply $\{B_3\}$, but Brown's information set is $\{B_3, B_4\}$.

Figure 2.4   Information sets and information partitions

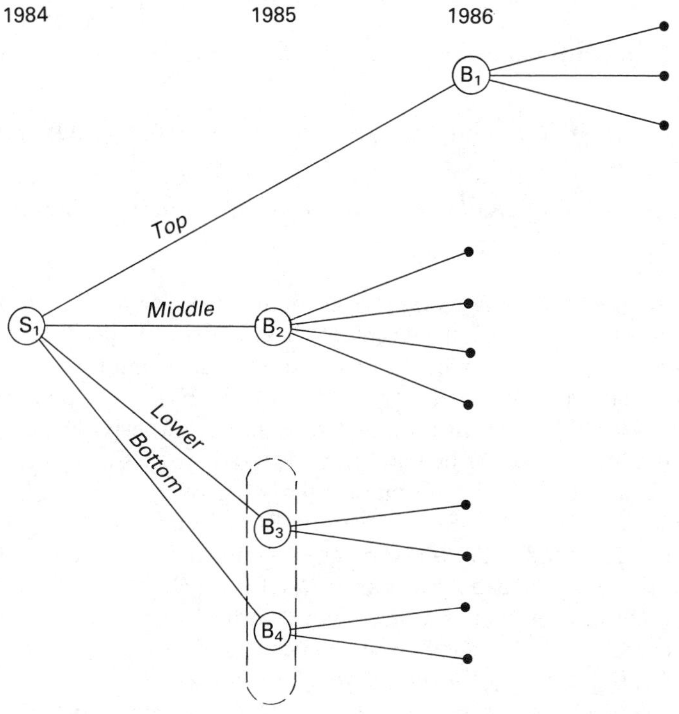

Players: Smith, Brown

One way to show information sets on a diagram is to put dotted lines around or between nodes in the same information set. The resulting diagrams can be very cluttered, so it is often more convenient to just dot the information sets of the player making the move at a node. The dotted lines in Figure 2.4 show that B_3 and B_4 are in the same information set for Brown, even though they are in different information sets for Smith.

If the nodes in one of Brown's information sets are nodes at which he moves, his action set must be the same at each node, because he knows his own action set (though his actions might differ later in the game depending on whether he advances from B_3 or B_4). Brown has the same action sets at Nodes B_3 and B_4, because if he had some different action available at B_3 he would know he was there and his information set would reduce to just $\{B_3\}$. We also require end nodes to be in different information sets for a player if they yield him different payoffs. With these exceptions, we do not include in the information structure of the game any information acquired by a player's rational deductions.

One node cannot belong to two different information sets of a single player. If

Node B_3 belonged to information sets $\{B_2, B_3\}$ and $\{B_3, B_4\}$ (unlike in Figure 2.4), then if the game reached B_3, Brown would not know whether he was at a node in $\{B_2, B_3\}$ or a node in $\{B_3, B_4\}$—which would imply that they were really the same information set.

Information sets also show the effects of unobserved moves by Nature. In Figure 2.4, if the initial move had been made by Nature instead of by Smith, Brown's information sets would be depicted the same way.

Player i's **information partition** *is a collection of his information sets such that*

(1) Each path is represented by one node in a single information set in the partition, and

(2) The predecessors of all nodes in a single information set are in one information set.

The information partition represents the different positions that the player knows he will be able to distinguish from each other at a given stage of the game, carving up the set of all possible nodes into the subsets called information sets. One of Smith's information partitions is $(\{B_1\},\{B_2\},\{B_3\},\{B_4\})$. The definition rules out information set $\{S_1\}$ being in that partition, because the path going through S_1 and B_1 would be represented by two nodes. Instead, $\{S_1\}$ is a separate information partition, all by itself. The information partition refers to a stage of the game, not chronological time. The information partition $(\{B_1\},\{B_2\},\{B_3, B_4\})$ includes nodes in both 1985 and 1986, but they are all immediate successors of Node S_1.

Brown has the information partition $(\{B_1\},\{B_2\},\{B_3, B_4\})$. There are two ways to see that his information is worse than Smith's: one of his information *sets*, $\{B_3, B_4\}$, contains *more* elements than Smith's; and one of his information *partitions*, $(\{B_1\},\{B_2\},\{B_3, B_4\})$, contains *fewer* elements.

We say that partition $(\{B_1\},\{B_2\},\{B_3, B_4\})$ is **coarser**, and $(\{B_1\},\{B_2\},\{B_3\},\{B_4\})$ is **finer**. A reduction of the number of information sets in a partition, which puts more nodes in each, is a **coarsening**; an increase is a **refinement**. The information partition $(\{B_1\},\{B_2\},\{B_3, B_4\})$ is a coarsening of $(\{B_1\},\{B_2\},\{B_3\},\{B_4\})$. The ultimate refinement is for each information set to be a **singleton**, containing one node. As in bridge, having a singleton can either help or hurt a player, as will be seen later.

Common Knowledge

We have been implicitly assuming that the players know what the game tree looks like. The assumption, in fact, is stronger: the players also know that the other players know what the game tree looks like. The term "common knowledge" is used to avoid spelling out the infinite recursion to which this leads, and the recursion is important, as will be seen in Section 5.3.

Information is **common knowledge** *if it is known to all players, each player knows that all of them know it, each of them knows that all of them know that all of them know it, and so forth ad infinitum.*

For clarity, models are set up so that information partitions are common knowledge. Every player knows how precise the other players' information is, however ignorant he himself may be of which node the game has reached. Modelled this way, the information partitions are independent of the equilibrium concept. Making the information partitions common knowledge is important for clear modelling, but it does not much restrict the kinds of games that can be modelled. This will be illustrated in Section 2.5 when the assumption will be imposed on a situation in which one player does not even know which of three games he is playing.

2.4   Perfect, Certain, Symmetric, and Complete Information

We categorize the information structure of a game in four different ways, so a particular game might have perfect, complete, certain, and symmetric information. The categories are summarized in Table 2.3.

Table 2.3 Information categories

Information category	Meaning
perfect	Each information set is a singleton.
certain	Nature does not move after any player moves.
symmetric	No player has information different from other players when he moves, or at the end nodes.
complete	Nature does not move first, or her initial move is observed by every player.

The first category divides games between those with perfect and those with imperfect information.

In a game of **perfect information** *each information set is a singleton. Otherwise the game is one of* **imperfect information**.

The strongest informational requirements are met by a game of perfect information, because in such a game each player always knows exactly where he is in the game tree. No moves are simultaneous, and all players observe Nature's moves. Pure Coordination is a game of imperfect information because of its simultaneous moves, but Follow the Leader I is a game of perfect information. Any game of incomplete or asymmetric information is also a game of imperfect information.

A game of **certainty** *has no moves by Nature after any player moves. Otherwise the game is one of* **uncertainty.**

The moves by Nature in a game of uncertainty may or may not be revealed to the players immediately. A game of certainty can be a game of perfect information if it has no simultaneous moves. The definition of "game of uncertainty" is new with this book, but I doubt it would surprise anyone. The only quirk in the definition is that it allows an initial move by Nature in a game of certainty, because in a game of incomplete information Nature moves first to select a player's "type." But most modellers do not think of this situation as uncertainty.

We have already talked about information in Pure Coordination, a game of imperfect, complete, and symmetric information with certainty. The Prisoner's Dilemma falls into the same categories. Follow the Leader I, which does not have simultaneous moves, is a game of perfect, complete, and symmetric information with certainty.

We can easily modify Follow the Leader to add uncertainty, creating the game Follow the Leader II. Imagine that if both players pick *Large* for their disks, the market yields either zero profits or very high profits, depending on the state of demand, but demand would not affect the payoffs in any other strategy combination. We can quantify this by saying that if (*Large*, *Large*) is picked, the payoffs are (10,10) with probability 0.2, and (0,0) with probability 0.8, as shown in Figure 2.5.

Figure 2.5   Follow the Leader II

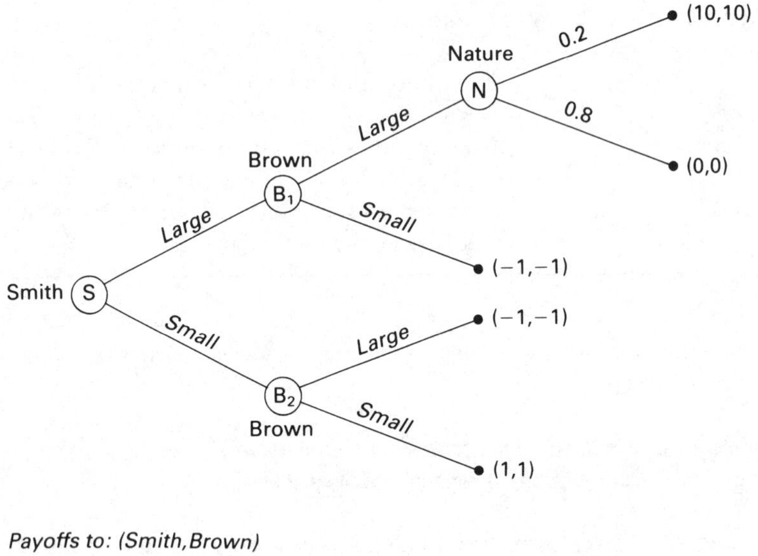

Payoffs to: (Smith, Brown)

When players face uncertainty, we need to specify how they evaluate their uncertain future payoffs. The obvious way to model their behavior is to say that the players maximize the expected values of their utilities. Players who behave in this way are said to have **von Neumann–Morgenstern utility functions,** because von Neumann & Morgenstern (1944) developed a rigorous justification for such behavior.

Maximizing their expected utilities, the players would behave exactly the same as in Follow the Leader I. Often a game of uncertainty can be transformed into a game of certainty without changing the equilibrium, by eliminating Nature's moves and changing the payoffs to their expected values based on the probabilities of Nature's moves. Here, we could eliminate Nature's move and replace the payoffs 10 and 0 with the single payoff 2 ($= 0.2[10] + 0.8[0]$). This cannot be done, however, if the actions available to a player depend on Nature's moves, or if information about Nature's move is asymmetric.

The players in Figure 2.5 might be either risk averse or risk neutral. Risk aversion is implicitly incorporated in the payoffs because they are in units of utility, not dollars. When players maximize their expected utility, they are not necessarily maximizing their expected dollars. Moreover, the players can differ in how they map money to utility. It could be that (0,0) represents ($0, $5,000), (10,10) represents ($100,000, $100,000), and (2,2), the expected utility, could here represent a non-risky ($3,000, $7,000).

In a game of **symmetric information**, *a player's information set at*

(1) any node where he chooses an action, or
(2) an end node

contains at least the same elements as the information sets of every other player. Otherwise the game is one of **asymmetric information.**

In a game of asymmetric information, the information sets of players differ in ways relevant to their behavior, or differ at the end of the game. Such games have imperfect information, since information sets which differ across players cannot be singletons. The definition of "asymmetric information" was invented for this book to try to capture the vague meaning commonly used today. The essence of asymmetric information is that some player has useful **private information**: an information partition that is different and not worse than another player's.

A game of symmetric information can have moves by Nature or simultaneous moves, but no player ever has an informational advantage. The one point at which information may differ is when the player *not* moving has superior information because he knows what his own move *was*; for example, if the two players move simultaneously. Such information does not help the informed player, since by definition it cannot affect his move.

A game also has asymmetric information if information sets differ at the end of the game. That is because we conventionally think of such games as ones in which information differs, even though no player takes an action after the end nodes. The principal–agent model that we will analyze in Chapter 6 is an example. The principal moves first, then the agent, and finally Nature. The agent observes the agent's move, but the principal does not, although he may be able to deduce it. This would be a game of symmetric information except that information continues to differ at the end nodes.

In a game of **incomplete information**, *Nature moves first and is unobserved by at least one of the players. Otherwise the game is one of* **complete information.**

A game with incomplete information also has imperfect information, because some player's information set includes more than one node. Two kinds of games have complete but imperfect information: games with simultaneous moves, and games where late in the game Nature makes moves not immediately revealed to all players.

Many games of incomplete information are games of asymmetric information, but the two concepts are not equivalent. If there is no initial move by Nature, but Smith takes a move unobserved by Brown and Smith moves again later in the game, then the game has asymmetric but complete information. The principal–agent games we will study in Chapter 6 are also examples: the agent knows how hard he worked, but his principal never learns, not even at the end nodes. A game can also have incomplete but symmetric information: let Nature, unobserved by either player, move first and choose the payoffs for $(Fink, Fink)$ in the Prisoner's Dilemma to be either $(-6, -6)$ or $(-100, -100)$.

A more interesting example of incomplete but symmetric information is in Harris & Holmstrom (1982), in which Nature assigns different abilities to workers, but when workers are young their ability is known neither to employers nor themselves. As time passes, the abilities become common knowledge, and if workers are risk averse and employers are risk neutral, the model shows that equilibrium wages are constant or rising over time.

Poker Examples of Information Classification

In the game of poker, the players make bets on who will have the best combination of cards in his hand, where a ranking of combinations has been pre-established. How would the following rules for behavior before betting be classified? (Answers are in note N2.4)

(1)   All cards are dealt face up.
(2)   All cards are dealt face down and a player cannot look even at his own cards before he bets.
(3)   All cards are dealt face down, and a player can look at his own cards.
(4)   All cards are dealt face up, but each player then scoops up his hand and secretly discards one card.
(5)   All cards are dealt face up, the players bet, and then each player receives one more card face up.
(6)   All cards are dealt face down, but then each player scoops up his cards without looking at them and holds them against his forehead so all the *other* players can see them (Indian poker).

2.5   Bayesian Games and the Harsanyi Transformation

The Harsanyi Transformation: Follow the Leader III

The term "incomplete information" is used in two quite different senses in the literature, usually without explicit definition. The definition in Section 2.4 is what economists commonly *use*, but if asked to *define* the term, they might come up with the following, older, definition.

Old definition

In a game of **complete information**, *all players know the rules of the game. Otherwise the game is one of* **incomplete information.**

The old definition is not meaningful, since the game itself is ill-defined if it does not specify exactly what the players' information sets are. Until 1967, game theorists only mentioned games of incomplete information to say that they could not be analyzed. Then Harsanyi pointed out that any game that had incomplete information under the old definition could be remodelled as a game of complete but imperfect information without changing its essentials, simply by adding an initial move in which Nature chose between different sets of rules. In the transformed game, all players know the new meta-rules, including the fact that Nature has made an initial move unobserved by them. Harsanyi's suggestion trivialized the definition of incomplete information, and people began using the term to refer to the transformed game instead. Under the old definition, a game of incomplete information was transformed to a game of complete information. Under the new definition, the original game is ill-defined, and the transformed version is a game of incomplete information.

Follow the Leader III serves to illustrate the Harsanyi transformation. Suppose that Brown does not know the game's payoffs precisely. He does have some idea of the payoffs, and we represent his beliefs by a subjective probability distribution. He places a 70 percent chance on the game being as depicted in Figure 2.6a (the same as Follow the Leader I), a 10 percent chance on Figure 2.6b, and a 20 percent on Figure 2.6c. In reality the game has a particular set of payoffs, and Smith knows what they are. This is a game of incomplete information (Brown does not know the payoffs), asymmetric information (when Smith moves, Smith knows something Brown does not), and certainty (Nature does not move after the players do).

The game cannot be analyzed in the form shown in Figure 2.6. The natural way to approach such a game is to use the Harsanyi transformation. We can remodel the game to look like Figure 2.7, in which Nature makes the first move and chooses the payoffs of either Figure 2.6a, 2.6b, or 2.6c, in accordance with Brown's subjective probabilities. Smith observes Nature's move, but Brown does not. Figure 2.7 depicts the same game as Figure 2.6, but now we can analyze it. Both Smith and Brown know the rules of the game, and the difference between them is that Smith has observed Nature's move. Whether Nature actually makes the moves with the indicated probabilities or Brown just imagines them is irrelevant, so long as Brown's initial beliefs or fantasies are common knowledge.

Often what Nature chooses at the start of a game is the strategy set, information partition, and payoff function of one of the players. We say that the player can be any of several "types," a term to which we will return in later chapters. When Nature moves, especially if she affects the strategy sets and payoffs of both players, it is often said that Nature has chosen a certain "state of the world." In Figure 2.7 Nature chooses the state of the world to be (a), (b), or (c).

A player's **type** *is the strategy set, information partition, and payoff function Nature chooses for him at the start of a game of incomplete information.*

A **state of the world** *is a move by Nature.*

Figure 2.6   Follow the Leader III: original

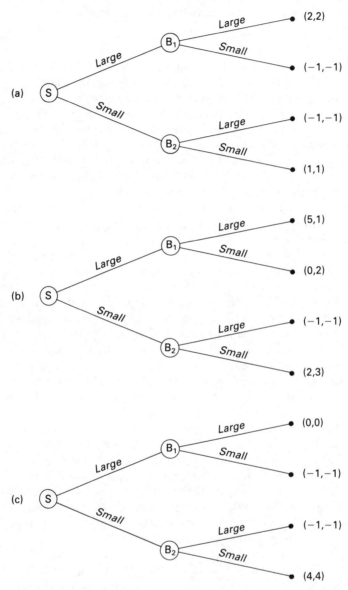

Payoffs to: (Smith, Brown)

Updating Beliefs with Bayes's Rule

When we classify a game's information structure we do not try to decide what a player can deduce from the other players' moves. Player Brown might deduce, upon seeing Smith choose *Large*, that Nature has chosen (a), but we do not draw Brown's information set in Figure 2.7 to take account of that. In drawing the game tree we

Figure 2.7   Follow the Leader III: after the Harsanyi transformation

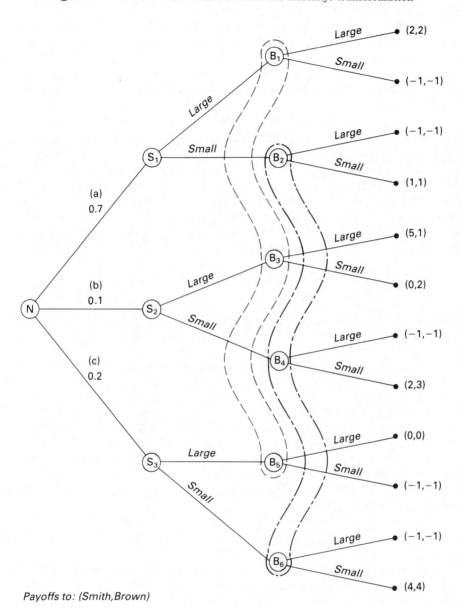

Payoffs to: (Smith,Brown)

want to illustrate only the exogenous elements of the game, uncontaminated by the equilibrium concept. But to find the equilibrium we do need to think about how beliefs change over the course of the game.

One part of the rules of the game is the collection of **prior beliefs** (or **priors**) held by the different players, beliefs that they update in the course of the game. A player holds prior beliefs concerning the types of the other players, and as he sees them take actions he updates his beliefs under the assumption that they are following equilibrium behavior.

Bayes's Rule is the rational way to update beliefs. Suppose, for example, that Smith starts with a particular prior belief, *Prob(event)*, about the probability that a certain event occurred. He then observes some data. The event might be *IBM was profitable* and the data might be *IBM's sales were high*. Seeing the data should make him update to the **posterior** belief, *Prob(event|data)*, where the symbol | denotes "conditional upon."

Bayes's Rule shows how to revise the prior belief in the light of data. It uses two pieces of information that Smith knows: the **likelihood** of seeing the data given that the event actually occurred, *Prob(data|event)*, and the likelihood of seeing the data given that the event did not occur, *Prob(data|not the event)*. If there were two alternatives to *event*, we could write *Prob(data|not the event) Prob(not the event) = Prob(data|first alternative) Prob(first alternative) + Prob(data|second alternative) Prob(second alternative)*. From these numbers, Smith can calculate *Prob(data)*, the **marginal likelihood** of seeing the data as a result of one or another of the possible events.

$$Prob(data) = Prob(data|event)Prob(event)$$
$$+ Prob(data|not\ the\ event)Prob(not\ the\ event). \quad (2.1)$$

We can relate all these probabilities using straightforward logic. The probability of both seeing the data and the event having occurred is

$$Prob(data, event) = Prob(event|data)Prob(data)$$
$$= Prob(data|event)Prob(event). \quad (2.2)$$

For example, if the event happened with probability 0.8 if the data was observed, and the marginal likelihood of the data is 0.5, then the probability of both seeing the data and the event having occurred is 0.4 (= [0.8][0.5]). If the likelihood that the data is observed if the event took place is 1, and the probability of the event is 0.4, then the probability of both the data and the event is also 0.4 (= [1][0.4]).

Since what the player is trying to calculate is *Prob(event|data)*, let us rewrite the last equation in (2.2) as

$$Prob(event|data) = \frac{Prob(data|event)Prob(event)}{Prob(data)}. \quad (2.3)$$

Our player needs to calculate his new belief—his posterior—using *Prob(data)*, which he calculates from his original knowledge using (2.1). Substituting the expression for *Prob(data)* from (2.1) into equation (2.3) we obtain our final result, a version of Bayes's Rule.

$$Prob(event|data) =$$
$$\frac{Prob(data|event)Prob(event)}{Prob(data|event)Prob(event) + Prob(data|not\ the\ event)Prob(not\ the\ event)} \quad (2.4)$$

Equation (2.5) is a verbal form of Bayes's Rule related to equation (2.3), which is useful for remembering the terminology.

$$(Posterior\ for\ event) = \frac{(Likelihood\ of\ data) \cdot (Prior\ for\ event)}{(Marginal\ likelihood\ of\ data)}. \qquad (2.5)$$

The reason for going through this derivation is to show that Bayes's Rule is not purely mechanical, but rather the only way to rationally update beliefs. The derivation is also worth understanding because Bayes's Rule is hard to memorize, but easy to rederive.

Updating Beliefs in Follow the Leader III

We will use Follow the Leader III as an example. Brown has a prior belief that the probability of event "Nature picks (a)" is 0.7 and he needs to update that belief on seeing the data "Smith picks *Small*." His prior is $Prob(a) = 0.7$, and we wish to calculate $Prob(a|Small)$.

To use Bayes's Rule from equation (2.4), we need the values of $Prob(Small|a)$, $Prob(Small|b)$, and $Prob(Small|c)$. These values depend on what Smith does in equilibrium, so we cannot calculate Brown's beliefs independently of the equilibrium. Instead, what we must do is propose an equilibrium and then use it to calculate the beliefs. Afterwards, we must check that the equilibrium strategies are indeed best responses given the beliefs they generate.

The term **Bayesian equilibrium** is used to refer to a Nash equilibrium when players update their beliefs according to Bayes's Rule. Since Bayes's Rule is the natural and standard way to handle imperfect information, the adjective "Bayesian" is really optional. But the two step procedure of checking a Nash equilibrium has now become a three-step procedure:

(1) Propose a strategy combination.
(2) See what beliefs the strategy combination generates.
(3) Check that given those beliefs and the strategies of the other players, each player is choosing a best response for himself.

A candidate for equilibrium in Follow the Leader III is for Smith to choose *Large* if the state is (a) or (b) and *Small* if it is (c); and for Brown to respond to *Large* with *Large* and to *Small* with *Small*. Let us test that this is an equilibrium, starting with the calculation of $Prob(a|Small)$. Using Bayes's Rule, equation (2.4) becomes

$$Prob(a|Small) = \frac{(0)(0.7)}{(0)(0.7) + (0)(0.1) + (1)(0.2)} = 0. \qquad (2.6)$$

Given that he believes the state is (c), Brown's best response to *Small* is *Small*, which agrees with our proposed equilibrium.

Suppose, on the other hand, that Brown observes *Large*. He can rule out (c), but he does not know whether the state is (a) or (b). Bayes's Rule is now truly useful. It tells us that the posterior probability of (a) is

$$Prob(a|Large) = \frac{(1)(0.7)}{(1)(0.7) + (1)(0.1) + (0)(0.2)} = 0.875, \qquad (2.7)$$

and the posterior probability of (b) is $1 - 0.875 = 0.125$, which could also be calculated from Bayes's Rule.

$$Prob(b|Large) = \frac{(1)(0.1)}{(1)(0.7) + (1)(0.1) + (0)(0.2)} = 0.125. \qquad (2.8)$$

Given that Brown believes that the state is (a) with probability 0.875 and (b) with probability 0.125, his best response is *Large*, even though he knows that if the state were actually (b) the better response would be *Small*. Given that he observes *Large*, Brown's expected payoff from *Small* is -0.625 ($= 0.875[-1] + 0.125[2]$), but from *Large* it is 1.875 ($= 0.875[2] + 0.125[1]$).

The calculations are relatively simple because Smith uses a nonrandom strategy in equilibrium, so, for instance, $Prob(Small|a) = 0$ in equation (2.6). Consider what happens if Smith uses a random strategy of picking *Large* with probability 0.2 in state (a), 0.9 in state (b), and 0 in state (c) (we will analyze such "mixed" strategies in Section 3.2). The equivalent of equation (2.7) is

$$Prob(a|Large) = \frac{(0.2)(0.7)}{(0.2)(0.7) + (0.9)(0.1) + (0)(0.2)} = 0.61 (rounded). \qquad (2.9)$$

If he sees *Large*, Brown's best guess is still that Nature chose (a), but the probability has fallen to 0.61 from his prior of 0.7.

2.6   An Example: the Png Settlement Game

We will use the Png (1983) model of out-of-court settlement as an example of a game with a fairly complicated extensive form.[1] The plaintiff alleges that the defendant was negligent in providing safety equipment at a chemical plant, a charge which is true with probability q. The plaintiff files suit, but the case is not decided immediately. In the meantime, the defendant and the plaintiff can settle out of court.

What are the moves in this game? It is really made up of two games: the one in which the defendant is liable for damages, and the one in which he is blameless. We therefore start the game tree with a move by Nature, who makes the defendant either liable or blameless. At the next node, the plaintiff takes an action: *Sue* or *No Action*. If he decides on *No Action* the game ends with zero payoffs for both players. If he decides to *Sue*, we go to the next node. The defendant then decides whether to *Hold Out* or *Offer* to settle. If the defendant chooses *Offer*, then the plaintiff can *Settle* or *Refuse*; if the defendant chooses to *Hold Out*, then the plaintiff can *Try* the case or *Drop* it.

The dotted lines in Figure 2.8 enclose the information sets of the plaintiff. His information partition is rather coarse—he must make decisions without knowing exactly which node the game has reached. Maybe the defendant is liable, but

---

[1]   "Png," by the way, is pronounced the same way it is spelled.

Figure 2.8 The Settlement Game

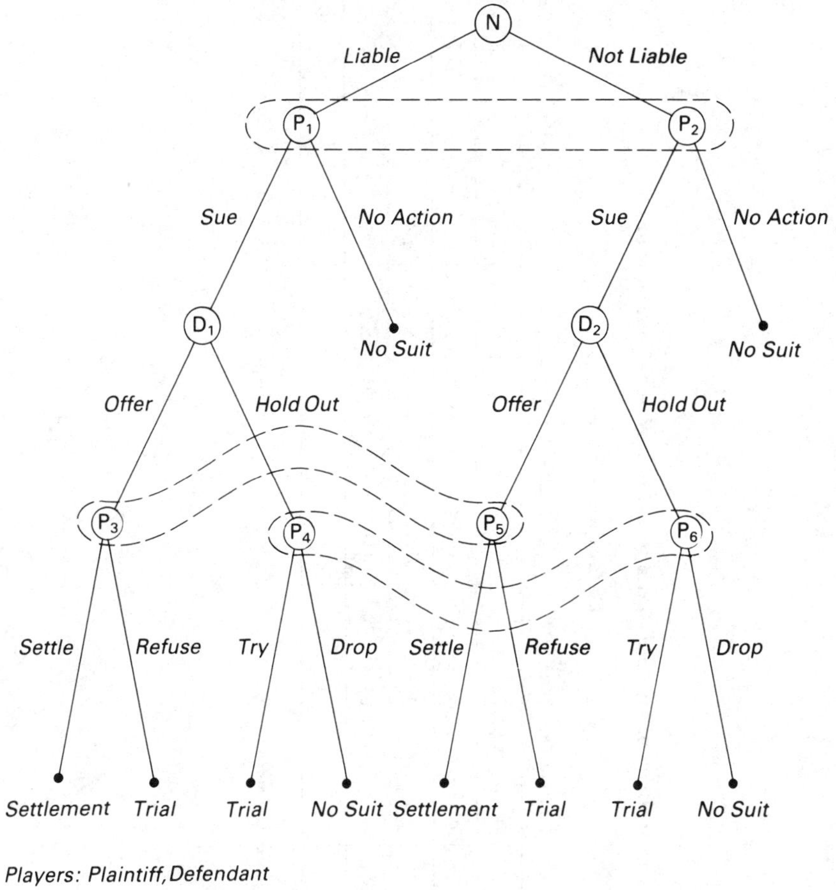

Players: Plaintiff, Defendant

Note: Dotted lines indicate the plaintiff's information sets.

maybe not. The plaintiff has three actions open to him, and there are three dotted ovals showing that his information remains coarse.

We assume that the settlement amount, S, and the amounts spent on legal fees are exogenous. Except in the infinitely long games without end nodes that will appear in Chapter 4, an extensive form should incorporate all costs and benefits into the payoffs at the end nodes, even if costs are incurred along the way. If the court required a $100 filing fee, we would subtract it from all the plaintiff's payoffs except when he chose *No Action*. Such consolidation makes it easier to analyze the game and would not affect the equilibrium strategies unless payments along the way revealed information, in which case what matters is the information, not the fact that payoffs change.

Table 2.4 The normal form of the Settlement Game: general parameters

Defendant's Strategy

Plaintiff's Strategy	(Offer, Offer)	(Hold Out, Hold Out)	(Offer, Hold Out)	(Hold Out, Offer)
(No Action)	0 $(0,0)$	0 $(0,0)$	0 $(0,0)$	0 $(0,0)$
(Sue; Settle, Try)	S $(-S,-S)$	$qW-\pi$ $(-W-\delta,-\delta)$	$qS-(1-q)\pi$ $(-S,-\delta)$	$q(W-\pi)+(1-q)S$ $(-W-\delta,-S)$
(Sue; Refuse, Try)	$qW-\pi$ $(-W-\delta,-\delta)$	$qW-\pi$ $(-W-\delta,-\delta)$	$qW-\pi$ $(-W-\delta,-\delta)$	$qW-\pi$ $(-W-\delta,-\delta)$
(Sue; Refuse, Drop)	$qW-\pi$ $(-W-\delta,-\delta)$	0 $(0,0)$	$q(W-\pi)$ $(-W-\delta,0)$	$-(1-q)\pi$ $(0,-\delta)$
(Sue; Settle, Drop)	S $(-S,-S)$	0 $(0,0)$	qS $(-S,0)$	$(1-q)S$ $(0,-S)$

Payoffs to: Plaintiff (Liable Defendant, Blameless Defendant)

We assume that if the case reaches the court, justice is done. In addition to his legal fees δ, the defendant pays the damages W only if he is liable. We also assume that players are risk neutral, so they only care about the expected dollars they will receive, not the variance. Without this assumption we would have to translate the dollar payoffs into utility, but the game tree would be unaffected.

This is a game of certain, asymmetric, imperfect, and incomplete information. We have assumed that the defendant knows whether he is liable, but we could modify the game by assuming that he has no better idea than the plaintiff of whether the evidence is sufficient to prove him so. The game would become one of symmetric information and we could reasonably simplify the extensive form by eliminating the initial move by Nature and setting the payoffs equal to the expected values. We cannot perform this simplification in the original game, because the fact that the defendant, and only the defendant, knows whether he is liable strongly affects the behavior of both players.

While the extensive form is the natural way to depict the game, we can also bypass it and associate strategies with payoffs using the normal form. This is a little different from the 2-by-2 normal forms, because the defendant gets a different payoff depending on whether he is really liable or not. We need three entries for each pair of strategies: for the Plaintiff, the Liable Defendant, and the Blameless Defendant. In complicated games like this one, the distinction between action and strategy is important. A typical action is *Settle*. A typical strategy is (*Sue*; *Settle*[if offered], *Try*[if the defendant holds out]). The strategies are shown in Table 2.4.

Using dominance we can rule out one of the plaintiff's strategies immediately— *No Action*—which is dominated by (*Sue, Settle, Drop*).

Whether a strategy combination is a Nash equilibrium depends on the parameters of the model—S, W, π, δ and q, which are the settlement amount, the damages, the court costs for the plaintiff and defendant, and the probability the defendant is liable. Depending on the parameter values, three outcomes are possible: settlement (if the settlement amount is low), trial (if expected damages are high and the plaintiff's court costs are low), and the plaintiff dropping the action (if expected damages minus court costs are negative). For some parameter values there are multiple Nash equilibria.

Consider the parameter values $S = 0.15, \delta = 0.2, W = 1, q = 0.13$, and $\pi = 0.1$. Table 2.5 (see p.64) shows the normal form in this case. The two Nash equilibria marked in boldface are both weak, and outcomes of either trial or settlement are possible in equilibrium. We will only trace through the outcome with settlement.

Consider the strategy combination {(*Sue, Settle, Try*), (*Offer, Offer*)}. The plaintiff sues, the defendant offers to settle (whether liable or not), and the plaintiff agrees to settle. Both players know that if the defendant did not offer to settle, the plaintiff would go to court and try the case. Such **out-of-equilibrium** behaviour is specified by the equilibrium, because the threat of trial is what induces the defendant to offer to settle, even though trials never occur in equilibrium. This strategy combination is a Nash equilibrium because given that the plaintiff chooses (*Sue, Settle, Try*) the defendant can do no better than (*Offer, Offer*), settling for a payoff of -0.15 whether he is liable or not; and given that the defendant chooses (*Offer, Offer*), the plaintiff can do no better than the payoff of 0.15 from (*Sue, Settle, Try*).

Table 2.5 The normal form of the Settlement Game: particular parameters

Defendant's Strategy

	(Offer, Offer)		(Hold Out, Hold Out)		(Offer, Hold Out)		(Hold Out, Offer)	
(No Action)	0	(0,0)	0	(0,0)	0	(0,0)	0	(0,0)
(Sue; Settle, Try)	**0.15**	**(−0.15, −0.15)**	0.03	(−1.2, −0.2)	−0.0675	(−0.15, −0.2)	0.2475	(−1.2, −0.15)
(Sue; Refuse, Try)	0.03	(−1.2, −0.2)	**0.03**	**(−1.2, −0.2)**	0.03	(−1.2, −0.2)	0.03	(−1.2, −0.2)
(Sue; Refuse, Drop)	0.03	(−1.2, −0.2)	0	(0,0)	0.117	(−1.2, 0)	−0.087	(0, −0.2)
(Sue; Settle, Drop)	0.15	(−0.15, −0.15)	0	(0,0)	0.0195	(−0.15, 0)	0.1305	(0, −0.15)

Plaintiff's Strategy

Payoffs to: Plaintiff (Liable Defendant, Blameless Defendant)

Recommended Reading

Png, Ivan (1983) "Strategic Behaviour in Suit, Settlement, and Trial" *Bell Journal of Economics.* Autumn 1983. 14, 2: 539–50.

Problem 2

Joint Ventures

Software Inc. and Hardware Inc. are in a joint venture together. Each can exert either high or low effort, which is equivalent to costs of 20 and 0. Hardware moves first, but Software cannot observe his effort. Profits are split equally at the end, and the two firms are risk neutral. If both firms exert low effort, total profits are 100. If the parts are defective, the total profit is 100; otherwise, if both exert high effort, profit is 200, but if only one player does, profit is 100 with probability 0.9 and 200 with probability 0.1. Before they start, both players believe that the probability of defective parts is 0.7. Hardware discovers the truth about the parts by observation before he chooses effort, but Software does not.

(1) Draw the extensive form and put dotted lines around the information sets of Software at any nodes where he moves.
(2) What is the Nash equilibrium?
(3) What is Software's belief in equilibrium as to the probability that Hardware chooses low effort?
(4) If Software sees that profit is 100, what probability does he assign to defective parts if he himself exerted high effort and he believes that Hardware chose low effort?

Notes

N2.2   The Normal and Extensive Forms of a Game

- The term "outcome matrix" is used in Shubik (1982, p. 70), but never formally defined there.
- In the terminology of Shubik (1982), what I call the "normal form" is the "strategic form," a more meaningful but less widespread term.
- The term "node" is sometimes defined to include only points at which a player or Nature makes a decision, which excludes the end points.

N2.3 Information Sets

- If you wish to depict a situation in which a player does not know whether the game has reached Node A_1 or A_2 and he has different action sets at the two nodes, restructure the game. If you wish to say that he has action set (X,Y,Z) at A_1 and (X,Y) at A_2, first add action Z to the information set at A_2. Then specify that at A_2, action Z simply leads to a new node, A_3, at which the choice is between X and Y.

- For rigorous but nonintuitive definitions of common knowledge, see Aumann (1976), Milgrom (1981a), and Tan & Werlang (unpub). Following Milgrom, let (Ω, p) be a probability space, let P and Q be partitions of Ω representing the information of two agents, and let R be the finest common coarsening of P and Q. Let ω be an event (an item of information) and $R(\omega)$ be that element of R which contains ω.

  An event A is **common knowledge** *at ω if $R(\omega) \subset A$.*

N2.4   Perfect, Certain, Symmetric, and Complete Information

- **Poker Classifications**. (1) Perfect, certain. (2) Incomplete, symmetric, certain. (3) Incomplete, asymmetric, certain. (4) Complete, asymmetric, certain. (5) Perfect, uncertain. (6) Incomplete, asymmetric, certain.
- For an explanation of von Neumann–Morgenstern utility, see p. 155 of Varian (1984). Machina (1982) is a well-written discussion of other ways of modelling choice between uncertain outcomes.
- Mixed strategies (Section 3.2) are allowed in a game of perfect information, because that is an aspect of the game's equilibrium, not of its exogenous structure.
- Although the word "perfect" appears in both "perfect information" (Section 2.4) and "perfect equilibrium" (Section 4.2), the concepts are unrelated.
- An unobserved move by Nature in a game of symmetric information can be represented in any of three ways: (1) As the last move in the game; (2) As the first move in the game; or (3) By replacing the payoffs with the expected payoffs and not using any explicit moves by Nature.

N2.5   Bayesian Games and the Harsanyi Transformation

- Mertens & Zamir (1985) probes the mathematical foundations of the Harsanyi transformation. The transformation requires the extensive form to be common knowledge, which raises subtle questions of recursion.
- A player always has some idea of what the payoffs are, so we can always assign him a subjective probability for each possible payoff. What would happen if he had no idea? Such a question is meaningless, because people always have some notion, and when they say they do not, they generally mean that their prior probabilities are low but positive for a great many possibilities. You, for instance, probably have as little idea as I do of how many cups of coffee I have consumed in my lifetime, but you would admit it to be a nonnegative number less than 3,000,000, and you could make a much more precise guess than that. On the topic of subjective probability, a classic reference is Savage (1954).
- If two players have common priors and their information partitions are finite, but they each have private information, then iterated communication between them leads to the adoption of a common posterior. This posterior is not always the posterior they would reach if they directly pooled their information, but it is almost always that posterior (Geanakoplos & Polemarchakis [1982]).
- In some contexts, "parameter" is a better word to use than "event" in applying Bayes's Rule. The various terms are listed in Table 2.6.

Table 2.6 Bayesian terminology

Name	Meaning
likelihood	$Prob(data|parameter)$
marginal likelihood	$Prob(data)$
conditional likelihood	$Prob(data\ X|data\ Y,\ parameter)$
prior	$Prob(parameter)$
posterior	$Prob(parameter|data)$

3 Mixed and Continuous Strategies

3.1 Introduction

So far, the games we have looked at have been simple in at least one respect: the number of moves in the action set has been finite. In this chapter we allow a continuum of moves, such as when a player chooses a price between 10 and 20 or a purchase probability between zero and one. Such games are important not only because continua are frequently more convenient to use than large finite numbers, but also because sometimes equilibria do not exist unless the strategy space is continuous. We begin in Section 3.2 with mixed strategies, using an example called the Welfare Game in which players choose to randomize their moves. Section 3.3 looks at two other games with mixed strategies, Chicken and the War of Attrition. Section 3.4 uses the Cournot Game to introduce a continuum of pure strategies and shows how the Cournot and Stackelberg equilibria are calculated.

3.2 Mixed Strategies: the Welfare Game

We invoked the concept of Nash equilibrium to provide predictions of outcomes when no dominant strategy equilibrium existed, but some games lack even a Nash equilibrium. It is often useful and realistic to expand the strategy space to include random strategies, in which case a Nash equilibrium exists in most games.

A **pure strategy** *maps each of a player's possible information sets to one action.* $s_i : \omega_i \rightarrow a_i$.

A **mixed strategy** *maps each of a player's possible information sets to a probability distribution over actions.*

$$s_i : \omega_i \rightarrow m(a_i), \quad \text{where } m \geq 0, \quad \int_{A_i} m(a_i) da_i = 1.$$

A **completely mixed** *strategy puts positive probability on every action, so* $m > 0$.

The version of a game expanded to allow mixed strategies is called the **mixed extension** *of the game.*

The pure strategy is a rule telling the player what action to choose, while the mixed strategy is a rule telling him what dice to throw to choose an action. If a player pursues a mixed strategy, he might choose any of several different actions in a given situation, an unpredictability that can be helpful to him. Do not be taken aback by the seeming artificiality of mixed strategies; they occur frequently in the real world. In American football games, for example, the offensive team has to decide whether to pass or to run. Passing generally gains more yards, but what is important is to choose the action not expected by the other team. Teams decide to run part of the time and pass part of the time in a way that seems random to observers, but rational to game theorists.

The Welfare Game

The Welfare Game models a government that wishes to aid a pauper if he searches for work but not otherwise, and a pauper who searches for work only if he cannot depend on government aid, and who may not succeed in finding a job even if he tries. Table 3.1 shows payoffs which represent this situation. Neither player has a dominant strategy, and with a little thought we can see that no Nash equilibrium exists in pure strategies either.

Table 3.1 The Welfare Game

Pauper

		Try to Work		*Be Idle*
	Aid	$3, 2$	$\rightarrow$	$-1, 3$
Government		$\uparrow$		$\downarrow$
	No Aid	$-1, 1$	$\leftarrow$	$0, 0$

Payoffs to: (Government, Pauper)

Each strategy combination must be examined in turn to check for Nash equilibria.

(1)   The strategy combination (*Aid, Try to Work*) is not a Nash equilibrium, because the pauper prefers to respond with *Be Idle* if the government picks *Aid*.
(2)   (*Aid, Be Idle*) is not Nash, because the government prefers *No Aid*.
(3)   (*No Aid, Be Idle*) is not Nash, because the pauper prefers *Try to Work*.
(4)   (*No Aid, Try to Work*) is not Nash, because the government prefers *Aid*, which brings us back to the first strategy combination.

The Welfare Game does have a mixed strategy Nash equilibrium, which we can calculate. Using von Neumann–Morgenstern utility (see Section 2.4), the players' payoffs are the expected values of the payments from Table 3.1. If the government plays *Aid* with probability P_a and the pauper plays *Try to Work* with probability P_w, the government's expected payoff is

$$
\begin{aligned}
E\pi_{Govt} &= P_a[3P_w + (-1)(1 - P_w)] + [1 - P_a][-1P_w + 0(1 - P_w)], \\
&= P_a[3P_w - 1 + P_w] - P_w + P_a P_w, \\
&= P_a[5P_w - 1] - P_w.
\end{aligned}
\tag{3.1}
$$

If only pure strategies are allowed, P_a equals zero or one, but in the mixed extension of the game, the government's action of P_a lies on the continuum from zero to one, the pure strategies being the extreme values. Following the usual procedure for solving a maximization problem, we differentiate the payoff function with respect to the choice variable to obtain the first order condition,

$$
0 = \frac{dE\pi_{Govt}}{dP_a} = 5P_w - 1,
\tag{3.2}
$$
$$
\Rightarrow P_w = 0.2.
$$

In the mixed strategy equilibrium, the pauper selects *Try to Work* 20 percent of the time. The way we obtained the number might seem strange: to obtain the pauper's strategy, we differentiated the government's payoff. Understanding why requires several steps.

(1) I assert that an optimal mixed strategy exists for the government.
(2) If the pauper selects *Try to Work* more than 20 percent of the time, then the government always selects *Aid*. If the pauper selects *Try to Work* less than 20 percent of the time, the government never selects *Aid*.
(3) If a mixed strategy is to be optimal for the government, the pauper must therefore select *Try to Work* with probability exactly 20 percent.

To obtain the probability of the government choosing *Aid*, we must turn to the pauper's payoff function, which is

$$
\begin{aligned}
E\pi_{Pauper} &= P_a(2P_w + 3[1 - P_w]) + (1 - P_a)(1P_w + 0[1 - P_w]), \\
&= 2P_a P_w + 3P_a - 3P_a P_w + P_w - P_a P_w, \\
&= -P_w(2P_a - 1) + 3P_a.
\end{aligned}
\tag{3.3}
$$

The first order condition is

$$
\frac{dE\pi_{Pauper}}{dP_w} = -(2P_a - 1) = 0,
\tag{3.4}
$$
$$
\Rightarrow P_a = 1/2.
$$

If the pauper selects *Try to Work* with probability 0.2, the government is indifferent among selecting *Aid* with probability 100 percent, 0 percent, or anything in between. If the strategies are to form a Nash equilibrium, however, the government must choose $P_a = 0.5$. In the mixed strategy Nash equilibrium, the government selects *Aid* with probability 0.5 and the pauper selects *Try to Work* with probability 0.2. The equilibrium outcome could be any of the four entries in the outcome matrix. The entries having the highest probability of occurence are *(No Aid, Be Idle)* and *(Aid, Be Idle)*, each with probability 0.4 ($= 0.5[1 - 0.2]$).

Interpreting Mixed Strategies

Mixed strategy equilibria are not so intuitive as pure strategy equilibria, and many modellers prefer to restrict themselves to pure strategy equilibria in games which have both. One objection to mixed strategies is that people in the real world do not take random actions. That is not a compelling objection, because, as in the football example, all that a model with mixed strategies requires to be a good description is that actions appear random to observers, even if the player himself has always been sure what action he would take. Moreover, explicitly random actions are not uncommon. When the Internal Revenue Service randomly selects which tax returns to audit, or the telephone company randomly monitors its operators' conversations to check their politeness, mixed strategies are being used.

A more troubling objection is that a player who selects a mixed strategy is always indifferent between two pure strategies. In the Welfare Game, the pauper is indifferent between his two pure strategies and a whole continuum of mixed strategies, given the government's mixed strategy. If the pauper were to decide not to follow the particular mixed strategy $P_w = 0.20$, the equilibrium would collapse because the government would change its strategy in response. Even a small deviation in the probability selected by the pauper, a deviation that does not change his payoff if the government does not respond, destroys the equilibrium completely because the government does respond. A mixed strategy Nash equilibrium is weak in the same sense as the (*North, North*) equilibrium in the Battle of the Bismarck Sea (Section 1.3): to maintain the equilibrium, a player who is indifferent between strategies must pick a particular strategy from among them.

One way to reinterpret the Welfare Game is to imagine that instead of one pauper there are many, with identical tastes and payoff functions, all of whom must be treated alike by the government. In the mixed strategy equilibrium, each of the paupers chooses *Try to Work* with probability 0.2, just as in the one-pauper game. But the many-pauper game has a pure strategy equilibrium: 20 percent of the paupers choose the pure strategy *Try to Work* and 80 percent choose the pure strategy *Be Idle*. The problem persists of how an individual pauper, indifferent between the pure strategies, chooses one or the other, but it is easy to imagine that individual characteristics outside the model could determine which actions are chosen by which paupers.

The number of players needed so that mixed strategies can be interpreted as pure strategies in this way depends on the equilibrium probability P_w, since we cannot speak of a fraction of a player. The number of paupers must be a multiple of five in the Welfare Game to use this interpretation, since the equilibrium mixing probability is a multiple of one–fifth. For the interpretation to apply no matter how we vary the parameters of a model we would need a *continuum* of players.

Another interpretation of mixed strategies, which works even in the single-pauper game, is that the pauper is drawn from a population of paupers, and the government does not know his characteristics. The government only knows that there are two types of paupers, in the proportions (0.2,0.8): those who pick *Try to Work* if the government picks $P_a = 0.5$, and those who pick *Be Idle*. A pauper drawn randomly from the population might be of either type.

3.3   Chicken, the War of Attrition, and Correlated Strategies

Chicken

The next game illustrates why we might decide that a mixed strategy equilibrium is best even if pure strategy equilibria also exist. In the game Chicken, the players are two Malibu teenagers, Smith and Brown. Smith drives a hot rod south down the middle of Route 1, and Brown drives north. When they approach each other, each has the choice to *Stay* in the middle or *Swerve*. If a player is the only one to *Swerve*, he loses face, but if neither player picks *Swerve* they are both killed, which has an even lower payoff. If a player is the only one to *Stay*, he is covered with glory, and if both *Swerve* they are both embarassed. Table 3.2 assigns numbers to these four outcomes.

Table 3.2 Chicken

Brown

		Stay	*Swerve*
	Stay	$-3, -3$	**2, 0**
Smith			
	Swerve	**0, 2**	1, 1

Payoffs to: (Smith, Brown)

Chicken has two pure strategy Nash equilibria, (*Swerve, Stay*) and (*Stay, Swerve*), but they have the defect of asymmetry. How do the players know which equilibrium is the one that will be played out? Even if they talked before the game started, it is not clear how they could arrive at an asymmetric result. We encountered the same dilemma in choosing an equilibrium for the Battle of the Sexes in Section 1.4. The mixed strategy equilibrium has the advantage of symmetry, which makes it a focal point of sorts and it does not require any differences between the players.

To calculate the mixing probability for Chicken, we use a method based on the logic in Section 3.2, but we can avoid the calculus of maximization. Chicken is simpler than the Welfare Game because it is symmetric, and we can find an equilibrium in which each player chooses the same mixing probability. In the mixed strategy equilibrium, Smith must be indifferent between *Swerve* and *Stay*. That requires that Brown's probability of *Stay*, which we denote by θ, be such that

$$\pi(Swerve) = (\theta) \cdot (0) + (1 - \theta) \cdot (1)$$
$$= (\theta) \cdot (-3) + (1 - \theta) \cdot (2) = \pi(Stay). \tag{3.5}$$

From equation (3.5) we can conclude that $1 - \theta = 2 - 5\theta$, so $\theta = 0.25$. To look at the question of most interest to their mothers, the two teenagers will survive with probability $1 - (\theta \cdot \theta) = 0.9375$.

The War of Attrition

The War of Attrition is a game something like Chicken stretched out over time, in which both players start with *Stay*, and the game ends when the first one picks *Swerve*. Until the game ends, both earn a negative amount per period, and when one exits, he earns zero and the other player earns some reward.

We will look at a war of attrition in discrete time. Two players, Smith and Brown, own firms in an industry that is a natural monopoly, so there is only enough demand for one of them to operate profitably. The actions are to *Exit* or to *Stay*. In each period that both *Stay*, they each earn -1. If a firm exits, its losses cease and the remaining firm obtains the value of the market's monopoly profit, which we set equal to 3. We will set the discount rate (which is akin to an interest rate and will be discussed in detail in Section 4.5) equal to $r > 0$, although that is inessential to the model, even if the possible length of the game is infinite.

The War of Attrition has a continuum of Nash equilibria. One simple equilibrium is for Smith to choose (*Stay* regardless of what Brown does) and for Brown to choose (*Exit* immediately), which are best responses to each other. But we will solve for a symmetric equilibrium in which each player chooses the same mixed strategy: a constant probability θ that he picks *Exit* given that the other player has not yet exited.

We can calculate θ as follows, adopting the perspective of Smith. Denote the expected discounted value of Smith's payoffs by V_{stay} if he stays and V_{exit} if he exits immediately. These two pure strategy payoffs must be equal in a mixed strategy equilibrium. If Smith exits, he obtains $V_{exit} = 0$. If Smith stays in, his payoff depends on what Brown does. If Brown stays in too, which has probability $(1 - \theta)$, Smith gets -1 currently and his expected value for the following period, which is discounted using r, is unchanged. If Brown exits immediately, which has probability θ, then Smith receives a payment of 3. In symbols,

$$V_{stay} = \theta \cdot (3) + (1 - \theta) \left(-1 + \left[\frac{V_{stay}}{1 + r} \right] \right), \tag{3.6}$$

which after solving for V_{stay} becomes

$$V_{stay} = \left(\frac{1 + r}{r + \theta} \right) (4\theta - 1). \tag{3.7}$$

Once we equate V_{stay} to V_{exit}, which equals zero, equation (3.7) tells us that $\theta = 0.25$ in equilibrium. In fact, it is easy to set $V_{Stay} = 0$ and see that $\theta = 0.25$ from equation (3.6) in this simple model, but usually the modeller must generate an equation something like (3.7) because when the equivalent of V_{stay} is not zero, equation (3.6) is harder to solve.

Returning from arithmetic to ideas, why does Smith *Exit* immediately with positive probability, given that if he waits long enough, Brown will exit first? The reason is that Brown might stay in for a long time and both players would earn -1 each period until Brown exited. The equilibrium mixing probability is calculated so that both of them are likely to stay in for a long enough time that their losses soak up the gain from being the survivor.

Correlated Strategies

One example of a war of attrition is setting up a market for a new security, which may be a natural monopoly for reasons explained in Section 8.5. Certain stock exchanges have avoided the destructive symmetric equilibrium by using lotteries to determine which of them would trade newly listed stock options, under a system similar to the football draft ("Big Board will Begin Trading of Options on 4 Stocks it Lists," *Wall Street Journal*, 4 October 1985, p. 15). Rather than waste resources fighting, they use the lottery as a coordinating device, even though it might not be a binding agreement.

Aumann (1974) has pointed out that it is often important whether players can use the same randomizing device for their mixed strategies. If they can, we refer to the resulting strategies as **correlated strategies**. Consider the game Chicken. The only mixed strategy equilibrium is the symmetric one in which each player chooses *Stay* with probability 0.25 and the expected payoff is 0.75. A correlated equilibrium would be for the two players to flip a coin, for Smith to choose *Stay* if it comes up heads, for Brown to choose *Stay* if it comes up tails, and for the player who loses the toss to choose the ignominious *Swerve*. Each player's strategy is a best response to the other's, the probability of each choosing *Stay* is 0.5, and the expected payoff for each is 1.0.

Usually the randomizing device is not modelled explicitly when a model refers to correlated equilibrium. If it is, uncertainty over variables that do not affect preferences, endowments, or production is called **extrinsic uncertainty.** Extrinsic uncertainty is the driving force behind **sunspot models**, so-called because the random appearance of sunspots might cause macroeconomic changes via correlated equilibria (Maskin & Tirole [1987]) or bets made between players (Cass & Shell [1983]).

One way to model correlated strategy outcomes explicitly is to specify a move by Nature that gives each player with equal probability the ability to commit first to an action such as *Stay*. This is often realistic, because it amounts to a zero probability of both players entering the industry at exactly the same time, but no one knowing in advance who will be the lucky starter. Neither firm has an *a priori* advantage, but the outcome is efficient.

The population interpretation of mixed strategies cannot be used for correlated strategies. In ordinary mixed strategies, the mixing probabilities are statistically independent, whereas in correlated strategies they are not. In Chicken, the usual mixed strategy can be interpreted as populations of Smiths and Browns, each population consisting of a certain proportion of pure swervers and pure stayers. The correlated equilibrium has no such interpretation.

Another coordinating device, useful in games that like the Battle of the Sexes have a coordination problem, is **cheap talk** (Crawford & Sobel [1982], Farrell [1987]). Cheap talk refers to costless communication before the game proper begins. In Pure Coordination, cheap talk instantly allows the players to make the desirable outcome a focal point. In Chicken, cheap talk is useless, because it is dominant for each player to announce that he will choose *Stay*. But in Battle of the Sexes, coordination and conflict are combined. Without communication, the only symmetric equilibrium is in mixed strategies. If both players know that making inconsistent announcements will lead to the wasteful mixed strategy outcome, then they are willing to mix announcing whether they will go to the ballet or the prize fight. With many periods of announcements before the final decision, their chances of coming to an agreement are high. Thus communication can help reduce inefficiency even if the two players are in conflict.

3.4   Continuous Strategies: the Cournot Game

In the games we have looked at so far, the strategies have been discrete: *Aid* or *No Aid*, *Fink* or *Not Fink*. Quite often when strategies are discrete no pure strategy Nash equilibrium exists. The only sort of compromise possible in the Welfare Game, for instance, was to choose *Aid* sometimes and *No Aid* sometimes, a mixed strategy. If *A Little Aid* were a possible action, maybe there would be a pure strategy equilibrium. The game we discuss next, the Cournot Game, has a continuous strategy space even without mixing, and it does have a pure strategy equilibrium.

If a game has a continuous strategy set, it is not always useful to depict the normal form and outcome matrix as tables, or the extensive form as a tree. The tables would require a continuum of rows and columns, the tree a continuum of branches. A new format for game descriptions of the players, actions, and payoffs will be used for the rest of the book. The new format is similar to the way OPEC Model II was presented in Section 1.1. Our example will be the Cournot model, in which two firms choose output levels in competition with each other.

The Cournot Game

Players

Firms Allied and Brydox.

Information

Imperfect, symmetric, complete, and certain.

Actions and Events

Allied and Brydox simultaneously choose quantities q_a and q_b.

Payoffs

Production costs are zero. Demand is a function of the total quantity sold, $q = q_a + q_b$.

$$P(q_a + q_b) = 120 - q_a - q_b. \tag{3.8}$$

Payoffs are profits, which are given by a firm's price times its quantity, i.e.,

$$\pi_{Allied} = 120q_a - q_a^2 - q_a q_b;$$
$$\pi_{Brydox} = 120q_b - q_a q_b - q_b^2. \tag{3.9}$$

The format first assigns the game a title, after which it lists the players, the information classification, the order of actions and events (together with who observes them), and the payoff functions. Listing the players and the information classification is redundant, strictly speaking, since they can be deduced from the Actions and Events section, but it is useful for letting the reader know what kind of model to expect. The format includes very little explanation; that is postponed, lest it obscure the raw description. This exact format is not standard in the literature, but every good article begins its technical section by specifying the same information, if perhaps in a less structured way, and the novice is strongly advised to use all the structure he can.

Cournot (1838) noted that this game has a unique equilibrium when demand curves are linear. If the game were cooperative (see Section 1.2), firms would end up producing somewhere on the 45° line in Figure 3.1 where total output is the monopoly output and maximizes the sum of the payoffs. More specifically, the monopoly output maximizes $Pq = (120 - q)q$ with respect to the total output of q, resulting in the first order condition

$$120 - 2q = 0, \tag{3.10}$$

which implies a total output of 60 and a price of 60. Deciding how much of that output of 60 should be produced by each firm—where to be on the 45° line—would be a zero-sum cooperative game, an example of bargaining. But since the Cournot Game is noncooperative, the strategy combinations such that $q_a + q_b = 60$ are not necessarily equilibria despite their Pareto optimality.[1]

To find the Nash equilibrium, we need to refer to the **best response functions** for the two players. If Brydox produced 0, Allied would produce the monopoly output of 60. If Brydox produced $q_b = 120$ or greater, the market price would fall to zero and Allied would choose to produce zero. The best response function is found by maximizing Allied's payoff, given in equation (3.9), with respect to his strategy q_a. This generates the first order condition $120 - 2q_a - q_b = 0$, or

$$q_a = 60 - q_b/2. \tag{3.11}$$

Another name for the best response function, the name usually used in the context

---

[1] Pareto-optimality is defined from the viewpoint of the players. When total output is 60 neither of them can be made better off without hurting the other. If consumers were added to the game and side payments were allowed from consumers to firms, the monopoly output of 60 would be inefficient.

of the Cournot Game, is the **reaction function**. Both names are somewhat mis-leading, since the players move simultaneously, with no chance to reply or react, but they are useful in imagining what a player would do if the rules of the game did allow him to move second. The reaction functions of the two firms are la-belled R_a and R_b in Figure 3.1. Where they cross, point C, is the **Cournot–Nash equilibrium**, which is simply the Nash equilibrium when the strategies consist of quantities. Algebraically, it is found by solving the two reaction functions for q_a and q_b, which generates the unique equilibrium, $(q_a = 40, q_b = 40)$. The equilibrium price is also 40, coincidentally.

Figure 3.1:   The Cournot Game: reaction functions

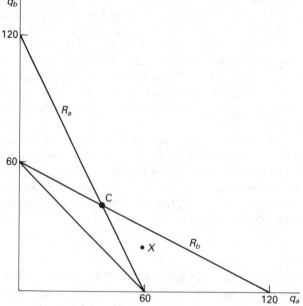

In the Cournot Game, the Nash equilibrium has the particularly nice property of **stability**: we can imagine how starting from some other strategy combination the players might reach the equilibrium. If the initial strategy combination is the point X in Figure 3.1, for example, Allied's best response is to decrease q_a and Brydox's is to increase q_b, which moves the combination closer to the equilibrium. But this is special to the Cournot Game, and Nash equilibria are not always stable in this way.

We might still be dissatisfied with the Cournot equilibrium. One problem is the assumption implicit in Nash equilibrium that Allied believes that changing q_a would leave q_b unchanged, an assumption which might be questioned. Another objection is that the strategy sets are oddly specified. If the strategies are prices, rather than quantities, the Nash equilibrium is much different. Both objections are discussed in Chapter 12.

Stackelberg Equilibrium

There are many ways to model duopoly, but while we will defer discussion of most of them to Chapter 12, we will make an exception for Stackelberg equilibrium. Stackelberg equilibrium differs from Cournot in that one firm gets to choose its quantity first. If Allied moved first, what output would it choose? Allied knows how Brydox will react to its choice, so it picks the point on Brydox's reaction curve that maximizes Allied's profit.

In a two-player game, the player moving first is the **Stackelberg leader** and the other player is the **Stackelberg follower.** The distinguishing characteristic of a Stackelberg equilibrium is that one player gets to commit himself first. In Figure 3.2, Allied moves first intertemporally. If moves were simultaneous, but Allied could commit himself to a certain strategy, the same equilibrium would be reached, as long as Brydox could not commit himself. Algebraically, since Allied forecasts Brydox's output to be $q_b = 60 - q_a/2$ from the analog of equation (3.11), Allied can substitute this into his payoff function in (3.9), obtaining

$$\pi_{\text{Allied}} = 120q_a - q_a^2 - q_a(60 - q_a/2). \tag{3.12}$$

Figure 3.2   Stackelberg equilibrium

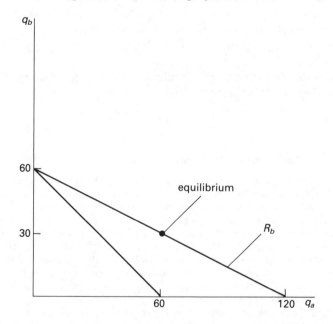

Maximizing with respect to q_a, we obtain the first order condition

$$120 - 2q_a - 60 + q_a = 0, \tag{3.13}$$

which generates $q_a = 60$. Once Allied chooses this output, Brydox chooses his output to be $q_b = 30$ (that Brydox chooses exactly half the monopoly output is accidental). The market price is 30 for both firms, so Allied has benefited from his status as Stackelberg leader.

Recommended Reading

Aumann, Robert (1974) "Subjectivity and Correlation in Randomized Strategies" *Journal of Mathematical Economics.* 1974. 1: 67–96.

Farrell, Joseph (1987) "Cheap Talk, Coordination, and Entry" *Rand Journal of Economics.* Spring 1987. 18, 1: 34–9.

Fudenberg, Drew & Jean Tirole (1986b) "A Theory of Exit in Duopoly" *Econometrica.* July 1986. 54, 4: 943–60.

Problem 3

Mixed Strategies in the Battle of the Sexes

Refer back to the Battle of the Sexes and Pure Coordination in Section 1.4, and denote the probabilities that the man and woman pick *Prize Fight* by p_m and p_w.

(1)   Find an expression for the man's expected payoff.
(2)   Find the first order condition for the man's choice of strategy.
(3)   What are the equilibrium values of p_m and p_w, and the expected payoffs?
(4)   Find the most likely outcome and its probability.
(5)   What is the equilibrium payoff in the mixed strategy equilibrium for Pure Coordination?
(6)   Why is the mixed strategy equilibrium a better focal point in Battle of the Sexes than in Pure Coordination?

Notes

N3.2   Mixed Strategies: the Welfare Game

- In this book we will always assume that players remember their previous moves. Without this assumption of **perfect recall**, the definition in the text is not that for a mixed strategy, but for a **behavior strategy**. As historically defined, a player pursues a mixed strategy when he randomly chooses between pure strategies at the starting node, but he plays a pure strategy thereafter. Under that definition, the modeller cannot talk of random choices at any but the starting node. Kuhn (1953) showed that the definition of mixed strategy given in the text is equivalent to the original definition if the game has perfect recall. Since all important games have perfect recall and the new definition of mixed strategy is better in keeping with the modern spirit of sequential rationality, I have abandoned the old definition.

  The classic example of a game without perfect recall is **bridge**, where the four players of the actual game can be cutely modelled as two players who forget what half their cards look like at any one time in the bidding. A more useful example is any game that has been simplified by restricting players to Markov strategies (see note N4.4), but usually the modeller sets up such a game with perfect recall and then rules out non-Markov equilibria after showing that the Markov strategies form an equilibrium for the general game.

- For more examples of calculating mixed strategy equilibria, see Sections 5.6, 10.5, 12.2, and 13.4.

- An equilibrium in mixed strategies depends heavily on the cardinal values of the payoffs, not just on their ordinal values like the pure strategy equilibria in other 2-by-2 games.
- For a careful interpretation of mixed strategy Nash equilibria as a result of players not knowing the parameters of the game with complete certainty, see Harsanyi (1973).
- It is *not* true that when two pure strategy equilibria exist a player would be just as willing to use a strategy mixing the two even though the other player is using a pure strategy. In the Battle of the Sexes, for instance, if the man knows the woman is going to the ballet he is not indifferent between the ballet and the prize fight.
- A continuum of players is useful not only because the modeller need not worry about fractions of players, but because he can use more modelling tools from calculus—taking the integral of the quantities demanded by different consumers, for example, rather than the sum. But using a continuum is also mathematically more difficult: see Aumann (1964a, 1964b).

N3.3   Chicken, the War of Attrition, and Correlated Strategies

- Papers on the war of attrition include Fudenberg & Tirole (1986b), Ghemawat & Nalebuff (1985), Maynard Smith (1974), Nalebuff & Riley (1985), and Riley (1980).
- The game of Chicken discussed in the text is simpler than in the movie *Rebel Without a Cause,* in which the players race towards a cliff and the winner is the player who jumps out of his car last. The pure strategy space in the movie game is continuous and the payoffs are discontinuous at the cliff's edge, which makes the game more difficult to analyze technically. (Looking ahead to Section 4.2, recall the importance of a "tremble" in the movie.)
- Technical difficulties arise in some models with a continuum of actions and mixed strategies. In the Welfare Game, the government chose a single number, a probability, on the continuum from zero to one. If we allowed the government to mix over a continuum of aid levels, it would choose a function, a probability density, over the continuum. The original game has a finite number of elements in its strategy set, so its mixed extension still has a strategy space in $\mathbf{R}^n$. But with a continuous strategy set extended by a continuum of mixed strategies for each pure strategy, the mathematics are difficult. A finite number of mixed strategies can be allowed without much problem, but usually that is not satisfactory.

  Games in continuous time, including the War of Attrition, frequently run into this problem. Sometimes it can be avoided by clever modelling, as in Fudenberg & Tirole's (1986b) version with asymmetric information. They specify as strategies the length of time firms would continue to *Stay* given their beliefs about the type of the other player, in which case there is a pure strategy equilibrium.

N3.4   Continuous Strategies: the Cournot Game

- An interesting class of simple continuous payoff games are the **Colonel Blotto games** (Tukey [1949]). In these games, two military commanders allocate their forces to m different battlefields, and a battlefield contributes more to the payoff of the commander with the greater forces there. A distinguishing characteristic is that player i's payoff increases with the value of player i's particular action relative to player j's, and i's actions are subject to a budget constraint. Except for the budget constraint, this is similar to the "tournament" that we will look at in Section 7.4.
- **Differential Games** are played in continuous time. The action is a function describing the value of a state variable at each instant, so the strategy maps the game's past history to such a function. Differential games are solved using dynamic optimization. A recent reference is Bagchi's 1984 book.

  A different class of games in continuous time are the **games of timing**, of which the **noisy duel** and the **silent duel** are the best known. In these games, the actions are discrete occurrences which the players locate at particular points in time. Two players with guns approach each other and must decide when to shoot. In the noisy duel, if a player shoots and misses, the other player observes the miss and can kill the first player at his leisure. A perfect equilibrium (see Section 4.2) exists in pure strategies. In a silent duel, a player does not know when the other player has fired, and the equilibrium is in mixed strategies. See Karlin (1959) for details.

- Fudenberg & Levine (1986) show circumstances under which the equilibria of games with infinite strategy spaces can be found as the limits of equilibria of games with finite strategy spaces.

- "Stability" is a word used in many different ways in game theory and economics. The natural meaning of a stable equilibrium is that it has dynamics that cause the system to return to that point after being perturbed slightly, and the discussion of the stability of Cournot equilibrium was in that spirit. The uses of the term by von Neumann & Morgenstern (1944) and Kohlberg & Mertens (1986) are completely different.

- The term "Stackelberg equilibrium" is not clearly defined in the literature. It is sometimes used to denote equilibria in which players take actions in a given order, but since that is just the perfect equilibrium (see Section 4.2) of a well-specified extensive form, I prefer to reserve it for the Nash equilibrium of the duopoly quantity game in which one player moves first, the context of Stackelberg (1934), though he himself did not use game theory.

  An alternative definition is that a Stackelberg equilibrium is a strategy combination in which players select strategies in a given order, and in which each player's strategy is a best response to the fixed strategies of the players preceding him and the yet-to-be-chosen strategies of players succeeding him, i.e., a situation in which players precommit to strategies in a given order. Such an equilibrium would not generally be either Nash or perfect.

- Stackelberg (1934) suggested that sometimes the players are confused about which of them is the leader and which the follower, resulting in the disequilibrium outcome called **Stackelberg Warfare**.

- With linear costs and demand, total output is greater in the Stackelberg equilibrium than in Cournot. The slope of the reaction curve is less than one, so Allied's output expands more than Brydox's contracts. Total output being greater, the price is less than in the Cournot equilibrium.

- A useful application of Stackelberg equilibrium is to an industry with a dominant firm and a **competitive fringe** of smaller firms that sell at capacity if the price exceeds their marginal cost. These smaller firms act as Stackelberg leaders (not followers), since each is small enough to ignore its effect on the behavior of the dominant firm. The oil market could be modelled this way with OPEC as the dominant firm and producers such as Britain on the fringe.

4 Dynamic Games with Symmetric Information

4.1 Introduction

In this chapter we will make heavy use of the extensive form to study games with a sequence of moves. We start in Section 4.2 with a refinement of the Nash equilibrium concept called perfectness that incorporates sensible implications of the order of moves. Perfectness is illustrated in Section 4.3 with a game of entry deterrence. Having established a way to deal with moves over time, we will analyze repeated games in Section 4.4 and use the Chainstore Paradox to show the perverse unimportance of repetition for the Prisoner's Dilemma. Section 4.5 shows how to model discounting: players who value future consumption less than present consumption. Neither discounting, probabilistic end dates, infinite repetitions, nor precommitment are satisfactory escapes from the Chainstore Paradox, and the Folk Theorem described in Section 4.6 explains why. Section 4.7 builds a framework for reputation models based on the Prisoner's Dilemma, and Section 4.8 presents one particular reputation model, the Klein–Leffler model of product quality.

4.2 Subgame Perfectness

The Perfect Equilibrium of Follow the Leader I

Subgame perfectness is an equilibrium concept based on the ordering of moves and the distinction between an equilibrium path and an equilibrium. The **equilibrium path** is the path through the game tree that is followed in equilibrium, but the equilibrium itself is a strategy combination, which includes the players' responses to other players' deviations from the equilibrium path. These off-equilibrium responses are very important to decisions on the equilibrium path. A threat, for example, is a promise to carry out a certain action if another player deviates from his equilibrium actions.

Perfectness is best introduced with an example. In Section 2.2, a flaw of Nash equilibrium was revealed in the game Follow the Leader I, which has three pure strategy Nash equilibria, only one of which is reasonable. The players are Smith and Brown, who choose disk sizes. Both their payoffs are greater if they choose

the same size, and greatest if they coordinate on the size *Large*. Smith moves
first, so his strategy set is {*Small, Large*}. Brown's strategy is more complicated,
because it must specify an action for each information set, and Brown's information
set depends on what Smith chose. A typical element of Brown's strategy set is
(*Large, Small*), which specifies that he chooses *Large* if Smith chose *Large*, and
Small if Smith chose *Small*. From the normal form we found the following three
Nash equilibria.

Equilibrium	Strategies	Outcome
X	{*Large, (Large, Large)*}	Both pick *Large*.
Y	{*Large, (Large, Small)*}	Both pick *Large*.
Z	{*Small, (Small, Small)*}	Both pick *Small*.

Only Equilibrium Y is reasonable, because the order of the moves should matter
to the decisions players make. The problem with the normal form, and thus with
simple Nash equilibrium, is that it ignores who moves first. Smith moves first, and
it seems reasonable that Brown should be allowed—in fact should be required—to
rethink his strategy after Smith moves.

Figure 4.1   Follow the Leader I

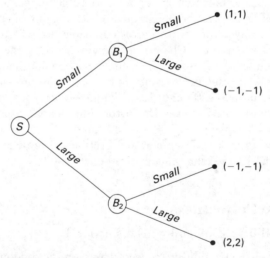

Payoffs to: (Smith,Brown)

Consider Brown's strategy of (*Small, Small*) in Equilibrium Z. If Smith deviated
from equilibrium by choosing *Large*, it would be unreasonable for Brown to stick
to the response *Small*. Instead, he should also choose *Large*. But if Smith expected
a response of *Large*, then he would indeed choose *Large*, and Z would not be an
equilibrium. A similar argument shows that (*Large, Large*) is an irrational strategy
for Brown, and we are left with Y as the unique equilibrium.

We say that Equilibria X and Z are Nash equilibria but not "perfect" Nash equilibria. A strategy combination is a perfect equilibrium if it remains an equilibrium

on all possible paths, both the equilibrium path and other paths which branch off into different "subgames."

> *A* **subgame** *is a game consisting of a node which is a singleton in every player's information partition, that node's successors, and the payoffs at the associated end nodes.*[1]

> *A strategy combination is a* **subgame perfect Nash equilibrium** *if (a) it is a Nash equilibrium for the entire game; and (b) its relevant action rules are a Nash equilibrium for every subgame.*

The extensive form of Follow the Leader I in Figure 4.1 (a reprise of Figure 2.1) has three subgames: (1) the entire game, (2) the subgame starting at Node B_1, and (3) the subgame starting at Node B_2. Strategy combination X is not a subgame perfect equilibrium, because it is only Nash in subgames (1) and (3), not in subgame (2). Strategy combination Z is not a subgame perfect equilibrium, because it is only Nash in subgames (1) and (2), not in subgame (3). But strategy combination Y is Nash in all three subgames.

One reason why perfectness (the word "subgame" is usually left off) is a good concept is because out-of-equilibrium behavior is irrational in a non-perfect equilibrium. A second justification is that a weak Nash equilibrium is not robust to small changes in the game. So long as he is certain that Smith will not choose *Large*, Brown is indifferent between the never-to-be-used responses (*Small* if *Large*) and (*Large* if *Large*). Equilibria X, Y, and Z are all weak because of this. But if there is even a small probability that Smith will choose *Large*—perhaps by mistake—then Brown would prefer the response (*Large* if *Large*), and equilibria X and Z are no longer valid. Perfectness is a way to eliminate some of the weak Nash equilibria. We call the small probability of a mistake a **tremble**, and we will return to this **trembling hand** approach in Section 5.1 as one way to extend the notion of perfectness to games of asymmetric information.

4.3   An Example of Perfectness: Entry Deterrence I

We turn now to a game in which perfectness plays a role just as important as in Follow the Leader I, but the players are in conflict. An old question in industrial organization is whether an incumbent monopolist can maintain his position by threatening to wage a price war against any new firm that enters the market. This idea was heavily attacked by Chicago School economists such as McGee (1958) on the grounds that a price war would hurt the incumbent more than colluding with the entrant. Game theory can present this reasoning very cleanly. Let us consider a single episode of possible entry and price warfare, which nobody expects to be repeated.

---

[1]   Technically, this is a *proper* subgame because of the information qualifier, but no economist is so ill-bred as to use any other kind of subgame.

Entry Deterrence I

Players

Two firms, the entrant and the incumbent.

Information

Perfect and certain.

Actions and Events

(1)   The entrant decides whether to *Enter* or *Stay Out*.
(2)   If the entrant enters, the incumbent can *Collude* with him, or *Fight* by cutting the price drastically.

Payoffs

Market profits are 100 at the monopoly price and 0 at the fighting price. Entry costs 10. Collusion shares the profits evenly.

The strategy sets can be discovered from the order of actions and events. They are {*Enter, Stay Out*} for the entrant, and {*Collude* if entry occurs, *Fight* if entry occurs} for the incumbent. The game has the two Nash equilibria indicated in boldface, (*Enter, Collude*) and (*Stay Out, Fight*). The equilibrium (*Stay Out, Fight*) is weak, because the incumbent would just as soon *Collude* given that the entrant is staying out.

Table 4.1 Entry Deterrence I

		Incumbent	
		Collude	*Fight*
	Enter	**40, 50**	−10, 0
Entrant			
	Stay Out	0, 100	**0, 100**

Payoffs to: (Entrant, Incumbent)

A piece of information has been lost by condensing from the extensive form, Figure 4.2, to the normal form, Table 4.1—the entrant gets to move first. Once he has chosen *Enter*, the incumbent's best response is *Collude*. The threat to fight is not credible and would be employed only if the incumbent could bind himself to fight, in which case he never does fight, because the entrant chooses to stay out. The equilibrium (*Stay Out, Fight*), is Nash but not subgame perfect, because if the game is started after the entrant has already entered, the incumbent's best

Figure 4.2   Entry Deterrence I

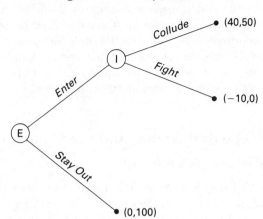

Payoffs to: (Entrant, Incumbent)

response is *Collude*. This does not prove that collusion is inevitable in duopoly—we will analyze many other duopoly models in Chapter 12 and 13—but collusion is the equilibrium for Entry Deterrence I.

The trembling hand interpretation of perfect equilibrium can be used here. So long as it is certain that the entrant will not enter, the incumbent is indifferent between *Fight* and *Collude*, but if there were even a small probability of entry—perhaps because of a lapse of good judgement by the entrant—the incumbent would prefer *Collude* and the Nash equilibrium would be broken.

Perfectness rules out threats that are not credible. Entry Deterrence I provides a good example, because if a communication move were added to the game tree, the incumbent might tell the entrant that entry would be followed by fighting, but the entrant would ignore this non-credible threat. If, however, some means existed by which the incumbent could precommit himself to fight entry, the threat would become credible.

Should the Modeller Ever Use Non-Perfect Equilibria?

A game in which a player can commit himself to a strategy can be modelled in two ways:

(1)   As a game in which non-perfect equilibria are acceptable, or
(2)   By changing the game to replace the action *Do X* with *Commit to Do X* at an earlier node.

An example of (2) in Entry Deterrence I is to remodel the game so that the incumbent moves first, deciding in advance whether or not to choose *Fight* before the entrant gets to move. Approach (2) is better than (1), because usually the modeller wants to let players commit to some actions and not others, and he can do this by carefully specifying the order of play. Allowing equilibria to be non-perfect forbids such discrimination and usually multiplies the number of equilibria.

Indeed, the problem with subgame perfectness is not that it is too restrictive, but that it still allows too many strategy combinations to be equilibria in games of asymmetric information. A subgame must start at a single node and not cut across any player's information set, so often the only subgame will be the whole game and imposing subgame perfectness does not restrict equilibrium at all. In Section 5.1, we will discuss perfect Bayesian equilibrium and other ways to extend the perfectness concept to games of asymmetric information.

4.4   Finitely Repeated Games and the Chainstore Paradox

The Chainstore Paradox (Selten [1978])

Suppose that we repeat Entry Deterrence I 20 times in the context of a chainstore that is trying to deter entry into 20 markets where it has outlets. We have seen that entry into just one market would not be deterred, but perhaps with 20 markets the outcome is different because the chainstore would fight the first entrant to deter the next 19. It turns out this is not the case.

The repeated game is much more complicated than the **one-shot game**, as we call the unrepeated version. A player's action is still to *Enter* or *Stay Out*, to *Fight* or *Collude*, but his strategy is a potentially very complicated rule telling him what action to choose depending on what actions both players took in each of the previous periods. Even the five-round repeated Prisoner's Dilemma has a strategy set for each player with over two billion strategies, and the number of strategy combinations is even greater (Sugden [1986], p. 108).

The obvious way to solve the game is from the beginning, where there is the least past history on which to condition a strategy, but that is not the easy way. We must follow Kierkegaard, who said, "Life can only be understood backwards, but it must be lived forwards." In picking his first action, a player looks ahead to its implications for all the future periods, so it is easiest to start by understanding the end of a multi-period game, where the future is shortest.

Consider the situation in which 19 markets have already been invaded (and maybe the chainstore fought, maybe not). In the last market, the subgame in which the two players find themselves is identical to the one-shot Entry Deterrence I, so the entrant will *Enter* and the chainstore will *Collude*, regardless of the past history of the game. Consider the next-to-last market. The chainstore can gain nothing from building a reputation for ferocity, because it is common knowledge that he will *Collude* with the last entrant anyway. So he might as well *Collude* in the 19th market. But we can say the same of the 18th market—and by continuing backward induction, of every market, including the first. This result is called the **Chainstore Paradox**.

Backward induction ensures that the strategy combination is a subgame perfect equilibrium. There are other Nash equilibria—(*Always Fight, Never Enter*), for example—but because of the Chainstore Paradox they are not perfect.

The Repeated Prisoner's Dilemma

The Prisoner's Dilemma of Section 1.2 is similar to Entry Deterrence I: the prisoners would like to commit themselves to *Cooperate*, but in the absence of commitment they *Fink*. The Chainstore Paradox can be applied to show that repetition does not induce cooperative behavior. Both prisoners know that in the last repetition, both will *Fink*. After 18 repetitions, they know that no matter what happens in the 19th, both will *Fink* in the 20th, so they might as well *Fink* in the 19th too. Building a reputation is pointless, because in the 20th period it is not going to matter. Proceeding inductively, both players *Fink* in every period, the unique perfect equilibrium outcome.

In fact, as a consequence of the fact that the one-shot Prisoner's Dilemma has a dominant strategy equilibrium, finking is the only Nash outcome for the repeated Prisoner's Dilemma, not just the only perfect outcome. The argument of the previous paragraph did not show that finking was the unique Nash outcome. To show subgame perfectness, we worked back from the end using longer and longer subgames. To show that finking is the only Nash outcome, we do not look at subgames, but instead rule out successive classes of strategies from being Nash. Consider the portions of the strategy which apply to the equilibrium path (that is, the portions directly relevant to the payoffs). No strategy in the class that calls for *Cooperate* in the last period can be a Nash strategy, because the same strategy with *Fink* replacing *Cooperate* would dominate it. But if both players have strategies calling for finking in the last period, then no strategy that does not call for finking in the next-to-last period is Nash, because a player should deviate by replacing *Cooperate* with *Fink* in the next-to-last period. The argument can be carried back to the first period, ruling out any class of strategies that does not call for finking everywhere along the equilibrium path.

The strategy of always finking is not a dominant strategy, as it is in the one-shot game, because it is not the best response to various suboptimal strategies such as (*Cooperate until the other player finks, then fink for the rest of the game*). Moreover, the uniqueness is only on the equilibrium path. Non-perfect Nash strategies could call for cooperation at nodes far away from the equilibrium path, since that action would never need to be taken. If Row has chosen (*Always Fink*), one of Column's best responses is (*Always fink unless Row has cooperated ten times; then always cooperate*).

4.5   Discounting

A model with many rounds must specify whether payments are valued less if they are made at later rounds, i.e., whether they are **discounted.** Discounting is measured by the discount rate or the discount factor.

> The **discount rate**, *r*, is the extra fraction of a payoff unit needed to compensate for delaying receipt by one period.

> The **discount factor**, *δ*, is the value in present payoff units of one payoff unit to be received one period from the present.

The discount rate is analogous to the interest rate, and in some models the interest rate determines the discount rate. The discount factor represents exactly the same idea as the discount rate, and $\delta = 1/(1+r)$. Models use r or δ depending on notational convenience. "No discounting" is equivalent to $r = 0$ and $\delta = 1$, so the notation includes no discounting as a special case.

Whether to put discounting into a model involves two questions. The first is whether the added complexity is accompanied by a change in the results or a surprising demonstration of no change in the results. If discounting makes no difference, it should not be included. If discounting does make a difference, the second question arises: do the events of the model occur in real time (Section 2.2), so that discounting is appropriate? The bargaining game Alternating Offers from Section 10.3 can be interpreted in two ways. One way, without discounting, is that the players make all their offers and counteroffers between dawn and dusk of a single day, so essentially no real time has passed. The other way, with discounting, is that each offer consumes a week of time, so that the delay before the bargain is reached is important to the players.

Discounting has two important sources: time preference and a probability that the game might end, represented by the rate of time preference, ρ, and the probability each period that the game ends, θ. It is usually assumed that ρ and θ are constant. If they both take the value zero, the player does not care whether his payments are scheduled now or ten years from now. Otherwise, a player is indifferent between $x/(1 + \rho)$ now and x guaranteed to be paid a period later. With probability $(1 - \theta)$ the game continues and the payment a period later is actually made, so the player is indifferent between $(1 - \theta)x/(1 + \rho)$ now and the promise of x to be paid a period later contingent upon the game still continuing. The discount factor is therefore

$$\delta = \frac{1}{1+r} = \frac{(1 - \theta)}{(1 + \rho)}. \tag{4.1}$$

Table 4.2 summarizes the implications of discounting for the value of payment streams of various kinds. We will not go into how these are derived, but they all stem from the basic fact that a dollar paid in the future is worth δ dollars now. Continuous time models usually refer to rates of payment rather than lump sums, so the discount factor is not so useful a concept, but discounting works the same

Table 4.2 Discounting

Payoff stream	Discounted value	
	r-notation	δ-notation
x at the end of one period	$\frac{x}{1+r}$	δx
x at the end of each period up through T	$\sum_{t=1}^{T} \frac{x}{(1+r)^t}$	$\sum_{t=1}^{T} \delta^t x$
x at the end of each period in perpetuity	x/r	$x\delta/(1 - \delta)$
x at the start of each period in perpetuity	$x + x/r$	$x/(1 - \delta)$
x at time t in continuous time	xe^{-rt}	—
a flow of x per period in continuous time up to T	$\int_0^T xe^{-rt}dt$	—
a flow of x per period in continuous time in perpetuity	x/r	—

way as in discrete time except that payments are continuously compounded. For a full explanation, see a finance text (e.g. Appendix A of Copeland & Weston [1983]).

4.6   Infinitely Repeated Games and the Folk Theorem

The contradiction between the Chainstore Paradox and what many people think of as real world behavior has been most successfully resolved by adding incomplete information to the model, as we will do in Section 5.4. Before we turn to the tools needed to handle incomplete information, however, we will explore certain other modifications. One idea is to repeat the Prisoner's Dilemma an infinite number of times instead of a finite number (after all, few economies, except perhaps Hong Kong, have a known end date). Without a last period, the inductive argument in the Chainstore Paradox fails. In fact, we can find a simple perfect equilibrium for the infinitely repeated Prisoner's Dilemma in which both players cooperate: both players adopt the "grim strategy."

Grim Strategy

(1) Start by choosing Cooperate.

(2) Continue to choose Cooperate unless some other player chooses Fink, in which case choose Fink forever thereafter.

If Column uses the grim strategy, the grim strategy is weakly Row's best response. If Row cooperates, he will continue to receive the high (*Cooperate, Cooperate*) payoff forever. If he finks, he will receive the higher (*Fink, Cooperate*) payoff once, but the best he can hope for thereafter is the (*Fink, Fink*) payoff.

Even in the infinitely repeated game, cooperation is not immediate, and not every strategy that punishes finking is perfect. A notable example is the strategy "tit-for-tat."

Tit-for-Tat

(1) Start by choosing Cooperate.

(2) Thereafter, in period n choose the action that the other player chose in period (n − 1).

If Column uses tit-for-tat, Row does not have an incentive to *Fink* first, because if Row cooperates he will continue to receive the high (*Cooperate, Cooperate*) payoff, but if he finks and then returns to tit-for-tat, the players alternate (*Fink, Cooperate*) with (*Cooperate, Fink*) forever. Row's average payoff from this alternation would be lower than if he had stuck to (*Cooperate, Cooperate*), and would swamp the one-time gain. But tit-for-tat is not perfect, because it is not rational for Column to punish Row's initial *Fink*. Adhering to tit-for-tat's punishments results in a miserable alternation of *Fink*s, so Column would rather ignore Row's first *Fink*. The deviation is not from the equilibrium path action of *Cooperate*, but from the off-equilibrium action rule of *Fink in response to a Fink*. Tit-for-tat, unlike the grim strategy, cannot enforce cooperation.

Unfortunately, although eternal cooperation is a perfect equilibrium outcome in the infinite game under at least one strategy, so is practically anything else, including eternal finking. The multiplicity of equilibria is summarized by the Folk Theorem, so called because no one remembers who should get credit for it. One version is stated below, and we will spend the rest of the section explaining it.

Theorem 4.1 (The Folk Theorem)

In an infinitely repeated n-person game with finite action sets at each repetition, any combination of actions observed in any finite number of repetitions is the unique outcome of some subgame perfect equilibrium given

Condition 1:   *The rate of time preference is zero, or positive and sufficiently small;*

Condition 2:   *The probability that the game ends at any repetition is zero, or positive and sufficiently small; and*

Condition 3 (Dimensionality):   *The set of payoff combinations that strictly Pareto-dominate the minimax payoff combinations in the mixed extension of the one-shot game is n-dimensional.*

What the Folk Theorem tells us is that claiming that particular behaviour arises in a perfect equilibrium is meaningless in an infinitely repeated game. This applies to any game that meets Conditions 1 to 3, not just the Prisoner's Dilemma. If an infinite amount of time always remains in the game, a way can always be found to make one player willing to punish another for the sake of a better future, even if the punishment currently hurts the punisher as well as the punished. Any finite interval of time is insignificant compared to eternity, so the threat of future reprisal makes the players willing to carry out the punishments needed.

We will next discuss Conditions 1 to 3.

Discounting

The Folk Theorem helps answer whether discounting future payments lessens the influence of the troublesome Last Period. Quite the opposite is true. With discounting, the present gain from finking is weighted more heavily and future gains from cooperation more lightly. If the discount rate is very high the game almost returns to being one-shot. When the real interest rate is 1,000 percent, a payment next year is little better than a payment a hundred years hence, so next year is practically irrelevant. Any model that relies on a large number of repetitions also relies on the discount rate not being too high.

Allowing a little discounting is nonetheless important to show there is no discontinuity at the discount rate of zero. If we come across an undiscounted infinitely repeated game with many equilibria, the Folk Theorem tells us that adding a small discount rate will not reduce the number of equilibria. This contrasts with the effect of changing the model by making the number of repetitions large but finite, which often eliminates all but one outcome by inducing the Chainstore Paradox.

A discount rate of zero supports many perfect equilibria, but if the rate is large enough, the only equilibrium outcome is eternal finking. We can calculate the critical value for given parameters. The grim strategy imposes the heaviest possible punishment for deviant behavior. Using the payoffs for the Prisoner's Dilemma from Table 4.3a in the next section, the equilibrium payoff from the grim strategy is the current payoff of 5 plus the value of the rest of the game, which from Table 4.2 is $5/r$. If Row deviated by finking, he would receive a current payoff of 10, but the value of the rest of the game would fall to 0. The critical value of the discount rate is found by solving the equation $5 + 5/r = 10 + 0$, which yields $r = 1$, a discount rate of 100 percent or a discount factor of $\delta = 0.5$. Unless the players are extremely impatient, finking is not much of a temptation.

Random Ending

Time preference is fairly straightforward, but what is surprising is that assuming that the game ends each period with probability θ does not make a drastic difference. In fact, we could even allow θ to vary over time, so long as it never became too large. If $\theta > 0$, the game ends with probability one; or, put less dramatically, the expected number of repetitions is finite, but it still behaves like a discounted infinite game, because the expected number of future repetitions is always large, no matter how many have already occurred. The game still has no Last Period, and it is still true that imposing one, no matter how far beyond the expected number of repetitions, would radically change the results.

I find it interesting that "(a) the game will end at some uncertain date before T," is different from "(b) a constant probability of the game ending." Under (a), the game is like a finite game, because as time passes the maximum amount of time still to run shrinks to zero. Under (b), even though the game will probably end by T, if it lasts until T the game looks exactly the same as at time zero.

The Dimensionality Condition

The "minimax payoff" mentioned in Theorem 4.1 is the payoff that results if all the other players pick strategies solely to punish player i, and he protects himself as best he can (see note N4.6). The dimensionality condition is needed only for games with three or more players. It is satisfied if for each player there is some payoff combination in which his payoff is greater than his minimax payoff but different from the payoff of every other player. This is satisfied by the n-person Prisoner's Dilemma in which a solitary finker gets a higher payoff than his cooperating fellow-prisoners. It is not satisfied by Pure Coordination, in which all the players have the same payoff. The condition is necessary because establishing the desired behavior requires some way for the other players to punish a deviator without punishing themselves.

Precommitment

What if we use metastrategies, abandoning the idea of perfectness by allowing players to commit at the start to a strategy for the rest of the game? We would still want to keep the game noncooperative by disallowing binding promises, but we could model it as a game with simultaneous choices by both players, or with one move each in sequence.

If precommitted strategies are chosen simultaneously, the equilibrium outcome of the finitely repeated Prisoner's Dilemma calls for always finking, because allowing commitment is the same as allowing equilibria to be non-perfect, in which case, as was shown earlier, the unique Nash outcome is always finking.

A different result is achieved if the players precommit to strategies in sequence. The outcome depends on the particular values of the parameters, but one possible equilibrium is the following: Row moves first and chooses the strategy (*Cooperate* until Column *Finks*; thereafter always *Fink*), and Column chooses (*Cooperate* until the last period; then *Fink*). The observed outcome would be for both players to cooperate until the last period, and then for Row to again cooperate, but be finked by Column. Row would submit to this because if he chose a strategy that initiated finking earlier, Column would choose a strategy of starting to fink earlier too. The game has a second-mover advantage.

4.7   Reputation: the One-Sided Prisoner's Dilemma

In Part II of this book we will look at moral hazard and adverse selection. Under moral hazard, a player wants to commit to high effort, but he cannot credibly do so. Under adverse selection, a player wants to prove he is high ability, but he cannot. In both, the problem is that the penalties for lying are insufficient. Reputation seems to offer a way out of the problem. If the relationship is repeated, perhaps a player is willing to be honest in early periods in order to establish a reputation for honesty valuable to himself later.

Reputation seems to play a similar role in making threats to punish credible. Usually punishment is costly to the punisher as well as the punished, and it is not clear why the punisher should not let bygones be bygones. Yet in 1988 we see the Soviet Union paying off 70-year-old debt to dissuade the Swiss authorities from blocking a mutually beneficial new bond issue ("Soviets Agree to Pay Off Czarist Debt to Switzerland," *Wall Street Journal*, 19 January 1988, p. 60). Why were the Swiss so vindictive towards Lenin?

The questions of why players do punish and do not cheat are really the same questions that arise in the repeated Prisoner's Dilemma, where only an infinite number of repetitions allows cooperation. That is the great problem of reputation: since everyone knows that a player will *Fink*, choose low effort, or default on debt in the last period, why do they suppose he will bother to build up a reputation in the present? Why should past behaviour be any guide to future behaviour?

To show how reputation works in a given situation, we must show why the Chainstore Paradox does not apply. The two approaches are to show that

(1)   There is cooperation in the last period, or
(2)   The early periods are different from the last period.

Not all reputation problems fit the basic Prisoner's Dilemma. Some, such as duopoly or the original Prisoner's Dilemma, are **two-sided** in the sense that each player has the same strategy set and the payoffs are symmetric. Others, such as product quality, are what we might call **one-sided Prisoner's Dilemmas**, which have properties similar to the Prisoner's Dilemma but do not fit the usual definition because they are asymmetric. Table 4.3 shows the normal forms for both the original Prisoner's Dilemma and the one-sided version. The important difference is that in the one-sided Prisoner's Dilemma at least one player really does prefer (*Cooperate, Cooperate*) to anything else. He finks defensively, rather than both offensively and defensively. The payoff (0,0) can often be interpreted as the refusal of one player to interact with the other: for example, the motorist who refuses to buy cars from Chrysler because he knows they once falsified odometers. Table 4.4 lists examples of both one-sided and two-sided games. Versions of the Prisoner's Dilemma with three or more players can also be classified as two-sided or one-sided, depending on whether all players find *Fink* a dominant strategy or not.

Table 4.3 Prisoner's Dilemmas

(a) Two-sided

		Column	
		Cooperate	*Fink*
Row	*Cooperate*	5, 5	−5, 10
	Fink	10, −5	**0, 0**

Payoffs to: (Row, Column)

(b) One-sided

	Consumer (Column)	
	Buy (*Cooperate*)	*Boycott* (*Fink*)
Honest (*Cooperate*)	5, 5	0, 0
Seller (Row) Cheat (*Fink*)	10, −5	**0, 0**

Payoffs to: (Seller, Consumer)

The Nash and iterated dominant strategy equilibria in the one-sided Prisoner's Dilemma are still (*Fink, Fink*), but it is not a dominant strategy equilibrium. Column does not have a dominant strategy, because if Row were to choose *Cooperate*, Column would also choose *Cooperate*, to obtain the payoff of five; but if Row chooses *Fink*, Column would choose *Fink* and obtain the payoff of zero. *Fink* is weakly dominant for Row, which makes (*Fink, Fink*) the iterated dominant strat-

Table 4.4 Repeated games in which reputation is important

Application	Sidedness	Players	Actions
Prisoner's Dilemma	two-sided	Row	*Cooperate/Fink*
		Column	*Cooperate/Fink*
Duopoly	two-sided	Firm	*Maintain price/Drop price*
		Firm	*Maintain price/Drop price*
Employer–Worker	two-sided	Worker	*Work/Slack off*
		Employer	*Bonus/No bonus*
Product Quality	one-sided	Firm	*High quality/Low quality*
		Customer	*High price/Low price*
Entry Deterrence	one-sided	Incumbent	$P < MC \ / \ P = MC$
		Entrant	*Not enter/Enter*
Financial Disclosure	one-sided	Firm	*Tell truth/Lie*
		Investor	*High price/Low price*
Borrowing	one-sided	Borrower	*Repay/Default*
		Lender	*Lend/Do not lend*

egy equilibrium. In both games, the players would like to persuade each other that they will cooperate, and devices that induce cooperation in the one-sided game will usually obtain the same result in the two-sided game.

4.8   Product Quality in an Infinitely Repeated Game

The Folk Theorem tells us that some perfect equilibrium of an infinitely repeated game—sometimes called an **infinite horizon model**—can generate any pattern of behavior observed over a finite number of periods. But since the Folk Theorem is no more than a mathematical result, the strategies that generate particular patterns of behavior may be unreasonable. The theorem's value is in provoking close scrutiny of infinite horizon models so that the modeller must show why his equilibrium is better than the multitude of others. He must go beyond satisfaction of the technical criterion of perfectness and justify the strategies on other grounds.

In the simplest model of product quality, a seller can choose between producing costly high quality or costless low quality, and the buyer cannot determine quality before he purchases. If the seller would produce high quality under symmetric information, we have a one-sided Prisoner's Dilemma, as in Table 4.3b. Both players are better off when the seller produces high quality and the buyer purchases the product, but the seller's weakly dominant strategy is to produce low quality, so the buyer will not purchase. This is also an example of moral hazard, the topic of Chapter 6.

A potential solution is to repeat the game, allowing the firm to choose quality at each repetition. If the number of repetitions is finite, however, the outcome stays the same because of the Chainstore Paradox. In the last repetition, the subgame is identical to the one-shot game, so the firm chooses low quality. In the next-to-last repetition, it is foreseen that the last period's outcome is independent of current actions, so the firm also chooses low quality, an argument that can be carried back to the first repetition.

If the game is repeated an infinite number of times, the Chainstore Paradox is inapplicable and the Folk Theorem says that a wide range of outcomes can be observed in equilibrium. Klein & Leffler (1981) construct a plausible equilibrium for an infinite period model. Their original article, in the traditional verbal style of UCLA, does not phrase the result in terms of game theory, but we will recast it here. In equilibrium, the firm is willing to produce a high quality product because it can sell at a high price for many periods, but consumers refuse to ever buy again from a firm that has once produced low quality. The equilibrium price is high enough that the firm is unwilling to sacrifice its future profits for a one-time windfall from deceitfully producing low quality and selling it at a high price. Although this is only one of a large number of subgame perfect equilibria, the consumers' behavior is simple and rational: no consumer can benefit by deviating from the equilibrium.

Product Quality

Players

An infinite number of potential firms and a continuum of consumers.

Information

Asymmetric, complete, and certain.

Actions and Events

(1) An endogenous number n of firms decide to enter the market at cost F.
(2) A firm that has entered chooses its quality to be *High* or *Low*, incurring the constant marginal cost c if it picks *High* and zero if it picks *Low*. The choice is unobserved by consumers. The firm also picks a price p.
(3) Consumers decide which firms (if any) to buy from, choosing firms randomly if they are indifferent. The amount bought from firm i is denoted q_i.
(4) All consumers observe the quality of all goods purchased.
(5) The game returns to (2) and repeats.

Payoffs

The total consumer benefit from a product believed to be low quality is zero, but consumers are willing to buy quantity $q(p) = \sum_{i=1}^{n} q_i$ for a product believed to be high quality, where $\frac{dq}{dp} < 0$.

If a firm stays out of the market, its payoff is zero.

If firm i enters, it receives $-F$ immediately. Its current period payoff is $q_i p$ if it produces *Low* quality and $q_i(p - c)$ if it produces *High* quality. The discount rate is $r \geq 0$.

That the firm can produce low-quality items at zero marginal cost is unrealistic, but it is only a simplifying assumption. By normalizing the cost of producing low quality to zero, we avoid having to carry an extra variable through the analysis without affecting the result.

The Folk Theorem tells us that this game has a wide range of perfect outcomes, including a large number with erratic quality patterns like (*High, High, Low, High, Low, Low*...). If we confine ourselves to pure strategy equilibria with the stationary outcome of constant quality and identical behavior by all firms in the market, then the two outcomes are low quality and high quality. Low quality is always an equilibrium outcome, since it is an equilibrium of the one-shot game. If the discount rate is low enough, high quality is also an equilibrium outcome, and this will be the focus of our attention. Consider the following strategy combination:

Firms. $\tilde{n}$ firms enter. Each produces high quality and sells at price $\tilde{p}$. If a firm ever deviates from this, it thereafter produces low quality and sells at price $\tilde{p}$. The values of $\tilde{p}$ and $\tilde{n}$ are given by equations (4.3) and (4.7) below.

Buyers.   Buyers start by choosing randomly among the firms charging $\tilde{p}$. Thereafter, they remain with their initial firm unless it changes its price or quality, in which case they switch randomly to a firm that has not changed its price or quality.

This strategy combination is a perfect equilibrium. Each firm is willing to produce high quality and refrain from price-cutting because otherwise it would lose all its customers. If it has deviated, it is willing to produce low quality because the quality is unimportant, given the absence of customers. Buyers stay away from a firm that has produced low quality because they know it will continue to do so, and they stay away from a firm that has cut the price because they know it will produce low quality. For this story to work, however, the equilibrium must satisfy three constraints that will be explained in more depth in Section 6.3: incentive compatibility, competition, and market clearing.

The incentive compatibility constraint says that the individual firm must be willing to produce high quality. Given the buyers' strategy, if the firm ever produces low quality it receives a one-time windfall profit, but loses its future profits. The tradeoff is represented by constraint (4.2), which is satisfied if the discount rate is low enough.

$$\textbf{(Incentive Compatibility)} \qquad \frac{q_i p}{1 + r} \leq \frac{q_i (p - c)}{r}. \qquad (4.2)$$

Inequality (4.2) determines a lower bound for the price, which must satisfy

$$\tilde{p} \geq (1 + r)c. \qquad (4.3)$$

We could write (4.3) as an equality rather than an inequality because any firm trying to charge a price higher than the quality-guaranteeing $\tilde{p}$ would lose all its customers and receive a payoff of $-F$.

The second constraint is that competition drives profits to zero, so firms are indifferent between entering and staying out of the market.

$$\textbf{(Competition)} \qquad \frac{q_i(p-c)}{r} = F. \qquad (4.4)$$

Treating (4.3) as an equation and using it to replace p in equation (4.4), we obtain

$$q_i = \frac{F}{c}. \qquad (4.5)$$

We have now determined p and q_i, and only n remains, which is determined by the equality of supply and demand. The market does not always clear in models of asymmetric information (see Stiglitz [1987]), and in this model each firm would like to sell more than its equilibrium output at the equilibrium price, but the market output must equal the quantity demanded by the market.

$$\textbf{(Market Clearing)} \qquad nq_i = q(p). \qquad (4.6)$$

Combining equations (4.3), (4.5), and (4.6) yields

$$\tilde{n} = \frac{cq(c+cr)}{F}. \qquad (4.7)$$

We have now determined the equilibrium values, the only difficulty being the standard existence problem caused by the requirement that the number of firms be an integer (see note N4.8).

The equilibrium price is fixed because F is exogenous and demand is not perfectly inelastic, which pins down the size of firms. If there were no entry cost, but demand were still elastic, then the equilibrium price would still be the unique p that satisfied constraint (4.3), and the market quantity would be determined by $q(p)$, but F and q_i would be undetermined. If consumers believed that any firm which might possibly produce high quality paid an exogenous dissipation cost F, the result would be a continuum of equilibria. The firms' best response would be for $\tilde{n}$ of them to pay F and produce high quality at price $\tilde{p}$, where $\tilde{n}$ is determined by the zero profit condition as a function of F. Klein & Leffler note this indeterminacy and suggest that the profits might be dissipated by some sort of brand-specific capital. The history of the industry may also explain the number of firms. Schmalensee (1982) shows how a pioneering brand can retain a large market share because consumers are unwilling to investigate the quality of new brands.

Recommended Reading

Fudenberg, Drew & Eric Maskin (1986) "The Folk Theorem in Repeated Games with Discounting or with Incomplete Information" *Econometrica*. May 1986. 54, 3: 533–54.

Klein, Benjamin & Keith Leffler (1981) "The Role of Market Forces in Assuring

Contractual Performance" *Journal of Political Economy.* August 1981. 89, 4: 615–41.

Selten, Reinhard (1978) "The Chain-Store Paradox" *Theory and Decision.* April 1978. 9, 2: 127–59.

Problem 4

Repeated Entry Deterrence

Consider two repetitions without discounting of the game Entry Deterrence I from Section 4.3. Assume that there is one entrant, who sequentially decides whether to enter two markets that have the same incumbent.

(1)   Draw the extensive form of this game.
(2)   What is one strategy combination for this game?
(3)   What is a subgame perfect equilibrium?
(4)   What is one of the non-perfect Nash equilibria?

Notes

N4.2   Subgame Perfectness

- Often "perfectness" is called "perfection." "Perfectness" is the term used in Selten (1975), and conveys an impression of completeness more appropriate to the concept than the goodness implied by "perfection".
- Perfectness is not the only way to eliminate weak Nash equilibria like *(Stay Out, Collude)*. In Entry Deterrence I, *(Enter, Collude)* is the only iterated dominant strategy equilibrium, because *Fight* is weakly dominated for the incumbent.
- The distinction between perfect and non-perfect Nash equilibria is like the distinction between **closed loop** and **open loop** trajectories in dynamic programming. Closed loop (or **feedback**) trajectories can be revised after they start, like perfect equilibrium strategies, while open loop trajectories are completely prespecified (though able to depend on state variables). In dynamic programming the distinction is not so important, because prespecified strategies do not change the behavior of other players. No threat, for example, is going to alter the pull of the moon's gravity on a rocket.
- A subgame can be infinite in length, and infinite games can have non-perfect equilibria. The infinitely repeated Prisoner's Dilemma is an example, in which every subgame looks exactly like the original game, but begins at a different point in time.
- **Perfectness in Macroeconomics.** In macroeconomics the requirement of **dynamic consistency** or **time consistency** is similar to perfectness. These terms are less precisely defined than perfectness, but they usually require only that strategies be best responses in subgames starting from nodes on the equilibrium path, instead of all subgames. Under this interpretation, time consistency is a less stringent condition than perfectness.

  The Federal Reserve would like to induce inflation to stimulate the economy, but the economy is stimulated only if the inflation is unexpected. If the inflation is expected, its effects are purely bad. Since members of the public know that the Fed would like to fool them, they disbelieve its claims that it will not generate inflation (see Kydland & Prescott [1977]). Likewise, the government would like to issue nominal debt, and promises lenders that it will keep inflation low, but once the debt is issued, the government has incentive to inflate its real value to zero. One reason that the Federal Reserve Board was established to be independent of Congress in the United States was to diminish this problem.

- In many situations, irrationality—behavior that is automatic rather than strategic—is an advantage. The Doomsday Machine of the movie *Dr Strangelove* is one example: it blows up the world if anyone explodes a nuclear bomb (although, as Dr Strangelove says, it is worse than useless if no one tells the bomb-owners about the machine).

  President Nixon reportedly told his aide H.R. Haldeman about a more complicated version of this strategy: "I call it the Madman Theory, Bob. I want the North Vietnamese to believe that I've reached the point where I might do *anything* to stop the war. We'll just slip the word to them that 'for God's sake, you know Nixon is obsessed about Communism. We can't restrain him when he's angry—and he has his hand on the nuclear button—and Ho Chi Minh himself will be in Paris in two days begging for peace,"(Haldeman & DiMona [1978] p. 83). The Gang of Four model in Section 5.4 tries to model a situation like this.

- The "lock-up agreement' (see Macey & McChesney [1985] p. 33) is an example of a credible threat: in a takeover defense, the threat to destroy the firm is made legally binding.

- Bernheim, Peleg, & Whinston (1987) and Bernheim & Whinston (1987) introduce the equilibrium concept of **coalition-proof Nash equilibrium**. In this refinement of Nash, a strategy combination is an equilibrium only if no coalition of players could form a self-enforcing agreement to deviate from it. The concept is easily extended to include perfectness.

N4.3   An Example of Perfectness: Entry Deterrence I

- The Stackelberg equilibrium of a duopoly game (Section 3.4) can be viewed as the perfect equilibrium of a Cournot Game modified so that one player moves first, a game similar to Entry Deterrence I. The player moving first is the Stackelberg leader and the player moving second is the Stackelberg follower. The follower could threaten to produce a large output, but he will not carry out his threat if the leader produces a large output first.

- Perfectness is not so desirable a property of equilibrium in biological games. The reason the order of moves matters is because the rational best reply depends on the node at which the game has arrived. In many biological games the players act by instinct and unthinking behavior is not unrealistic (see Section 5.6).

- Reinganum & Stokey (1985) is a clear presentation of the implications of perfectness and commitment illustrated with the example of natural resource extraction.

N4.4   Finitely Repeated Games and the Chainstore Paradox

- The Chainstore Paradox does not apply to all games as neatly as to Entry Deterrence and the Prisoner's Dilemma. If the one-shot game has only one Nash equilibrium, the perfect equilibrium of the finitely repeated game is unique and has that same outcome. But if the one-shot game has multiple Nash equilibria, the perfect equilibrium of the finitely repeated game can have not only the one-shot outcomes, but others besides. See Benoit & Krishna (1985), Harrington (1987), and Moreaux (1985).

- The quotation is attributed to Soren Kierkegaard's *Life* in John Bartlett, *Familiar Quotations*, 14th edition, Boston: Little, Brown, and Co., 1968, p. 676. Bartlett's reference is vague and should be treated skeptically.

- The peculiarity of the unique Nash equilibrium for the Repeated Prisoner's Dilemma was noticed long before Selten (1978) (see Luce & Raiffa [1957] p. 99), but the term "Chainstore Paradox" is now generally used for all unravelling games of this kind.

- *An* **epsilon equilibrium** *is a strategy combination s^* such that no player has more than an ϵ incentive to deviate from his strategy given that the other players do not deviate. Formally,*

$$\forall i, \quad \pi_i(s_i^*, s_{-i}^*) \geq \pi_i(s_i', s_{-i}^*) - \epsilon, \quad \forall s_i' \in S_i. \tag{4.8}$$

Radner (1980) has shown that cooperation can arise as an ϵ-equilibrium of the finitely repeated Prisoner's Dilemma. Fudenberg & Levine (1986) compare the ϵ-equilibria of finite games with the Nash equilibria of infinite games. Other concepts besides Nash can also use the ϵ-equilibrium idea.

- A general way to decide whether a mathematical result is a trick of infinity is to see if the same result is obtained as the limit of results for longer and longer finite models. Applied to games, a good criterion for picking among equilibria of an infinite game is to select one which is the limit of the equilibria for finite games as the number of periods gets longer. Fudenberg & Levine (1986) show under what conditions one can find the equilibria of infinite horizon games by this process. For the Prisoner's Dilemma, (*Always Fink*) is the only equilibrium in all finite games, so it uniquely satisfies the criterion.

- A **Markov strategy** *is a strategy that at each node chooses the action independently of the history of the game except for the immediately preceding action (or actions, if they were simultaneous) and except as the history determines the action set.*

- Markov strategies are memoryless. If the players in an undiscounted, infinitely repeated Prisoner's Dilemma are restricted to Markov strategies, the strategy set is reduced to the eight strategies of the form (*Begin with X; pick Y if the other player finked, pick Z if the other player cooperated*). If we consider only Markov pure strategies, the unique iterated dominant strategy equilibrium is for both players to choose tit-for-tat (Aumann [1981]). See Section 12.4 for another application of Markov strategies, to customers who have switching costs.

- There are two ways to use Markov strategies: (1) just look for equilibria that use Markov strategies, and (2) disallow non-Markov strategies and then look for equilibria. Because the first way does not disallow non-Markov strategies, the equilibrium must be such that no player wants to deviate by using any other strategy, Markov or not. This is just a way of eliminating possible multiple equilibria by discarding ones that use non-Markov strategies. The second way is much more dubious, because it requires the players not to use non-Markov strategies, even if they are best responses.

- Defining payoffs in games that last an infinite number of periods presents the problem that the total payoff is infinite for any positive payment per period. Ways to distinguish one infinite amount from another include

  (1) Use an **overtaking criterion**. Payoff stream π is preferred to $\tilde{\pi}$ if there is some time T^* such that for every $T \geq T^*$,

$$\sum_{t=1}^{T} \delta^t \pi_t > \sum_{t=1}^{T} \delta^t \tilde{\pi}_t.$$

  (2) Specify that the discount rate is strictly positive, and use the present value. Since payments in faroff periods count for less, the discounted value is finite unless the payments are growing faster than the discount rate.

  (3) Use the average payment per period, a tricky method since some sort of limit needs to be taken as the number of periods averaged goes to infinity.

  Whatever the approach, game theorists assume that the payoff function is **additively separable** over time, which means that the total payoff is based on the sum or average, possibly discounted, of the one-shot payoffs. Macroeconomists worry about this assumption, which rules out, for example, a player whose payoff is very low if any of his one-shot payoffs dips below a certain subsistence level. The issue of separability will arise again in Section 12.5 when we discuss durable monopoly.

- Ending in finite time with probability one means that the limit of the probability the game has ended by date t goes to one as t goes to infinity; the probability that the game lasts till infinity is zero. Equivalently, the expectation of the end date is finite, which it could not be were there a positive probability of an infinite length.

- A realistic expansion of a game's strategy may eliminate the Chainstore Paradox. Hirshleifer & Rasmusen (forth), for example, show that allowing the players in a multi-person, finitely repeated Prisoner's Dilemma to ostracize offenders can enforce cooperation even if there are economies of scale in the number of players who cooperate and are not ostracized.

N4.6   Infinitely Repeated Games and the Folk Theorem

- References on the Folk Theorem include Aumann (1981), Fudenberg & Maskin (1986), and Rasmusen (unpub). The most commonly cited version of the Folk Theorem says that if Conditions 1 to 3 are satisfied, then

Any payoff combination that strictly Pareto-dominates the minimax payoff combinations in the mixed extension of an n-person one-shot game with finite action sets is the average payoff in some perfect equilibrium of the infinitely repeated game.

- **Trigger strategies** are an important kind of strategy for repeated games. Consider the oligopolist facing uncertain demand (as in Stigler [1964]). He cannot tell whether the low demand he observes facing him is due to Nature or price cutting by his fellow oligopolists. Two things that could trigger him to cut his own price in retaliation are a series of periods with low demand or one period of especially low demand. Finding an optimal trigger strategy is a difficult problem (see Porter [1983a]). Trigger strategies are usually not subgame perfect unless the game is infinitely repeated, in which case they are a subset of the equilibrium strategies.

  Empirical work on trigger strategies includes Porter (1983b), who examines price wars between railroads in the 19th century, and Slade (1987) who concluded that price wars among gas stations in Vancouver used small punishments for small deviations rather than big punishments for big deviations.

- The reason why games with a constant probability of ending are like infinite games is nicely pointed out by a verse from *Amazing Grace*:

  When we've been there ten thousand years,
  Bright shining as the sun,
  We've no less days to sing God's praise
  Than when we'd first begun.

- A macroeconomist's technical note related to the similarity of constant-ending games and infinite games is Blanchard (1979), which discusses speculative bubbles.

- In the repeated Prisoner's Dilemma, if the end date is infinite with positive probability and only one player knows it, cooperation is possible for reasons explained in Section 5.4.

- An equilibrium concept quite different from Nash or dominance is based on maximin strategies.

The strategy s_i^* is a **maximin strategy** *for player i if, given that the other players pick strategies to make his payoff as low as possible, s_i^* gives him the highest possible payoff. In our notation, s_i^* solves*

$$\underset{s_i}{Maximize} \ \underset{s_{-i}}{Minimum} \ \pi_i(s_i, s_{-i}). \tag{4.9}$$

The maximin strategy need not be unique, and it can be in mixed strategies. Since maximin behavior can also be viewed as minimizing the maximum loss that might be suffered, decision theorists refer to such a policy as a **minimax criterion,** a catchier phrase (Luce & Raiffa [1957] p. 279).

Maximin strategies have very little justification. They are not simply the optimal strategies for risk-averse players, because risk aversion is accounted for in the utility payoffs. The players' implicit beliefs can be inconsistent in a maximin equilibrium, and a player must believe that his opponent would choose the most harmful strategy out of spite rather than self-interest.

In the same vein, the minimax strategy has also been defined (the term is used differently by game theorists and decision theorists).

*The strategy s^{i*}_{-i} is a set of $(n-1)$ **minimax strategies** chosen by all the players except i to keep i's payoff as low as possible, no matter how he responds. In our notation, s^{i*}_{-i} solves*

$$\underset{s_{-i}}{Minimize}\; \underset{s_i}{Maximum}\; \pi_i(s_i, s_{-i}). \tag{4.10}$$

As an example, consider a two player game with player 1 as i.

$$\text{Maximin:}\quad \underset{s_1}{Maximize}\; \underset{s_2}{Minimum}\; \pi_1.$$

$$\text{Minimax:}\quad \underset{s_2}{Minimize}\; \underset{s_1}{Maximum}\; \pi_1.$$

In the Prisoner's Dilemma, the minimax and maximin strategies are both *Fink*. Although the Welfare Game (Table 3.1) has only a mixed strategy Nash equilibrium, if we restrict ourselves to the pure strategies the pauper's maximin strategy is *Try to Work*, which guarantees him at least 1, and his strategy for minimaxing the government is *Be Idle*, which prevents the government from getting more than 0.

Under minimax, player 2 is purely malicious but must move first (at least in choosing a mixing probability) in his attempt to cause player 1 the maximum pain. Under maximin, player 1 moves first, in the belief that player 2 is out to get him. In non-zero-sum games, minimax is for sadists and maximin for paranoids. In zero-sum games, the players are merely neurotic: minimax is for optimists and maximin for pessimists.

- The **Minimax Theorem** (von Neumann [1928]) says that a minimax equilibrium exists in pure or mixed strategies for every two-person zero-sum game, and is identical to the maximin equilibrium. This theorem, while famous, is not useful in economics.

- An alternative to Condition 3 (dimensionality) in the Folk Theorem is

   Condition 3': *The repeated game has a "desirable" subgame-perfect equilibrium in which the strategy combination $\bar{s}$ played each period gives player i a payoff that exceeds his payoff from some other "punishment" subgame-perfect equilibrium in which the strategy combination $\underline{s}^i$ is played each period:*

$$\exists \bar{s} : \forall i,\ \exists \underline{s}^i : \pi_i(\underline{s}^i) < \pi_i(\bar{s}). \tag{4.11}$$

   Condition 3' is useful because sometimes it is easy to find a few perfect equilibria. To enforce the desired pattern of behavior, use the "desirable" equilibrium as a carrot and the "punishment" equilibrium as a self-enforcing stick. (See Rasmusen [unpub].)

- Any Nash equilibrium of the one-shot game is also a perfect equilibrium of the finitely or infinitely repeated game.

N4.7   Reputation: the One-Sided Prisoner's Dilemma

- *A game that is repeated an infinite number of times without discounting is called a* **supergame**.

   There is no connection between the terms "supergame" and "subgame."

- The terms "one-sided" and "two-sided" Prisoner's Dilemma are new with this book. Only the two-sided version is a true Prisoner's Dilemma according to the definition of note N1.2.

- Empirical work on reputation is scarce. One worthwhile effort is Jarrell & Peltzman (1985), which finds that product recalls inflict costs greatly in excess of the measurable direct costs of the operations. The sociologist Macaulay (1963) is much cited and little imitated. He notes that reputation seems to be more important than the written details of business contracts.

- **Vengeance and Gratitude.** Most models have excluded these feelings (although see J. Hirshleifer [1987]), which can be modelled in two ways:

   (1)   A player's current utility from *Fink* or *Cooperate* depends on what the other player has played in the past; or

(2)   A player's current utility depends on current actions and the other players' current utility in a way that changes with past actions of the other player.

The two approaches are subtly different in interpretation. In (1), the joy of revenge is in the action of finking. In (2), the joy of revenge is in the discomfiture of the other player. Especially if the players have different payoff functions, these two approaches can lead to different results.

N4.8   Product Quality in an Infinitely Repeated Game

- The product quality game may also be viewed as a principal agent model of moral hazard (see Chapter 6). The seller (an agent), takes the action of choosing quality that is unobserved by the buyer (the principal), but which affects the principal's payoff, an interpretation used in much of the Stiglitz (1987) survey of the links between quality and price.

  The intuition behind the Klein & Leffler model is similar to the explanation for high wages in the Shapiro & Stiglitz (1984) model of involuntary unemployment (Section 7.3). Consumers, seeing a low price, realize that with a price that low the firm cannot resist lowering quality to make short-term profits. A large margin of profit is needed for the firm to decide to continue to produce high quality.

- A paper related to Klein & Leffler (1981) is Shapiro (1983), which reconciles a high price with free entry by requiring that firms price under cost during the early periods to build up a reputation. If consumers believe, for example, that any firm charging a high price for any of the first five periods has produced a low quality product, but any firm charging a high price thereafter has produced high quality, then firms behave accordingly and the beliefs are confirmed. That the beliefs are self-confirming does not make them irrational; it only means that many different beliefs are rational in the many different equilibria.

- An equilibrium exists in the product quality model only if the entry cost F is just the right size to make n an integer in equation (4.7). Any of the usual assumptions to get around the integer problem could be used: allowing potential sellers to randomize between entering and staying out; assuming that for historical reasons n firms have already entered; or assuming that firms lie on a continuum and the fixed cost is a uniform density across firms that have entered.

5 Dynamic Games with Asymmetric Information

5.1 Perfect Bayesian Equilibrium: Entry Deterrence II and III

In Chapter 4 we tried various ways to steer between the Scylla of the Chainstore Paradox and the Charybdis of the Folk Theorem. We found that uncertainty did not make a big difference, but we did not make information incomplete. One might imagine that if the players did not know each others' types the resulting confusion might allow cooperation. To model this idea, we need to begin by redefining the equilibrium concept to fit games of asymmetric information. The basic concept, perfect Bayesian equilibrium, is introduced in Section 5.1 and criticized in Section 5.2. Section 5.3 applies it to show that a player's ignorance can benefit him, and to show that common knowledge means more than just that all the players observe something happen. Section 5.4 returns to the Chainstore Paradox, and shows how the Gang of Four model generates cooperation in a finitely repeated game of incomplete information. We then turn to quite different approaches to repeated games: the experimental tournament of Axelrod (Section 5.5) and the evolutionary equilibrium of Maynard Smith (Section 5.6). Section 5.7 closes Part I of the book with a look at theorems for the existence of equilibrium.

We still require that an equilibrium be subgame perfect in a game of asymmetric information, but the mere forking of the game tree might not be relevant to a player's decision. Player Smith might know he is at one of two different nodes of the game tree, depending on whether Brown has high or low production costs, but if Smith does not know the exact node, the "subgames" starting at each node are irrelevant to his decisions. In fact, they are not even subgames as we have defined them, because they cut across Smith's information sets.

Subgame Perfectness is Not Enough

To see how asymmetric information makes subgame perfectness unsatisfactory, let us look at an asymmetric information version of Entry Deterrence I (see Section 4.3). Suppose that with 50 percent probability the incumbent's payoff from *Fight* is x, rather than the 0 in Figure 4.2, and Nature's choice of 0 or x is unobserved by the incumbent. We model this as an initial move by Nature, who chooses between $(-10, 0)$ (*Zero*) and $(-10, x)$ (*X*) for the payoffs from (*Enter, Fight*). Figure 5.1 shows the extensive form of this game, which we will call Entry Deterrence II, III, or IV depending on the value of x.

Figure 5.1   Entry Deterrence II, III, IV

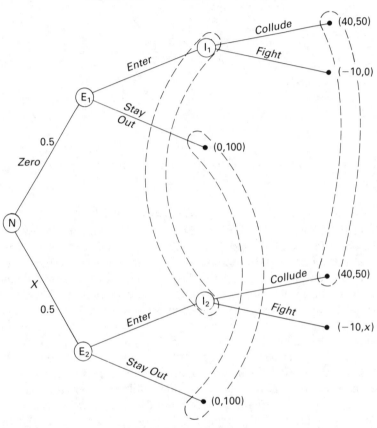

Payoffs to: (Entrant, Incumbent)

For Entry Deterrence II, assume $x = 1$, so information is not very asymmetric; it is common knowledge that the incumbent never benefits from *Fight*, even though his exact payoff is unknown. Unlike in Entry Deterrence I, subgame perfectness does not rule out any Nash equilibria in Entry Deterrence II, because the only subgame is that starting at Node N, the entire game. A subgame cannot start at Nodes E_1 or E_2, because neither of those nodes are singletons in the information partitions. Thus the bad Nash equilibrium, (*Stay Out, Fight*), escapes elimination by a technicality.

Two general approaches can be taken to refining the equilibrium concept to eliminate bad equilibria: introducing small "trembles" into the game, and requiring that strategies be best responses given rational beliefs. The first approach takes us to "trembling hand perfect" equilibrium, while the second takes us to "perfect

Bayesian" and "sequential" equilibrium. The results are similar whichever approach is taken.

Trembling Hand Perfectness

Trembling hand perfectness is an equilibrium concept introduced by Selten (1975) which says that for a strategy to be part of an equilibrium it must continue to be optimal for the player even if there is a small chance that the other player will pick some out-of-equilibrium action (that the other player's hand will "tremble"). Trembles certainly do occur in the real world. In an SEC filing, Crazy Eddie Inc, an electronics retail chain, reported that sales from stores open a year or more were up only 5 percent. That was a mistake: the true figure was 15 percent, and the understatement hurt the company because analysts steered investors away (*Wall Street Journal*, 11 August 1987, p. 27). It turned out that the company really was in trouble, for reasons that the filing did not reveal. What should investors have thought if they found out that Crazy Eddie underreported, to its own detriment? Was a badly run company more likely to tremble?

Trembling hand perfectness is defined for games with finite action sets as follows.

> *The strategy combination s^* is a **trembling hand perfect** equilibrium if for any ϵ there is a vector of positive numbers $\delta_1, \ldots, \delta_n$ and a vector of completely mixed strategies $\sigma_1, \ldots, \sigma_n$ such that the perturbed game where every strategy is replaced by $(1 - \delta_i)s_i + \delta_i\sigma_i$ has a Nash equilibrium in which every strategy is within distance ϵ of s^*.*

Every trembling hand perfect equilibrium is subgame perfect; indeed, in Section 4.2 we justified subgame perfectness using a tremble argument. Unfortunately, it is often hard to tell whether a strategy combination is trembling hand perfect, and the concept is undefined for games with continuous strategy spaces because it is hard to work with mixtures of a continuum (see note N3.3). Moreover, the equilibrium depends on which trembles are chosen, and rationalizing why one tremble is chosen over another may be difficult.

Perfect Bayesian Equilibrium and Sequential Equilibrium

The second approach to asymmetric information, introduced by Kreps & Wilson (1982b) in the spirit of Harsanyi (1967), is to start with prior beliefs common to all players that specify the probabilities with which Nature chooses the types of the players at the beginning of the game. Some of the players observe Nature's move and update their beliefs, while other players can update their beliefs only by deductions they make from observing the actions of the informed players.

The deductions used to update beliefs are based on the actions specified by the equilibrium. When players update their beliefs, they assume that the other players are following the equilibrium strategies, but since the strategies themselves depend on the beliefs, an equilibrium can no longer be defined based on strategies alone. Under asymmetric information, an equilibrium is a strategy combination and a set of beliefs such that the strategies are best responses. The combination of beliefs and strategies is called an **assessment**.

On the equilibrium path, all that the players need to update their beliefs are their priors and Bayes's Rule, but off the equilibrium path this is not enough. Suppose that in equilibrium, the entrant always enters. If, for whatever reason, the impossible happens and the entrant stays out, what is the incumbent to think about the probability that Nature chose *Zero*? Bayes's Rule does not help, because when $Prob(data) = 0$, as for data that like *Stay Out* is never observed in equilibrium, the posterior belief cannot be calculated using Bayes's Rule. From Section 2.5,

$$Prob(Zero|Stay\ Out) = \frac{Prob(Stay\ Out|Zero)Prob(Zero)}{Prob(Stay\ Out)}. \qquad (5.1)$$

But the posterior $Prob(Zero|Stay\ Out)$ is undefined, because (5.1) requires dividing by zero. Does the incumbent decide that a *Zero*-entrant is more likely to tremble than an X-entrant? A natural way to define equilibrium is as a strategy combination consisting of best responses given that equilibrium beliefs follow Bayes's Rule and out-of-equilibrium beliefs follow a pattern that does not contradict Bayes's Rule.

A **perfect Bayesian equilibrium** *is a strategy combination* s *and a set of beliefs* μ *such that at each node of the game:*

(1) The strategies for the remainder of the game are Nash given the beliefs and strategies of the other players.

(2) The beliefs at each information set are rational given the evidence appearing that far in the game (meaning that they are based, if possible, on priors updated by Bayes's Rule given the observed actions of the other players under the hypothesis that they are in equilibrium).

Kreps & Wilson (1982b) use the same idea to introduce their equilibrium concept of **sequential equilibrium**, but they impose a third condition to restrict beliefs a little further. Their third condition is defined only for games with discrete strategies. This adds:

(3) The beliefs are the limit of a sequence of rational beliefs, i.e., if (μ^*, s^*) *is the equilibrium assessment, then some sequence of rational beliefs and completely mixed strategies converges to it:*

$$(\mu^*, s^*) = Lim_{n \to \infty}(\mu^n, s^n) \text{ for some sequence } (\mu^n, s^n) \text{ in } \{\mu, s\}.$$

The third condition makes sequential equilibrium close to trembling hand perfect equilibrium, but it adds more to the concept's difficulty than to its usefulness. If players are using the sequence of completely mixed strategies s^n, then every action is taken with some positive probability, so Bayes's Rule can be applied to form the beliefs μ^n after any action is observed. Condition (3) says that the equilibrium assessment has to be the limit of some such sequence (though not of every such sequence). For the rest of the book we will simply use perfect Bayesian equilibrium, and dispense with condition (3), although it usually can be satisfied.

Sequential equilibria are always subgame perfect (condition (1) takes care of that). Every trembling hand perfect equilibrium is a sequential equilibrium. "Almost every" sequential equilibrium is trembling hand perfect. Every sequential equilibrium is Bayesian perfect, but not every Bayesian perfect equilibrium is sequential.

An Equilibrium for Entry Deterrence II

Armed with the concept of the perfect Bayesian equilibrium, we can find a sensible equilibrium assessment for Entry Deterrence II.

> (*Enter* if Nature chose *Zero*), (*Enter* if Nature chose *X*).
> (*Collude*).
> *Prob*(*Zero* | *Stay Out*) = 0.4.

In this assessment, the entrant enters whether he observes Nature choosing *Zero* or *X*. The incumbent's strategy is *Collude*, which is not conditioned on Nature's move, since he does not observe it. Because the entrant enters regardless of Nature's move, an out-of-equilibrium belief for the incumbent if he should observe *Stay Out* must be specified, and this belief is arbitrarily chosen to be that the probability Nature chose *Zero* is 0.4 if the entrant deviated from equilibrium. Given this assessment, neither player has incentive to change his strategy.

There is no perfect Bayesian equilibrium in which the entrant chooses *Stay Out*. Perfect Bayesian equilibrium is not defined structurally, like subgame perfectness, but rather in terms of optimal responses. *Fight* is a bad response even under the most optimistic possible belief: that Nature chose *X* with probability one.

Finding the perfect Bayesian equilibria of a game, like finding the Nash equilibria, requires intelligence; there is no algorithm. To find a Nash equilibrium, the modeller thinks about his game, picks a plausible strategy combination, and tests whether the strategies are best responses to each other. To make it a perfect Bayesian equilibrium, he notes which actions are never taken in equilibrium and specifies the beliefs players use to interpret those actions. He then tests whether each player's strategies are best responses given his beliefs at each node, checking in particular whether any player would like to take an out-of-equilibrium action in order to set in motion the other players' out-of-equilibrium beliefs and strategies. Note that this process does not involve testing whether a player's beliefs are beneficial to him, because a player cannot choose his beliefs; the priors and out-of-equilibrium beliefs are exogenously specified by the modeller.

One might wonder why the beliefs have to be specified in Entry Deterrence II. Doesn't the game tree specify the probability of *Zero*? And what difference does it make if the entrant stays out? Admittedly, Nature does choose each type with probability 0.5, so if the incumbent had no other information than this prior, that would be his belief. But the entrant's action might convey additional information. The concept of perfect Bayesian equilibrium leaves the modeller free to specify how the players form beliefs from that additional information, so long as the beliefs do not violate Bayes's Rule. (A technically valid choice of beliefs by the modeller might still be met with scorn, however, as with any silly assumption.) Here, the equilibrium says that if the entrant stays out the incumbent believes that Nature chose *Zero* with probability 0.4 and *X* with probability 0.6, beliefs that are arbitrary, and might be scorned, but do not contradict Bayes's Rule.

Whether beliefs matter after the entrant has stayed out is a different question, and in this example they do not. In this section, perfect Bayesian equilibrium has been introduced only as a way out of a technical problem. In the next section, out-of-equilibrium beliefs will be crucial to which strategy combinations are equilibria.

5.2   Refining Perfect Bayesian Equilibrium: PhD Admissions

Entry Deterrence III: a Good Equilibrium

Now assume $x = 60$, not $x = 1$, so fighting is profitable for the incumbent if Nature chooses X. The entrant observes Nature's move, and hence knows the values of the payoffs, but the incumbent remains ignorant. The following is a perfect Bayesian equilibrium in which the incumbent's out-of-equilibrium belief is that if the entrant stays out, Nature chose *Zero* with probability 0.5 and X with probability 0.5.

(*Enter* if Nature chose *Zero*), (*Enter* if Nature chose X).
(*Collude*).
Prob(*Zero* | *Stay Out*) = 0.5.

In choosing whether to enter, the entrant must predict the incumbent's behavior. If the probability of each (*Enter, Fight*) payoff is 0.5, the expected payoff for the incumbent from choosing *Fight* is 30 ($= 0.5[0] + 0.5[60]$), which is less than the *Collude* payoff of 50. The incumbent will collude, so the entrant enters. The entrant may know that the incumbent's payoff is actually 60, but that is irrelevant to the incumbent's behavior.

The out-of-equilibrium belief does not matter to this equilibrium. Retaining the prior, which in this game is *Prob*(*Zero*) = 0.5, is a convenient way to form beliefs that is called **passive conjectures.**

Entry Deterrence III: a Bad Equilibrium

While beliefs in a perfect Bayesian equilibrium must follow Bayes's Rule, that puts very little restriction on how players interpret out-of-equilibrium behaviour. Out-of-equilibrium behaviour is "impossible," so when it does occur there is no obvious way the player should react. Some beliefs may seem more reasonable than others, however, and Entry Deterrence III has another equilibrium with the outcome that the entrant stays out:

(*Stay Out* if Nature chose *Zero*), (*Stay Out* if Nature chose X).
(*Fight*).
Prob(*Zero*|*Enter*) = 0.1.

This is an equilibrium because if the entrant were to enter, the incumbent would calculate his payoff from fighting to be 54 ($= 0.1[0] + 0.9[60]$), which is greater than the *Collude* payoff of 50. The entrant would therefore stay out.

The beliefs in the bad equilibrium are different and less reasonable than the beliefs in the good equilibrium. The beliefs do not violate Bayes's Rule, but they have no justification: why should the incumbent believe that entrants would enter mistakenly nine times as often when the incumbent benefits from choosing *Fight*?

The reasonableness of the beliefs is important here, because if the incumbent uses passive conjectures, the bad equilibrium breaks down. With passive conjectures, the incumbent would want to change his strategy to *Collude*, because the expected payoff from *Fight* would be less than 50. The bad equilibrium is less robust with respect to beliefs than the good equilibrium, and it requires beliefs that are hard to justify.

PhD Admissions

Passive conjectures may not always be the most satisfactory belief, as the next example shows. Suppose that the university knows that 90 percent of the population hates economics, and hence would be unhappy in its PhD program, but 10 percent love economics, and would do well. The university cannot observe the type of an applicant. If the university rejects an application, its payoff is 0 and the applicant's, because of his expenditure on postage, is -1. If the university accepts the application of someone who hates economics, their payoffs are both -10, but if the applicant loves economics, their payoffs are both $+20$. Figure 5.2 shows this extensive form, in which the population proportions are represented by a node at which Nature chooses the student to be a *Lover* or a *Hater*.

PhD Admissions is a signalling game of the kind we will look at in Chapter 9. It has various perfect Bayesian equilibria, depending on the beliefs, but they can be divided into **pooling equilibria**, in which neither type of student applies, and the **separating equilibrium**, in which the lovers of economics apply and the haters do not.

Separating Equilibrium

> (*Apply* if Nature chose *Lover*), (*Do Not Apply* if Nature chose *Hater*).
> (*Admit*).

The separating equilibrium does not need to specify out-of-equilibrium beliefs, because when both of the possible actions *Apply* and *Do Not Apply* could occur in equilibrium, Bayes's Rule can be used.

Pooling Equilibrium

> (*Do Not Apply* if Nature chose *Lover*), (*Do Not Apply* if Nature chose *Hater*).
> (*Reject*).
> $Prob(Hater \mid Apply) = 0.9$ (passive conjectures).

Passive conjectures supports the pooling equilibrium, in which the students refrain from applying, believing they would be rejected and receive a payoff of -1 if they applied; and the university is willing to reject any student who foolishly applied, believing that an applicant has a 90 percent chance of being a *Hater*.

Figure 5.2   PhD Admissions

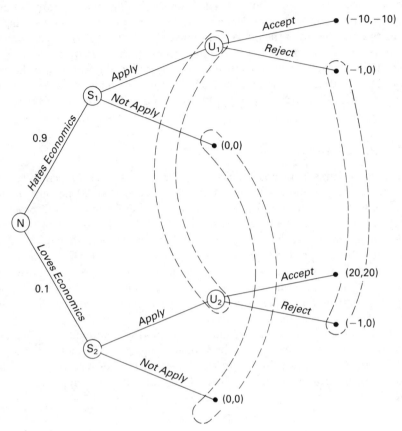

Payoffs to: (Student,University)

Because the perfect Bayesian equilibrium concept imposes no restrictions on out-of-equilibrium beliefs, researchers starting with McLennan (1985) have come up with a variety of what I call "exotic refinements" of the equilibrium concept. Note N5.2 lists a number of these, and we will return to the problem of beliefs again when we look at signalling models of education in Section 9.2. Let us consider whether various beliefs would support a pooling equilibrium in PhD Admissions.

Passive Conjectures

$$Prob(Hater \mid Apply) = 0.9.$$

This is the belief specified above, under which the prior beliefs are unchanged by out-of-equilibrium behavior. The argument for this belief is that applying is a mistake, and both types are equally likely to make mistakes, although *Haters* are more common in the population. This supports the pooling equilibrium.

An Ad Hoc Argument

$Prob(Hater \mid Apply) = 1.$

Sometimes the modeller can justify beliefs by the circumstances of the particular game. Here, we could argue that anyone so foolish as to apply when they know the university will reject them could not possibly have the good taste to love economics. This supports the pooling equilibrium also.

The Intuitive Criterion

$Prob(Hater \mid Apply) = 0.$

Under the Intuitive Criterion of Cho & Kreps (1987), if there is a type of informed player who could not benefit from the out-of-equilibrium action no matter what beliefs were held by the uninformed player, the uninformed player's belief must put zero probability on that type. Here, the *Hater* could not benefit from applying under any possible beliefs of the university, so the university puts zero probability on an applicant being a *Hater*. This argument will not support the pooling equilibrium, because if the university holds this belief, it will want to admit anyone who applies.

Complete Robustness

$Prob(Hater \mid Apply) = m, \forall m, 0 \leq m \leq 1.$

Under this approach, the equilibrium strategy combination must consist of responses that are best given any and all out-of-equilibrium beliefs. Our equilibrium for Entry Deterrence II satisfied this requirement. Complete robustness rules out a pooling equilibrium in PhD Admissions, because a belief like $m = 0$ makes accepting applicants a best response, in which case only the *Lovers* will apply.

An alternative approach to the problem of out-of-equilibrium beliefs is to remove its origin by building a model in which every outcome is possible in equilibrium, because different types of players take different equilibrium actions. In PhD Admissions, we could assume that there are a few students who both love economics and actually enjoy writing applications. Those students would always apply in equilibrium, so there would never be a pure pooling equilibrium in which nobody applied, and Bayes's Rule could always be used. In equilibrium, the university would always accept someone who applied, because applying is never out-of-equilibrium behavior and it always indicates that the applicant is a *Lover*.

These arguments can also be applied to the beliefs in Entry Deterrence III, which had two different pooling equilibria and no separating equilibrium. We used passive conjectures in the "good equilibrium". The intuitive criterion would not restrict beliefs at all, because in Entry Deterrence III both types would enter if the incumbent's beliefs were such as to make him collude, and both would stay out if they made him fight. Complete robustness would rule out the equilibrium in which the entrant stays out, because the optimality of that strategy depends on the beliefs, but it would support the equilibrium in which the entrant enters and out-of-equilibrium beliefs do not matter.

5.3   The Importance of Common Knowledge: Entry Deterrence IV and V

To demonstrate the importance of common knowledge (Section 2.3), let us consider two more versions of Entry Deterrence. We will use passive conjectures in both. In Entry Deterrence III, the incumbent was hurt by his ignorance. Entry Deterrence IV will show how he can benefit from ignorance, and Entry Deterrence V will show what can happen when the incumbent has the same information as the entrant, but the information is not common knowledge.

Entry Deterrence IV: the Incumbent Benefits from his Ignorance

To construct Entry Deterrence IV, assume $x = 300$ in Figure 5.1, so fighting is even more profitable than in Entry Deterrence III but the game is otherwise the same: the entrant observes Nature's move, but the incumbent is ignorant. The following is a perfect Bayesian equilibrium:

> (*Stay Out* if Nature chose *Zero*), (*Stay Out* if Nature chose *X*).
> (*Fight*).
> *Prob*(*Zero* | *Enter*) = 0.5. (passive conjectures).

In Entry Deterrence IV, the incumbent's expected payoff from *Fight* is 150 (= $0.5[0] + 0.5[300]$), which is greater than the *Collude* payoff of 50. In contrast to Entry Deterrence III, the incumbent benefits from his own ignorance, because he would always fight entry, even if the payoff were (unknown to himself) just 0. The entrant would very much like to communicate the costliness of fighting, but the incumbent would not believe him, so entry never occurs.

Entry Deterrence V: Lack of Common Knowledge of Ignorance

In Entry Deterrence V, it is possible that both the entrant and the incumbent know the payoff from (*Enter*, *Fight*), but the entrant does not know whether the incumbent knows; the information is not common knowledge. Figure 5.3 depicts this somewhat complicated situation. If we ignore the dotted lines of the entrant's information sets, the subgame starting at Node N_2 is the same as Entry Deterrence IV, in which only the entrant knows the payoffs, and entry is deterred. But now we allow for the possibility that the incumbent might know the payoffs. In Figure 5.3 this takes the form of an even earlier move by Nature at N_1, where Nature chooses between the game in which only the entrant knows the payoffs and the game in which both players know them, assigned probabilities 0.9 and 0.1. We model the choice by Nature not in terms of players "knowing" this or that, but as whether Nature tells the incumbent or not. We assume further that the entrant

Figure 5.3   Entry Deterrence V

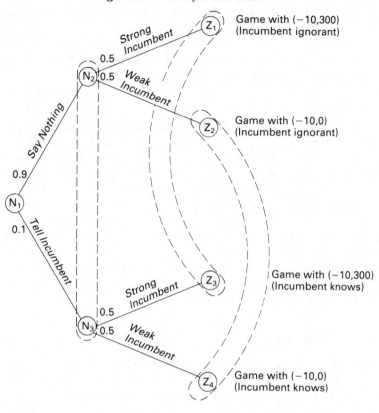

Payoffs to: (Entrant, Incumbent)

does not know whether Nature has told the incumbent anything, so that Nodes N_2 and N_3 are in the same information set for the entrant and further down the tree the partition is ($\{Z_1, Z_3\}, \{Z_2, Z_4\}$).

A perfect Bayesian equilibrium for Entry Deterrence V is

> (*Stay Out* if Nature chose *Zero*), (*Stay Out* if Nature chose *X*).
> (*Fight* if Nature told the incumbent *X*), (*Collude* if Nature told the incumbent *Zero*), (*Fight* if Nature said nothing).
> *Prob(Zero | Enter*, Nature said nothing) = 0.5. (passive conjectures).

Since the entrant puts a high probability on the incumbent not knowing, the entrant should stay out, because the incumbent will fight for either of two reasons: with probability 0.9 Nature has said nothing and the incumbent calculates his expected payoff from *Fight* to be 150, and with probability 0.05 (= 0.1[0.5]) Nature has told the incumbent that his payoff from *Fight* is 300. Even if the true payoff from *Fight* were 0 and the incumbent was told this by Nature, the entrant would choose *Stay Out*, because he does not know that the incumbent knows, and his expected payoff from *Enter* would be -7.5 ($= [0.9 + 0.05][-10] + 0.05[40]$).

5.4   Incomplete Information in the Repeated Prisoner's Dilemma: the Gang of Four Model

A possible escape route from the unhappy *Fink* equilibrium of the repeated Prisoner's Dilemma is to add incomplete information and look at the perfect Bayesian equilibria. A bad way to use incomplete information is to assume that a large number of players are of an irrational type. The incomplete information is that with high probability Row is a player who blindly follows the strategy tit-for-tat described in Section 4.6. If Column thinks he is playing against a tit-for-tat player, his optimal strategy is to *Cooperate* until near the last period (how near depends on the parameters), and then *Fink*. If he were not certain, but the probability was high that he faced a tit-for-tat player, Row would also choose that strategy. Such a model begs the question, because it is not the incompleteness of the information that drives the model, but the high probability that one player is a tit-for-tat player. Tit-for-tat is not a rational strategy, and to put lots of tit-for-tat players in the model is to assume away the problem.

The Gang of Four Model

One of the most important explanations of reputation is that of Kreps, Milgrom, Roberts, & Wilson (1982), hereafter referred to as the Gang of Four. In their model, a few firms are genuinely unable to play any strategy but tit-for-tat, and many firms pretend to be that type. The beauty of the model is that it requires only a little incomplete information: a small probability γ that player Row is a tit-for-tat player. It is not unreasonable to suppose that a world which contains Neo-Ricardians and McGovernites contains a few mildly irrational tit-for-tat players, and such behavior is especially plausible among consumers, who are subject to less evolutionary pressure than firms. The irrational players have a small direct influence, but they matter because other players imitate them. Even if Column knows that with high probability Row is just pretending to be tit-for-tat, Column does not care what the truth is so long as Row keeps on pretending. Not only is hypocrisy the tribute vice pays to virtue; it's just as good for maintaining social order.

Theorem 5.1: The Gang of Four Theorem

In the T-stage, repeated Prisoner's Dilemma without discounting but with a probability γ of a tit-for-tat player, in any perfect Bayesian equilibrium the number of stages in which either player chooses Fink is less than some number M that depends on γ but not on T.

The Gang of Four Theorem characterizes the equilibrium outcome rather than the equilibrium. Finding perfect Bayesian equilibria is difficult and tedious, since the modeller must check all the out-of-equilibrium subgames, as well as the equilibrium path. Modellers usually content themselves with describing important characteristics of the equilibrium strategies and payoffs.

The significance of the Gang of Four Theorem is that while the players do resort to *Fink* as the last period approaches, the number of periods during which they *Fink* is independent of the total number of periods. Perhaps $M = 2,500$, a very large number. If $T = 2,500$, there might be a *Fink* every period. But if $T = 10,000$ then there are 7,500 periods without a *Fink*. In applications, the model can look quite impressive. Wilson (unpub) has set up an entry deterrence model in which the incumbent fights entry (the equivalent of *Cooperate* above) up to seven periods from the end, although the probability the entrant is an irrational type is only 0.008.

The Gang of Four Theorem provides a way out of the Chainstore Paradox, but it creates a problem of multiple equilibria in much the same way as the infinitely repeated game. For one thing, if the asymmetry is two-sided, so that both players might be tit-for-tat players, the equilibrium is indeterminate. Also, a version of the Folk Theorem has been proven that says the modeller can make the average payoffs take any particular values by making the game long enough and choosing the form of the irrationality carefully:

Theorem 5.2: The Incomplete Information Folk Theorem (Fudenberg & Maskin [1986] p. 547)

For any two-person repeated game without discounting, the modeller can choose a form of irrationality such that for any probability $\epsilon > 0$ there is some finite number of repetitions such that with probability $(1 - \epsilon)$ a player is rational and the average payoffs in some sequential equilibrium are closer than ϵ to any desired payoffs greater than the minimax payoffs.

5.5   The Axelrod Tournament

Another way to approach the repeated Prisoner's Dilemma is through experiments such as the round-robin Prisoner's Dilemma tournament that political scientist Robert Axelrod describes in his 1984 book. Contestants submitted strategies for a 200-repetition game. Since the strategies could not be updated during play, players could precommit, but the strategies could be as complicated as they wished. If a player wanted to specify a strategy which simulated subgame perfectness by adapting to past history just as a non-committed player would, he was free to do so, but he could also submit a non-perfect strategy such as tit-for-tat or the slightly more forgiving tit-for-two-tats. Strategies were submitted in the form of computer programs that were matched with each other and played automatically. In Axelrod's first tournament, 14 programs were submitted as entries. Every program played every other program, and the winner was the one with the greatest sum of payoffs over all the plays.

The Axelrod approach is useful for showing which strategies are robust against a variety of other strategies in a game with given parameters. It is quite different from trying to find a Nash equilibrium, because it is not common knowledge what the equilibrium is in such a tournament. The situation could be viewed as a game of incomplete information in which Nature chooses the number and cognitive abilities of the players, and their priors regarding each other. Another source of confusion is that if there are multiple Nash equilibria, the players may use conflicting Nash strategies, a problem discussed in Section 1.4.

The winner of Axelrod's first tournament was Anatol Rapoport, whose strategy was tit-for-tat. After the results were announced, Axelrod ran a second tournament, adding a probability $\theta = 0.00346$ that the game would end each round so as to avoid the Chainstore Paradox. The winner among the 62 entrants was again Anatol Rapoport, and again he used tit-for-tat.

Rapoport had written an entire book on the Prisoner's Dilemma in analysis, experiment, and simulation (Rapoport & Chammah [1965]) before choosing his tournament strategy. Why did he choose tit-for-tat? Axelrod points out that tit-for-tat has three strong points:

(1)   It never initiates finking (**niceness**).
(2)   It retaliates instantly against finking (**provokability**).
(3)   It forgives a finker who goes back to cooperating (it is **forgiving**).

Care must be taken in interpreting the results of the tournament. For a variety of reasons it does not prove that tit-for-tat is the best strategy, and it certainly does not prove that collusion is inevitable.

(1)   Tit-for-tat never beats any other strategy in a one-on-one contest. It won the tournament by piling up points through cooperation, having lots of high-score plays and very few low-score plays. In an elimination tournament, tit-for-tat would be eliminated very early, because it scores *high* payoffs but never the *highest* payoff.
(2)   The other players' strategies are important. In neither tournament were the strategies a Nash equilibrium: if a player knew what strategies he was facing, he would want to revise his own. Some of the strategies submitted in the second tournament would have won the first, but they did poorly because the environment had changed. Other programs, designed to try to probe the strategies of their opposition, wasted too many finks on the learning process, but if the games had lasted a thousand repetitions they might have done better.
(3)   In a game in which players occasionally finked "accidentally," two tit-for-tat players facing each other would do very badly.

Optimality depends on the environment. When players are rational and that rationality is common knowledge, finking is the only equilibrium, but if most players are irrational, or most players think most players are irrational, the optimal strategy is different. Tit-for-tat is suboptimal for any given environment, but it is robust across environments, and that is its advantage.

5.6   Evolutionary Equilibrium: the Hawk–Dove Game

For most of this book we have been using the Nash equilibrium concept or refinements of it based on information and sequentiality, but in biology such concepts are often inappropriate. The lower animals are less likely than humans to think about the strategies of their opponents at each stage of a game. Their strategies are more likely to be pre-programmed and their strategy sets more restricted than the businessman's, if perhaps not his customer's. In addition, behavior evolves, and any equilibrium must take account of the possibility of odd behavior caused by the occasional mutation. That the equilibrium is common knowledge, or that players cannot precommit to strategies, are not compelling assumptions.

Game theory has grown to some importance in biology, but the style is different than in economics. The goal is not to explain how players would rationally pick actions in a given situation, but to explain how behavior evolves or persists over time under exogenous shocks. Both approaches end up defining equilibria to be strategy combinations that are best responses in some sense, but biologists care much more about the stability of the equilibrium and how strategies interact over time. In Section 3.4, we touched briefly on the stability of the Cournot equilibrium, but economists view stability as a pleasing byproduct of the equilibrium rather than its justification. For biologists, stability is the point of the analysis.

Consider a game with identical players who engage in pairwise contests. In this special context, it is useful to think of an equilibrium as a strategy combination such that no player with a new strategy can enter the environment (**invade**) and receive a higher expected payoff than the old players. Moreover, the invading strategy should continue to do well even if it plays itself with finite probability, or its invasion could never grow to significance. In the commonest model in biology, all the players adopt the same strategy, called an evolutionarily stable strategy, in equilibrium. John Maynard Smith originated this idea, which is somewhat confusing because it really aims at an equilibrium concept, and only applies to games with pairwise interactions and identical players.

*A strategy s_i^* is an **evolutionarily stable strategy** or **ESS** if, using the notation $\pi_i(s_i, s_{-i})$ for player i's payoff when his opponent uses strategy s_{-i}, for every other strategy s_i' either*

$$\pi_i(s_i^*, s_i^*) > \pi_i(s_i', s_i^*) \tag{5.2}$$

or

$$\text{if } \pi_i(s_i^*, s_i^*) = \pi_i(s_i', s_i^*), \text{ then } \pi_i(s_i^*, s_i') > \pi_i(s_i', s_i'). \tag{5.3}$$

If Condition (5.2) holds, then a population of players using s_i^* cannot be invaded by a deviant using s_i'. If Condition (5.3) holds, then s_i' does well against s_i^*, but badly against itself, so that if more than one player tried to use s_i' to invade a population using s_i^*, the invaders would fail.

We can interpret ESS in terms of Nash equilibrium. Condition (5.2) says that s^* is a strong Nash equilibrium (although not every strong Nash strategy is an ESS). Condition (5.3) says that if s_i^* is only a weak Nash strategy, the weak alternative

s_i' is not a best response to itself. The ESS is a refinement of Nash, narrowed by the requirement that the ESS be not only a best response, but (1) have the highest payoff of any strategy (which rules out equilibria with asymmetric payoffs) and (2) be a best response to itself. The motivations behind the two equilibrium concepts are quite different, but the similarities are useful because even if the modeller prefers ESS to Nash, he can start with the Nash strategies in his efforts to find an ESS.

An Example of ESS: the Hawk–Dove Game

The best-known illustration of the ESS is the Hawk–Dove Game. Imagine that we have two populations of birds, and each bird in each population can behave as an aggressive Hawk or a pacific Dove. A resource worth $V = 2$ "fitness units" is at stake when two birds meet. If they both fight, the loser incurs a cost of $C = 4$, which means that the expected payoff when two Hawks meet is -1 ($= 0.5[2] + 0.5[-4]$) for each of them. When two Doves meet, they split the resource, for a payoff of 1 apiece. When a Hawk meets a Dove, the Dove flees with a payoff of 0, leaving the Hawk with a payoff of 2. Table 5.1 summarizes this.

Table 5.1 The Hawk–Dove Game

		Beta	
		Hawk	*Dove*
	Hawk	$-1, -1$	$2, 0$
Alpha			
	Dove	$0, 2$	$1, 1$

Payoffs to: (Alpha, Beta)

The Hawk–Dove Game is Chicken (Section 3.3) with new feathers. The two games have the same ordinal ranking of payoffs, as can be seen by comparing Table 5.1 with Table 3.2, and their equilibria are the same except for the mixing parameters. The Hawk–Dove Game has no symmetric pure strategy Nash equilibrium, and hence no pure strategy ESS, since in the two asymmetric Nash equilibria the payoffs from *Hawk* are greater than from *Dove*. Neither Hawks nor Doves can completely take over the environment. If the population consisted entirely of Hawks, a Dove could invade and obtain a one-round payoff of 0 against a Hawk, compared to the -1 that a Hawk obtains against itself. If the population consisted entirely of Doves, a Hawk could invade and obtain a one-round payoff of 2 against a Dove, compared to the 1 that a Dove obtains against itself.

In the mixed strategy ESS, the equilibrium strategy is to be a Hawk with probability 0.5 and a Dove with probability 0.5, which can be interpreted as a population 50 percent Hawks and 50 percent Doves. As in the mixed strategy equilibria we computed in Chapter 3, the players are indifferent as to their strategies. The expected payoff from being a Hawk is the $0.5(2)$ from meeting a Dove plus the $0.5(-1)$ from meeting another Hawk, a sum of 0.5; the expected payoff from being a Dove is the $0.5(1)$ from meeting another Dove plus the $0.5(0)$ from meeting a Hawk, also a sum of 0.5. Moreover, the equilibrium is stable in a sense similar to the Cournot equilibrium. If 60 percent of the population were Hawks, for example, a bird would do better as a Dove. If "doing better" means being able to reproduce faster, the number of Doves increases and the proportion returns to 50 percent over time.

The ESS depends on the strategy sets allowed the players. If two birds can base their behavior on commonly observed random events such as which bird arrives at the resource first, and $V > C$ (as specified above), then a strategy called the **bourgeois strategy** is an ESS. Under this strategy, the bird respects property rights like a good bourgeois; it behaves as a Hawk if it arrives first, and a Dove if it arrives second, where we assume the order of arrival is random. The bourgeois strategy has an expected payoff of 1 from meeting itself, and behaves exactly like a 50:50 randomizer when it meets a strategy that ignores the order of arrival, so it can successfully invade a population of 50:50 randomizers. But the bourgeois strategy is a correlated strategy (see Section 3.3), and requires something like the order of arrival to decide which of two identical players will play Hawk.

The evolutionary approach can also be applied to the repeated Prisoner's Dilemma. Boyd & Lorberbaum (1987) show that no pure strategy, including tit-for-tat, is evolutionarily stable in a population-interaction version of the Prisoner's Dilemma. J. Hirshleifer & Martinez-Coll (1988) have found that tit-for-tat is no longer part of an ESS in an evolutionary Prisoner's Dilemma if (1) more compli-cated strategies have higher computation costs; or (2) sometimes a *Cooperate* is observed to be a *Fink* by the other player.

The ESS approach is suited to games in which all the players are identical and interacting in pairs. It does not apply to games with non-identical players—wolves who can be wily or big and deer who can be fast or strong. The approach follows three steps, specifying first the initial population proportions and the probabilities of interactions, second the pairwise interactions, and third the dynamics by which players with higher payoffs increase in number in the population. Economic games generally use only the second step, which describes the strategies and payoffs from a single interaction.

The third step, the evolutionary dynamics, is especially foreign to the style of game theory used in economics. In specifying dynamics, the modeller must specify a difference equation (for discrete time) or differential equation (for continuous time) that describes how the strategies employed change over iterations, whether because players differ in the number of their descendants or because they learn to change their strategies over time. In economic games, the adjustment process is usually degenerate: the players jump instantly to the equilibrium.

5.7   Existence of Equilibrium

As a coda to Part I of the book, I would like to discuss the issue of the existence of equilibrium. We have seen that often, as in the Welfare Game of Section 3.2, no pure strategy equilibrium exists. Sometimes the modeller cannot find a particular equilibrium easily, but he can show what characteristics an equilibrium must have if it exists, and he would like to prove that it does exist. For any reader interested in doing research using game theory, the list of theorems in this section will be quite useful. Others should read the paragraphs below but skip the list.

A little general discussion of existence would be helpful for everybody. One useful feature is continuity of the payoffs in the strategies. A payoff is continuous in a player's strategy if a small change in his strategy causes a small (or zero) change in the payoff. With continuity, the players' strategies can adjust more finely and come closer to being best responses to each other. In the Welfare Game, the payoffs are discontinuous in pure strategies, so no compromise between the players is possible and no pure strategy equilibrium exists. Once mixed strategies are incorporated, the payoffs are continuous in the mixing probability, so an equilibrium does exist.

A second feature promoting existence is a closed and bounded strategy set. Suppose in a stock market game that Smith can borrow money and buy as many shares of stock as he likes, so his strategy set, the amount of stock he can buy, is $[0, \infty)$, which is unbounded above (we assume he can buy fractional shares but cannot sell short). If Smith knows that the price today is lower than it will be tomorrow, he wants to buy an infinite number of shares, which is not an equilibrium purchase. If the amount he buys is restricted to be strictly less than 1,000, then the strategy set is bounded (by 1,000), but not closed, and no equilibrium purchase exists, because he wants to buy 999.999 ... shares. If the strategy set is closed and bounded, the extreme solutions are well-defined: if Smith can buy no more than 1,000 shares, that is his equilibrium purchase. Sometimes, as in the Cournot Game, the unboundedness of the strategy sets does not matter because the optimum is an interior solution, but it is an easy matter to close and bound the strategy sets—just require that $0 \leq q_a \leq 10,000$.

When an economist has proved that an equilibrium exists, and he is trying to characterize it, he often looks at the **comparative statics**: how the equilibrium changes when values of exogenous parameters are changed. No general results on comparative statics will be presented here, but the reader is warned that they can be especially tricky in game theory. If a parameter is changed, and one player changes his strategy in response, that change in turn induces a change in the strategies of the other players, so when the game settles down to its new equilibrium, the strategies might be quite different (Lippman, Mamer, & McCardle [1987] is a recent paper on the subject).

Existence Theorems

These theorems use many mathematical terms with which the average reader will not be familiar, but which are listed in the Mathematical Appendix at the end of the book.

Theorem 5.3: Nash Existence Theorem (Nash [1951])

Every game with a finite number of pure strategies has at least one Nash equilibrium, possibly in mixed strategies.

Theorem 5.4: Odd Number Theorem (R. Wilson [1971])

Generically, every game with a finite number of pure strategies has a positive and odd number of Nash equilibria, including mixed strategy equilibria.

Observation on Theorem 5.4.   The meaning of "generically" depends on the context, which can be found for this theorem in R. Wilson (1971). If two pure strategy equilibria have been found for a game such as the Battle of the Sexes, Theorem 5.4 says that unless the game is very special, at least one other equilibrium remains to be found (see note N3.2).

Theorem 5.5: Glicksberg Theorem (Glicksberg [1952], Dasgupta & Maskin [1986a] p. 4)

If individual players' pure strategy sets are compact, and every player's payoff function is continuous in the pure strategies of all the players, then a Nash equilibrium exists, possibly in mixed strategies.

Observation on Theorem 5.5.   Player i might have severally equally good best replies to a given set of strategies for the other players, so the best reply mapping is sometimes a correspondence rather than a function. Allowing mixed strategies makes the set of best replies convex and the best reply correspondence upper semi-continuous, which permits use of the Kakutani fixed point theorem (Theorem A.2). The Glicksberg Theorem is unusual in that it refers to mixed strategies of games with a continuous pure strategy space, which makes the mixed strategy space hard to define (see note N3.3).

Theorem 5.6: Reaction Function Existence Theorem

If the strategy set for each player is compact and convex, and all best reply correspondences are continuous, then a Nash equilibrium exists.

Observation on Theorem 5.6.   The best reply correspondences are continuous functions in the linear Cournot game of Section 3.4. They are discontinuous in the Bertrand game with capacity constraints and the three-player Hotelling Location Game in Sections 12.2 and 12.3. Mixed strategy equilibria exist for those games, but that can be concluded only from Theorem 5.8 below, not Theorem 5.6.

Theorem 5.7: Quasi-Concavity Existence Theorem (Dasgupta & Maskin [1986a] p. 4)

If pure strategy sets are compact and convex, and every player's payoff function is continuous in the pure strategies of all players and quasi-concave in his own pure strategies, then there exists a pure strategy Nash equilibrium.

Observation on Theorem 5.7.   The theorem of Nikaido & Isoda (1955) is a weaker version of Theorem 5.7 which requires that payoffs be concave in own-strategies and that the sum of all the players' payoffs be a continuous function of the strategies.

Theorem 5.8: Discontinuity Existence Theorem (Dasgupta & Maskin [1986a] p. 24)

For each player i (i = 1,...,n) let the action set $A_i \in R^m$ be convex and compact. Let the payoff function $\pi_i(a_1,...,a_n)$ be continuous except on a subset A_i^* consisting of action combinations such that for some $j \neq i$, some action k, and some integer d, there is some continuous one-to-one function $f_{ij}^d : R^1 \rightarrow R^1$ such that

$$f_{ij}^d = (f_{ji}^d)^{-1}, \quad and \quad a_{jk} = f_{ij}^d(a_{ik}). \tag{5.4}$$

Suppose that $\sum \pi_i(a)$ is upper semicontinuous, and for all i the payoff is bounded and is weakly lower semicontinuous in a_i. Then the game has a mixed strategy Nash equilibrium.

Observation on Theorem 5.8.   This is a hard theorem, but it has been put to practical use (Dasgupta & Maskin [1986b]). It is applicable when the payoff function is discontinuous only at a few points that are special in the following way: if a' is an action combination at which the payoff function is discontinuous, then if a'' is an action combination exactly the same except for one player's action, there is no discontinuity at a''.

Theorem 5.8 can be applied to the Bertrand Game with capacity constraints of Section 12.2. Two sellers, Allied and Brydox, simultaneously choose prices for the same good. If their prices are equal, they split the market; otherwise, the seller with the lower price gets all the customers. The payoffs are discontinuous at $a' = (p_a = 10, p_b = 10)$, but they are continuous at $a'' = (p_a = 10, p_b = 9.99)$. Theorem 5.8 says that a Nash equilibrium exists.

Theorem 5.8 would not be applicable to the Cournot Game modified by requiring firms to pay a set-up cost for outputs greater than zero but not outputs of zero. We could let $a' = (q_a = 0, q_b = 4)$ and $a'' = (q_a = 0, q_b = 3)$, and at both action combinations the payoff for Allied would be discontinuous because on increasing q_a even the slightest bit he would incur the set-up cost.

Theorem 5.9: Perfect Information Existence Theorem

Generically, a finite game of perfect information has a unique perfect Nash equilibrium in pure strategies.

Observation on Theorem 5.9.   By a finite game is meant a game of finite length in which the number of moves is finite (not a continuum). The proof is not difficult. A game of perfect information has no simultaneous moves or moves by Nature. At the node before each end node, there is some best move for the last player (because the number of moves is finite), and only one move is best, for almost all sets of payoffs (which is what "generically" means here). Go back one move. Anticipating what the last move will be, the player making the next-to-last move has a best move. We can work all the way back to the starting node that way.

Theorem 5.10: Symmetric Equilibrium Theorem (Dasgupta & Maskin [1986a] p. 18)

Every symmetric game that has an equilibrium has a symmetric equilibrium; that is, every game in which players' action sets are identical at each point in time has an equilibrium in which the mixing probabilities (perhaps equal to one) are identical.

Observation on Theorem 5.10.   Note that "symmetric game" as used in Theorem 5.10 is different and much stronger than "game of symmetric information." Not only must the information be symmetric, but the action sets too, and all moves must be simultaneous.

Theorem 5.11: Asymmetric Information Existence Theorem (Selten [1975])

Every game with perfect recall and a finite action set has at least one trembling hand perfect equilibrium.

Observation on Theorem 5.11.   "Trembling hand perfect" equilibrium is defined in Section 5.1. Theorem 5.11 also implies that the less restrictive perfect Bayesian and sequential equilibria exist.

Recommended Reading

Dasgupta, Partha & Eric Maskin (1986a) "The Existence of Equilibrium in Discontinuous Economic Games, I: Theory" *Review of Economic Studies.* January 1986. 53(1), 172: 1–26.

Dasgupta, Partha & Eric Maskin (1986b) "The Existence of Equilibrium in Discontinuous Economic Games, II: Applications" *Review of Economic Studies.* January 1986. 53(1), 172: 27–41.

Hofstadter, Douglas (1983) "Computer Tournaments of the Prisoner's Dilemma Suggest how Cooperation Evolves" *Scientific American.* May 1983. 248, 5: 16–26.

Kreps, David, Paul Milgrom, John Roberts, & Robert Wilson (1982) "Rational Cooperation in the Finitely Repeated Prisoners' Dilemma" *Journal of Economic Theory.* August 1982. 27, 2: 245–52.

Kreps, David & Robert Wilson (1982b) "Sequential Equilibria" *Econometrica.* July 1982. 50, 4: 863–94.

Nash, John (1951) "Non-Cooperative Games" *Annals of Mathematics.* September 1951. 54, 2: 286–95.

Problem 5

Limit Pricing

Firm Allied is a monopolist in the local computer market, a natural monopoly. In the initial period, Allied can price *Low*, earning 1 in profits, or *High*, earning 5. Allied knows its own operating cost, which is 20 with probability 0.75 and 30 with probability 0.25. Brydox knows those probabilities, but not Allied's exact cost. Brydox can enter at a cost of 40, and its cost of 25 is common knowledge. Profits for both firms are zero during the period of entry (instead of the 5 that Allied would otherwise get). Then the firm with the highest cost drops out, and the survivor earns monopoly profit equal to 125 minus its operating costs.

(1)   Draw the extensive form for this game.
(2)   Why is there no perfect bayesian equilibrium in which Allied prices *Low* only when its costs are 20? (no "separating" equilibrium)
(3)   What is a perfect bayesian equilibrium for this game?
(4)   What is a set of out-of-equilibrium beliefs that do not support a pooling equilibrium at the *Low* price?

Notes

N5.2   Refining Perfect Bayesian Equilibrium: PhD Admissions

- For discussions of the appropriateness of different equilibrium concepts in actual economic models see Rubinstein (1985b) on bargaining, Shleifer & Vishny (1986) on greenmail, and D. Hirshleifer & Titman (unpub) on tender offers. See also Van Damme (1987).
- **Exotic Refinements.** Binmore (1987) is a lengthy discussion of rationality and equilibrium refinements. New-and-improved equilibrium concepts refining perfect Bayesian equilibrium have been suggested by McLennan (1985), Cho & Kreps (1987) and Cho (1987) (**the intuitive criterion**), Banks & Sobel (1987) (**divinity**), Grossman & Perry (1986) (**perfect sequential equilibrium**), and Farrell (unpub) (**credible neologisms**).

  Two refinements somewhat different in flavor are stability and properness.

  Stability. (Kohlberg & Mertens [1986]) This concept is axiomatic, rather than based on beliefs or trembles. Its main requirements are that an equilibrium exist, that it be subgame perfect, that game trees that are essentially the same generate the same equilibria, and that deleting dominated strategies does not change the equilibrium. These requirements are so stringent that for many games no stable equilibrium exists. (The use of the term "stability" is unrelated to its use by von Neumann-Morgenstern or in connection with the Cournot equilibrium.)

  Proper Equilibrium. (Myerson [1978]) Properness is different from most refinements in being tremble-based, not belief-based. The idea is that "worse" trembles are less likely. Take the trembling hand perfect equilibrium concept and set the trembles so that if pure strategy s_i' is a worse reply than pure strategy s_i'', the probability weight put by the tremble's mixed strategy on s_i' is less than ϵ times the weight on s_i''. Take the limit as ϵ goes to zero, and a proper equilibrium results.

N5.5   The Axelrod Tournament

- Hofstadter (1983) is a nice discussion of the Prisoner's Dilemma and the Axelrod tournament by an intelligent computer scientist who came to the subject untouched by the preconceptions or training of economics. It is useful for elementary economics classes. Axelrod's 1984 book provides a fuller treatment.

N5.6   Evolutionary Equilibrium: the Hawk–Dove Game

- See Axelrod & Hamilton (1981) for a short article on biological applications of the Prisoner's Dilemma, Hines (1987) for a survey, and Maynard Smith (1982) for a book. J. Hirshleifer (1982) compares the approaches of economists and biologists. Among economists, Cornell & Roll (1981) and Riley (1979a) have used the ESS. Jacquemin's 1985 industrial organization text is sympathetic to the evolutionary approach, and Boyd & Richerson (1985) uses it to examine cultural transmission, which has important differences from purely genetic transmission. Sugden (1986) has written a book using game theory with the biological approach to look at social theory.

N5.7   Existence of Equilibrium

- Despite being so mathematical a discipline, noncooperative game theory has not generated many citable theorems. The books by J. Friedman (1986), Szep & Forgo (1985), and Tirole (1988) all have short sections on existence. Conditions for uniqueness are even less accessible. Rosen (1965) is one of the few papers on the subject.
- Theorem 5.6 is an adaption of Theorem A.2, the Kakutani Fixed Point Theorem (see the Mathematical Appendix). In Theorem 5.6 the correspondence must be continuous, not just upper semicontinuous, so that the image set is convex, as required by A.2. I do not know of a published source for Theorem 5.6, which is not original with me, but it follows immediately from A.2.
- Theorem 5.8 is not the main theorem in the body of Dasgupta & Maskin (1986a), but the more general theorem from the appendix to their article. Simon (1987) has been able to slightly relax the conditions that Dasgupta & Maskin use to characterize discontinuities.
- An early and very important application of a Nash equilibrium existence theorem (from Debreu [1952]) is by Arrow & Debreu (1954) to prove the existence of equilibrium prices in a competitive economy. They model the economy as a game between consumers, firms, and a Walrasian auctioneer whose payoff function is designed so that supply equals demand in the equilibrium of the game.

Part II

Asymmetric Information

6 Moral Hazard: Hidden Actions

6.1 Categories of Asymmetric Information Models

It used to be that an economist's generic answer to someone who brought up peculiar behavior that seemed to contradict basic theory was "It must be some kind of price discrimination." Today, we have a new answer: it must be some kind of asymmetric information. In a game of asymmetric information, player Smith knows something that player Brown does not. This covers a broad range of models (including price discrimination), so perhaps it is not surprising that so many situations come under its rubric. We will divide games of asymmetric information into five categories, to be studied in four chapters.

(1) **Moral Hazard with Hidden Actions** (Chapter 6)

Smith and Brown begin with symmetric information and agree to a contract, but then Smith takes an action unobserved by Brown. Information is complete.

(2) **Moral Hazard with Hidden Information** (Chapter 7)

Smith and Brown begin with symmetric information and agree to a contract, but then Nature takes a move observed by Smith but not Brown. Information is complete.

(3) **Adverse Selection** (Chapter 8)

Nature begins the game by choosing Smith's type (his payoff and strategies), unobserved by Brown. Smith and Brown then agree to a contract. Information is incomplete.

(4, 5) **Signalling** and **Screening** (Chapter 9)

Nature begins the game by choosing Smith's type, unobserved by Brown. To demonstrate his type, Smith takes actions that Brown can observe. If Smith takes the action before they agree to a contract, we say he is signalling; if he takes it afterwards, he is being screened. Information is incomplete.

Signalling and screening are special cases of adverse selection, and all three are forms of hidden information (or **hidden knowledge**, as it is sometimes called). Information is complete in either kind of moral hazard, and incomplete in adverse selection, signalling, and screening.

We will make heavy use of the "principal–agent model" to analyze asymmetric information. The two players are the principal and the agent, who are usually representative individuals. The principal hires an agent to perform a task, and the agent acquires an informational advantage about his type, his actions, or the outside world at some point in the game.

> The **principal** *(or **uninformed player**) is the player who has the coarser information partition.*

> The **agent** *(or **informed player**) is the player who has the finer information partition.*

Figure 6.1 shows the game trees for five principal-agent problems corresponding to the categories listed above. In each model the principal (P) offers the agent (A) a contract, which he accepts or rejects. In some, Nature (N) makes a move or the agent chooses an effort level, message, or signal. The moral hazard models (Figures 6.1a and 6.1b), are games of complete information with uncertainty. The principal offers a contract, and after the agent accepts, Nature adds noise to the task being performed. In moral hazard with hidden actions (Figure 6.1a), the agent moves before Nature, and in moral hazard with hidden information (Figure 6.1b), the agent moves after Nature and conveys a "message" to the principal about Nature's move.

Adverse selection models have incomplete information, so Nature moves first and picks the type of the agent, generally his ability to perform the task. In the simplest model, Figure 6.1c, the agent simply accepts or rejects the contract. If the agent can send a "signal" to the principal, as in Figures 6.1d and 6.1e, the model is signalling if he sends the signal before the principal offers a contract, and screening otherwise. A "signal" is different from a "message" in that it is not a costless statement, but a costly action. Some adverse selection models, like those of Figures 6.1c, 6.1d, and 6.1e, are games of certainty, but others are games of uncertainty.

As an example, consider the problem of an employer (the principal) hiring a worker (the agent), which we will consider later in more detail. If the employer knows the worker's ability, but not his effort level, the problem is moral hazard with hidden actions. If neither player initially knows the worker's ability, but after the worker accepts a contract he discovers it, the problem is moral hazard with hidden information. If the worker knows his ability from the start, but the employer does not, the problem is adverse selection. If in addition to the worker knowing his ability from the start, he can acquire education observable by the employer before they make a contract, the problem is signalling. If the worker acquires his education in response to the contract offered by the employer, the problem is screening.

Figure 6.1   Categories of asymmetric information models

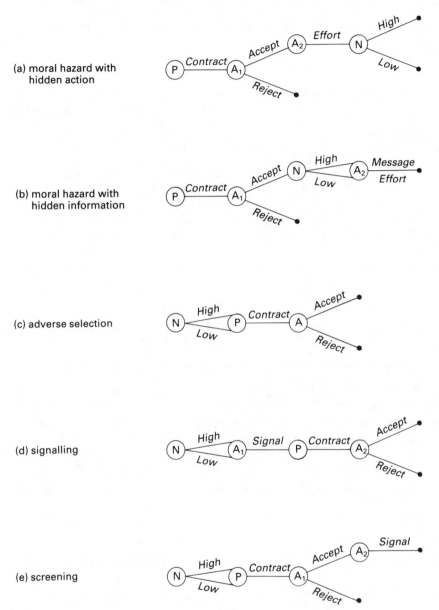

(a) moral hazard with
hidden action

(b) moral hazard with
hidden information

(c) adverse selection

(d) signalling

(e) screening

The five categories are only gradually rising from the swirl of the literature on
agency models, and the definitions are not well established. In particular, some
would argue that what I have called moral hazard with hidden information is really
a form of adverse selection, and that what I have called screening is indistinguishable
from simple adverse selection. Others do not realize that screening and signalling
are different models, and use the terms interchangeably. Most researchers have not

thought very hard about any of the definitions. The importance of the distinctions will become clearer as we explore the properties of the models. For readers whose minds are more synthetic than analytic, Table 6.1 may be more help than anything else to clarify the categories.

Table 6.1 Applications of the principal–agent model

	Principal	Agent	Effort or Type / Signal
moral hazard with hidden action	insurance company	policyholder	care to avoid theft
	insurance company	policyholder	health precautions
	plantation owner	sharecropper	farming effort
	bondholders	stockholders	riskiness of company projects
	owner	renter	upkeep of building
	renter	owner	upkeep of building
	society	criminal	refraining from crime
moral hazard with hidden information	shareholders	top manager	investment decisions
	FDIC	bank	safety of loans
adverse selection	insurance company	policyholder	infection with AIDS virus
	stockholders	bondholders	riskiness of company projects
	employer	worker	skill
signalling or screening	employer	worker	skill / education
	buyer	seller	durability / warranty
	investor	stock-issuer	stock value / percentage retained

Section 6.2 discusses the roles of uncertainty and asymmetric information in a principal-agent model of moral hazard with hidden actions, and Section 6.3 shows how various constraints are satisfied in equilibrium. Section 6.4 uses diagrams to approach the classical problem of moral hazard: lack of effort by the insured party to avoid accidents. Section 6.5 collects several unusual contracts produced under moral hazard and discusses the properties of optimal contracts. The chapter ends with Section 6.6, which talks less formally about the institutional implications of agency theory.

6.2   A Principal–Agent Model: the Production Game

In the archetypal principal–agent model, the principal is a manager and the agent is a worker. Under moral hazard, it is easier for the manager to observe the worker's output than his effort, so he offers a contract to pay the worker based on output, which depends on the worker's effort. We will work through three production games, the last of which is the standard principal–agent model.

Denote the monetary value of output by $q(e)$, which is increasing in effort, e. The agent's utility function $U(e, w)$ is decreasing in effort and increasing in the wage w, while the principal's utility $V(q - w)$ is increasing in the difference between output and the wage.

Production Game I: a Flat Wage under Certainty

Players

The principal and the agent.

Information

Asymmetric, complete, and certain.

Actions and Events

(1) The principal offers the worker a wage w.
(2) The agent decides whether to accept. If he does, he exerts effort e, which is unobserved by the principal.
(3) Output equals $q(e)$, observed by both players.

Payoffs

If the agent rejects the contract, then $\pi_{agent} = \bar{U}$ and $\pi_{principal} = 0$.

If the agent accepts the contract, then $\pi_{agent} = U(e, w)$ and $\pi_{principal} = V(q - w)$.

An assumption common to most principal–agent models is that either the principal or the agent is one of many perfect competitors. In the background other principals compete to employ the agent, so the principal's equilibrium profit equals zero; or many agents compete to work for the principal, so the agent's equilibrium utility equals the **reservation utility** $\bar{U}$, the minimum that induces him to work. In Production Games I, II, and III, by specifying that the principal moves first, we will implicitly assume that many agents compete for one principal.

The outcome of Production Game I is simple and inefficient. If the wage is nonnegative, the agent accepts the job and exerts zero effort, and the principal's best response is to offer a wage of zero. But Production Game I is a strange model, because even though the principal cannot observe effort, the output, which he does observe, maps one-to-one to effort. Expanding the strategy space to allow contracts based on output, we obtain Production Game II.

Production Game II:
an Output-Based Wage under Certainty

Players

The principal and the agent.

Information

Asymmetric, complete, and certain.

Actions and Events

(1)   The principal offers the worker a wage function $w(q)$.
(2)   The agent decides whether to work. If he does, he exerts effort e, which is unobserved by the principal.
(3)   Output equals $q(e)$, observed by both players.

Payoffs.

If the agent rejects the contract, then $\pi_{agent} = \bar{U}$ and $\pi_{principal} = 0$.

If the agent accepts the contract, then $\pi_{agent} = U(e, w)$ and $\pi_{principal} = V(q - w)$.

The principal's problem is more complicated in Production Game II. He must pick not a number w, but a function $w(q)$, so the strategy space is harder to define. This principal's decision can be simplified, however, because with his monopsony power he collects all the gains from trade, so he wishes to make the agent pick the effort level e^* that generates the efficient output level, q^*. The contract must provide the agent with utility $\bar{U}$ in equilibrium, but any wage function such that $U(e^*, w(q^*)) = \bar{U}$ and $U(e, w(q)) < \bar{U}$ for $e \neq e^*$ will make the agent pick $e = e^*$. Such a contract is called a **forcing contract**, because it forces the agent to pick a particular effort level. It achieves efficiency despite the lack of direct observability of effort.

In Production Game III, however, the outcome is not efficient. We now let Nature move, so that effort does not map cleanly to the observed output, which might have been produced by any of several different effort levels. Information is asymmetric in an important way.

Production Game III:
an Output-Based Wage Under Uncertainty

Players

The principal and the agent.

Information

Asymmetric, complete, and uncertain.

Actions and Events

(1)   The principal offers the worker a wage function $w(q)$.

(2)   The agent decides whether to accept. If he does, he exerts effort e, which is unobserved by the principal.

(3)   Unobserved by the principal, but observed by the agent, Nature chooses the state of the world $\theta \in \mathbf{R}$, according to the probability density $f(\theta)$. Output equals $q(e, \theta)$.

Payoffs

If the agent rejects the contract,
   then $\pi_{agent} = \bar{U}$ and $\pi_{principal} = 0$.

If the agent accepts the contract,
   then $\pi_{agent} = EU(e, w)$ and $\pi_{principal} = EV(q - w)$.

In Production Game III, the great variety of possible wage functions matters. The principal cannot just choose an output level and tell the agent to produce it, because unlike in Production Game II, the principal cannot deduce that $e = e^*$ just by looking at the output, which is $q(e, \theta)$, not just $q(e)$. The optimal wage contract might specify the highest wage for q^*, but not necessarily, and it will usually not specify a zero wage for a slightly lower output, because Nature might be to blame. If the agent is risk averse, as is usually assumed, then keeping the agent's expected utility equal to $\bar{U}$ is more expensive when he bears risk, and the principal wants to insure the agent by keeping the risk imposed on him low. The tradeoff is between incentives and insurance.

Since the space of all possible functions is more difficult to work with than $\mathbf{R}^n$, the space of n-vectors, the modeller may wish to restrict the principal's strategy set to

$$(\alpha, \beta) \in \mathbf{R}^2 \text{ where } w = \alpha + \beta q \textbf{ (Linear Contract)}$$

or

$$(\alpha, \beta, \bar{q}) \in \mathbf{R}^3 \text{ where } w = \begin{cases} \alpha, & \text{if } q < \bar{q} \\ \beta, & \text{if } q \geq \bar{q} \end{cases} \textbf{ (Threshold Contract)}$$

Arbitrarily restricting the strategy space is generally not good modelling practice, but sometimes tractability demands it, and we hope that the equilibrium is not affected much. Simplification might be justified by the administrative convenience of simple contracts in the real world, where complicated contracts incur transaction and computation costs. They may also have the advantage of robustness, but this is hard to establish without a specific context.

Usually the model loses very little generality if the modeller sets $\alpha = 0$ in either the linear or the threshold contract, since what matters is the difference in expected utility between slacking and slaving, not the absolute levels. An exception is if the agent's initial wealth is important to the model, in which case α should not be normalized to zero.

However specified, the wage function cannot be a direct function of either the agent's effort or the state of the world, since the principal observes neither of these (if the principal knew the state, he could deduce the effort, as in Production Game II). But the source of moral hazard is not unobservability, but the fact that the contract cannot be conditioned on effort. Effort is **noncontractible**. Production Game III applies even if the principal can see very well that the agent is slacking, but he cannot prove it in court. We have gone through Production Games I, II, and III to illustrate what is at the heart of the principal–agent problem: setting the wage without exactly knowing the agent's effort. The rest of the chapter will tackle the problem of how to best compensate the agent under these conditions.

6.3   The Self-Selection, Participation, and Competition Constraints

There are two approaches to finding equilibria in asymmetric information models. One is the straight game theory approach: search for Nash strategies. The other is the economist's standard approach: set up a maximization problem and solve using calculus. The drawback of game theory is that it is not mechanical: the modeller must make a guess that some strategy combination is an equilibrium before he tests it. Some economists are quite frustrated when they learn there is no general way to make the initial guess. The maximization approach is so straightforward, on the other hand, that we could let a trained ape do it. Just set up a payoff function with some constraints, take derivatives, and solve the first order conditions. That approach is less appropriate to game theory, because the players must solve their optimization problems jointly: Smith's strategy affects Brown's maximization problem and vice versa. In sequential games, however, the maximization approach can be used, with some care, if the last player's maximization problem is embedded in the first player's problem as a constraint. We did this to find the Stackelberg solution in Section 3.4, and we will discuss it for the principal–agent model here.

The principal's problem in Production Game III is to maximize his utility, knowing that the agent chooses effort after accepting the contract and he is free to reject the contract entirely.

$$\underset{w(\cdot)}{Maximize} \ EV(q(\tilde{e}, \theta) - w(q(\tilde{e}, \theta))) \tag{6.1}$$

subject to

$$\tilde{e} = \underset{e}{argmax} \ EU(e, w(q(e, \theta))) \ \textbf{(Incentive Compatibility Constraint)} \tag{6.2}$$

$$EU(\tilde{e}, w(q(\tilde{e}, \theta))) \geq \bar{U} \ \textbf{(Participation Constraint)} \tag{6.3}$$

The incentive compatibility constraint takes account of the fact that the agent moves second, so the contract must induce him to voluntarily pick the desired effort. The participation constraint, also called the **reservation utility** or **individual rationality** constraint, requires that the worker prefer the contract to unemployment.

Expression (6.1) is the way an economist instinctively sets up the problem, but setting it up is often as far as he can get with this **first order condition approach**. The difficulty is not just that the maximizer is choosing a wage function instead of a number (control theory or calculus of variations can solve such problems). Rather, it is that the constraints are nonconvex.

A different approach, developed by Grossman & Hart (1983) and called the **Three-Step Procedure** by Fudenberg & Tirole (unpub), is to focus on contracts that induce the agent to pick a particular action, rather than to directly attack the problem of maximizing profits. The first step in this procedure is to find for each possible effort level the set of wage contracts that induce the agent to choose that effort level. The second step is to find the contract which supports that effort level at the lowest cost to the principal. The third step is to choose the effort level that maximizes profits, given the necessity to support that effort with a costly wage contract.

To support the effort level e, the wage contract $w(\cdot)$ must satisfy the incentive compatibility and participation constraints. Mathematically, the problem of finding the least cost $C(\tilde{e})$ of supporting the effort level $\tilde{e}$ combines steps one and two.

$$C(\tilde{e}) = \underset{w(\cdot)}{Minimum} \; Ew(q(\tilde{e}, \theta)) \tag{6.4}$$

subject to

(6.2) $\tilde{e} = \underset{e}{argmax} \; EU(e, w(q(e, \theta)))$ (Incentive Compatibility Constraint)

(6.3) $EU(\tilde{e}, w(q(\tilde{e}, \theta))) \geq \bar{U}$ (Participation Constraint)

Step three takes the principal's problem of maximizing his payoff, (6.1), and restates it as

$$\underset{\tilde{e}}{Maximize} \; EV(q(\tilde{e}, \theta) - C(\tilde{e})). \tag{6.5}$$

After finding which contract most cheaply induces each effort, he discovers the optimal effort by solving problem (6.5). Breaking the problem into two parts makes it easier to solve.

Models with Competing Principals: the Competition Constraint

In most principal–agent problems, the interesting interactions are between a single principal and a single agent, but one side of the market is constrained by the presence of competitors. In Production Game III, agents compete for a principal. In the next section, insurance companies (principals) compete for a customer (the agent). We will model a competitive insurance industry by specifying that there are two insurance companies offering contracts, a method which explicitly puts competition into the game.

A second approach to modelling competition is to change the order of moves. The questions of how moves are ordered and which players have market power are closely related. Instead of the firm offering a contract and the agent accepting or rejecting, let the agent offer a contract and the firm accept or reject (after which the agent chooses his effort). This approach generates the same outcome as using two principals, because the agent offers a contract which maximizes his own utility, subject to the constraint that the principal makes non-negative profits. This approach was used in Production Game III to represent competing agents. The issue will arise again in Chapter 10 when we look at bargaining models.

A third method, which is the most common, is to model the game with just one principal and one agent, but to constrain the principal: he must offer the contract that maximizes the utility of the agent as well as meeting the other constraints we have discussed. We call the new constraint the **competition constraint.** The competition constraint is often simplified to say that profits equal zero and called the **zero-profit constraint**, an ambiguous name since it fits the firm's participation constraint equally well. Loosely speaking, competition requires profits to be non-positive and participation requires them to be non-negative. That is true only loosely because zero profits are neither necessary nor sufficient to satisfy the competition constraint. It is not logically necessary, because the constraint just says that no principal can offer a more attractive contract. In markets with asymmetric information, we cannot rule out the possibility that profits are positive even under competition—recall the Klein–Leffler product quality model of Section 4.8 and consider its implications. Nor are zero profits sufficient to show that a market satisfies the competition constraint: profits could be zero because all of the firms offer equally inefficient contracts, even though a profit opportunity remains.

The agent's participation constraint is likely not to be binding in a model with competing principals, because the agent will be able to collect the rents from the transaction and achieve more than his reservation utility. Only rarely, in fact, will both players' participation constraints be binding in the same model.

6.4   State-Space Diagrams: Insurance Games I and II

The preceding section was about ways to find equilibria using algebra. Another approach, especially useful when the strategy space is continuous, is to use diagrams of a kind we will use again in Sections 7.2, 8.4, and 9.4. The term "moral hazard" comes from the insurance industry. Suppose Mr Smith (the agent) is considering buying theft insurance for a car with a value of 12. Figure 6.2 is an example of a **state-space diagram**, a diagram whose axes measure the values of one variable in two different states of the world. Before Smith buys insurance, his dollar wealth is 0 if there is a theft and 12 otherwise, depicted as his endowment, $\omega = (12, 0)$. The point (12,0) indicates a wealth of 12 in one state and 0 in the other, while the point (6,6) indicates a wealth of 6 in each state.

One cannot tell the probabilities of each state just by looking at the state-space diagram. For our model we specify that if Smith is careful where he parks, the state *Theft* occurs with probability 0.5, but if he is careless, the probability rises to 0.75. He is risk averse, and, other things equal, he mildly prefers to be careless, a preference worth only ϵ to him. Other things are not equal, however, and he would choose to be careful were he uninsured, because of the high correlation of carlessness with carelessness.

The insurance company (the principal) is risk neutral, perhaps because it is owned by diversified shareholders. We assume that no transaction costs are incurred in providing insurance and the market is competitive, a switch from Production Game III where the principal collected all the gains from trade. If the insurance company can require Smith to park carefully, it offers him insurance at a premium of 6, with a payout of 12 if theft occurs, leaving him with an allocation of $C_1 = (6, 6)$. This satisfies the competition constraint because it is the most attractive contract any company can offer without making losses. Smith, whose allocation is 6 no matter what happens, is **fully insured**. In state-space diagrams, allocations like C_1 which fully insure one player are on the 45° line through the origin, the line along which his allocations in the two states are equal.

Figure 6.2   Insurance Game I

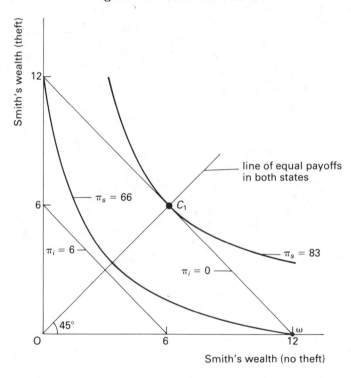

The game is described below in a specification that includes two insurance companies to simulate a competitive market. For Smith, who is risk averse, we must distinguish between dollar *allocations* such as (12,0) and utility *payoffs* such as $0.5U(12) + 0.5U(0)$. The curves in Figure 6.2 are labelled in units of utility for Smith and dollars for the insurance company.

Insurance Game I: Observable Care

Players

Smith and two insurance companies.

Information

Perfect and uncertain.

Actions and Events

(1)   Smith chooses to be either *Careful* or *Careless*, observed by the insurance company.
(2)   Each insurance company offers a contract (x, y), where Smith pays premium x and receives compensation y if there is a theft.
(3)   Smith picks a contract.
(4)   Nature chooses whether there is a theft, with probability 0.5 if Smith is *Careful* or 0.75 if Smith is *Careless*.

Payoffs

The insurance company not picked by Smith has a payoff of zero.

Smith is risk averse, so his utility function U is such that $U' > 0$ and $U'' < 0$. If Smith picks contract (x, y), the payoffs are:

If Smith chooses *Careful,*

$$\pi_{Smith} = 0.5U(12 - x) + 0.5U(0 + y - x).$$
$$\pi_{company} = 0.5x + 0.5(x - y), \text{ for his insuror.}$$

If Smith chooses *Careless,*

$$\pi_{Smith} = 0.25U(12 - x) + 0.75U(0 + y - x) + \epsilon.$$
$$\pi_{company} = 0.25x + 0.75(x - y), \text{ for his insuror.}$$

In the equilibrium of Insurance Game I, Smith chooses to be *Careful* because he foresees that otherwise his insurance would be more expensive. Figure 6.2 is the corner of an Edgeworth box which shows the indifference curves of Smith and his insurance company given that Smith's care keeps the probability of a theft down to 0.5. The company is risk neutral, so its indifference curve, $\pi_i = 0$, is a straight line, with slope $-1/1$. Its payoffs are higher on indifference curves such as $\pi_i = 6$ that are closer to the origin and thus have smaller expected payouts to Smith. The insurance company is indifferent between ω and C_1, at both of which its profits are zero. Smith is risk averse, so his indifference curves are closest to the origin on the 45° line, where his wealth in the two states is equal. I have labelled his original indifference curve $\pi_s = 66$ and drawn the preferred indifference curve $\pi_s = 83$ through the equilibrium contract C_1 (the numbers 66 and 83 are picked out of thin air to provide labels). The equilibrium contract is C_1, which satisfies the competition constraint by generating the highest expected utility for Smith that allows nonnegative profits for the company.

Insurance Game I is a game of symmetric information. But suppose that

(1) The company cannot observe Smith's action; or

(2) The state insurance commission does not allow contracts to require Smith to be careful; or

(3) A contract requiring Smith to be careful is impossible to enforce because of the cost of proving he was careless.

In each case, Smith's action is a noncontractible variable, so we model all three the same way: by putting Smith's move second. The new game is like Production Game III, with uncertainty, unobservability, and two levels of output, *Theft* and *No Theft*. The insurance company may not be able to directly observe Smith's action, but his dominant strategy is to be *Careless*, so the company knows the probability of a theft is 0.75.

Insurance Game II: Unobservable Care

(The same as Insurance Game I except)

Information

Asymmetric, complete, and uncertain.

Actions and Events

(1) Each insurance company offers a contract (x, y), where Smith pays premium x and receives compensation y if there is a theft.

(2) Smith picks a contract.

(3) Smith chooses either *Careful* or *Careless*.

(4) Nature chooses whether there is a theft, with probability 0.5 if Smith is *Careful* or 0.75 if Smith is *Careless*.

Smith's dominant strategy is *Careless*, but if he is careless, the insurance company must change the contract to a premium of 9 and a payout of 12 to prevent losses, which leaves Smith with an allocation of $C_2 = (3, 3)$. Making thefts more probable reduces the slopes of both players' indifference curves, because it decreases the utility of points to the southeast of the 45° line and increases utility to the northwest. In Figure 6.3, the insurance company's isoprofit curve swivels to the dotted line $\tilde{\pi}_i = 0$. Smith's indifference curve also swivels, to the dotted curve $\tilde{\pi}_s = 66 + \epsilon$, where the ϵ difference appears because Smith chooses the action *Careless*, which he prefers.

Figure 6.3 shows that no full insurance contract will be offered. The contract C_1 is acceptable to Smith, but not to the insurance company (it earns negative profits), and the contract C_2 is acceptable to the insurance company, but Smith prefers ω, where he achieves at least his reservation utility. Smith would like to commit himself to being careful, but he cannot make his commitment credible. If the means existed to prove his honesty, he would use them even if they were costly.

Figure 6.3   Insurance Game II: full and partial insurance

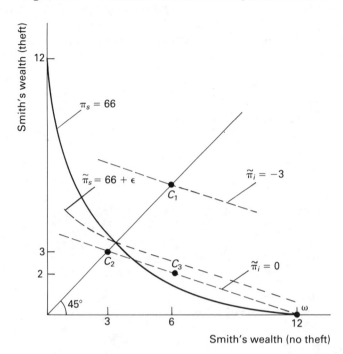

He might, for example, agree to buy off-street parking even though locking his car would be cheaper, if verifiable.

Although no full insurance contract such as C_1 or C_2 is mutually agreeable, other contracts can be used. Suppose, for a moment, that Smith chooses *Careful* if his insurance policy has incomplete coverage. Figure 6.3 shows a third contract, C_3, which has a premium of 6 and a payout of 8. Smith prefers C_3 to his endowment of $\omega = (12, 0)$. We can think of C_3 in two ways:

(1)   Full insurance except for a **deductible** of four. The insurance company pays for all losses in excess of four.
(2)   Insurance with a **coinsurance** rate of one third. The insurance company pays two-thirds of all losses.

The outlook is brighter, because Smith chooses *Careful* under partial insurance. The moral hazard is "small" in the sense that Smith barely prefers *Careless*. With even a small deductible, Smith would choose *Careful* and the probability of theft would fall to 0.5, allowing the company to provide much more generous insurance. The solution of full insurance is "almost" reached. In reality, we rarely observe truly full insurance, because insurance contracts repay only the price of the car and not the bother of replacing it, which is great enough to deter owners from leaving their cars unlocked.

Figure 6.4 illustrates partial insurance. Smith has a choice between dotted indifference curves (*Careless*) and solid ones (*Careful*). To the southeast of the 45° line, the dotted indifference curve for a particular utility level is always above that utility's solid indifference curve. Offered contract C_4, Smith chooses *Careful*, remaining on the solid indifference curve, so C_4 yields zero profit to the insurance company. In fact, the competing insurance companies will offer contract C_5 in equilibrium, which is almost full insurance, but just almost, so that Smith will choose *Careful* to avoid the small amount of risk he still bears.

Figure 6.4   Insurance Game II: more on partial insurance

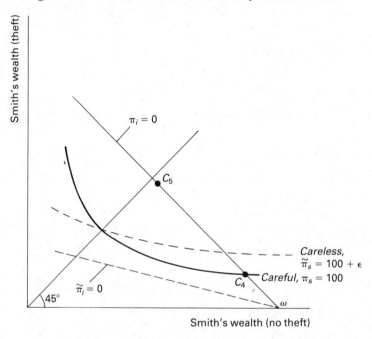

Even when the ideal of full insurance cannot be reached, there exists some best choice like C_5 in the set of feasible contracts. We use the terms "first-best" and "second-best" to distinguish two kinds of optimality.

> A **first-best contract** *achieves the same allocation as the contract optimal when the principal and the agent have the same information set and all variables are contractible.*

> A **second-best contract** *is Pareto-optimal given information asymmetry and constraints on writing contracts.*

6.5   Optimal Contracts: the Broadway Game

The Relation Between Output and Compensation

The next game, inspired by Mel Brooks's film *The Producers*, illustrates some peculiarities of optimal contracts. Investors advance funds to a producer to produce a Broadway show which may succeed or fail. The producer has a choice between embezzling and not embezzling the funds advanced, with a direct gain to himself of 50 if he embezzles. If the show is a success, the revenue is 500 if he did not embezzle, and 100 if he did. If the show is a failure, revenue is -100 in either case; extra expenditure on a fundamentally flawed show is useless.

The Broadway Game

Players

The producer and the investors.

Information

Asymmetric, complete, and uncertain.

Actions and Events

(1)   The investors offer a wage contract $w(q)$ as a function of revenue q.
(2)   The producer accepts or rejects the contract.
(3)   The producer chooses to *Embezzle* or *Be Honest*.
(4)   Nature picks the state of the world to be *Success* or *Failure* with equal probability. The resulting revenue q is shown in Table 6.2.

Payoffs.

The producer's payoff is $U(20)$ if he rejects the contract, where $U' > 0$ and $U'' < 0$, and the investors' payoff is 0. Otherwise,

$$\pi_{producer} = \begin{cases} U(w(q) + 50) & \text{if he embezzles;} \\ U(w(q)) & \text{if he is honest.} \end{cases}$$

$$\pi_{investors} = q - w(q).$$

Table 6.2 The Broadway Game: profits from the show

State of the World (equally probable)

		Failure	Success
	Be Honest	-100	$+500$
Effort			
	Embezzle	-100	$+100$

Another way to tabulate payoffs, shown in Table 6.3, is to put the probabilities of outcomes in the boxes, with effort in the rows and output in the columns. The Broadway Game illustrates that optimal contracts do not always pay higher wages for better performance. An optimal contract here is the **Boiling-in-Oil Contract**:

$w(+500) = 20.$
$w(-100) = 20.$
$w(+100) = -10,000.$

In this model, investors are risk neutral and the producer is risk averse, so the producer should bear as little risk as is consonant with providing his incentives. The boiling-in-oil contract is an application of the **sufficient statistic condition**, which states that for incentive purposes, if the agent's utility function is separable in effort and money, wages should be based on whatever evidence best indicates effort, and only incidentally on output (see Holmstrom [1979] and note N6.4). In the spirit of the Three-Step Procedure, what the principal wants is to induce the agent to choose the appropriate effort, *Be Honest*, and his data on what the agent chose is the output. In equilibrium (though not out of it), the datum $q = +500$ contains exactly the same information as the datum $q = -100$. Both lead to the same posterior probability that the agent chose *Be Honest*, so the wages conditioned on each datum should be the same. We need to insert the qualifier "in equilibrium," because to form the posterior probabilities the principal needs to have some beliefs as to the agent's behavior. Otherwise, he could not interpret $q = -100$ at all.

Table 6.3 The Broadway Game: probabilities of profits from the show

		Profit		
		−100	+100	+500
	Be Honest	0.5	0	0.5
Effort				
	Embezzle	0.5	0.5	0

If both the producer and the investors were risk averse, risk sharing would change the part of the contract that applies in equilibrium. The optimal contract would then provide for $w(-100) < w(+500)$ to share the risk. The principal would have a lower marginal utility of wealth when output was +500, so he would be better able to pay an extra dollar of wages in that state than when output was −100.

The Broadway Game is special because the first-best can be attained by a contract that is the same regardless of the agent's level of risk aversion. If the producer is paid enough when output is −100 or +500 that his expected utility equals his reservation expected utility $\bar{U}$, and if he is boiled in oil (that is, forced down to an arbitrarily low level of utility) if output is +100, then he refrains from embezzlement. If he refrains, the probability of being boiled in oil is zero, so the participation constraint is satisfied and he is willing to accept the contract.

A less gaudy name for this kind of contract is the alliterative "**shifting support scheme**," so named because the contract depends on the support of the output distribution being different when effort is optimal than when effort is anything else. In other words, the set of possible outcomes under optimal effort is different from under any other effort. As a result, certain outputs show without any doubt that the producer embezzled. Very heavy punishments inflicted only for those outputs achieve the first-best, because a non-embezzling producer has nothing to fear.

Figure 6.5 shows shifting supports in a model where output can take not three,

Figure 6.5   Shifting supports in an agency model

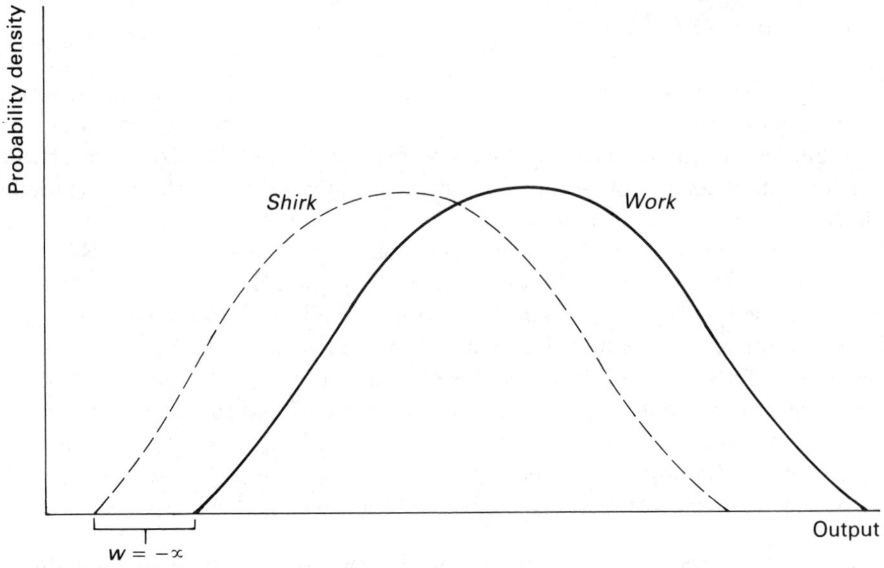

but a continuum of values. If the agent shirks instead of working, certain low outputs become possible and certain high outputs become impossible. In a case like this, when the support of the output shifts when behavior changes, boiling-in-oil contracts are useful: the wage is $-\infty$ for the low outputs possible only under shirking. When there is a limit to the amount the agent can be punished, or the support is the same under all actions, the threat of boiling in oil might not achieve the first-best, but similar contracts can still be useful. The conditions favoring boiling-in-oil contracts are

(1)   The agent is not very risk averse.
(2)   There are outcomes with high probability under shirking that have low probability under optimal effort.
(3)   The agent can be severely punished.
(4)   It is credible that the principal will carry out the severe punishment.

Selling the Store

Another first-best contract that can sometimes be used is **selling the store**. Under this arrangement, the agent buys the entire output for a flat fee paid to the principal, becoming the **residual claimant**, since he keeps every additional dollar of output that his extra effort produces. This is equivalent to fully insuring the principal.

In the Broadway Game, selling the store takes the form of the producer paying the investors 200 ($= 0.5[-100] + 0.5[+500]$) and keeping all the profits for himself. The drawbacks are that (1) the producer might not be able to afford to pay the investors the flat price of 200; and (2) the producer might be risk averse and incur a heavy utility cost in bearing the entire risk. These two drawbacks are why producers go to investors in the first place.

Public Information that Hurts the Principal and the Agent

We can use a modified Broadway Game to show how having more public information available can hurt both players. This will also provide a little practice in using information sets. Let us split *Success* into two states of the world.

Table 6.4 The Broadway Game: profits from the show with three states

		State		
		Failure	*Minor Success*	*Big Success*
Effort	*Do Not Embezzle*	−100	+50	+500
	Embezzle	−100	−100	+100

Under the optimal contract,

$$w(-100) = w(+50) = w(+500) > w(+100) \tag{6.6}$$

because only the datum $q = +100$ is proof that the producer embezzled. Were (6.6) the contract, the agent would not embezzle. But consider what happens when the information set is refined so that before the agent takes his action both he and the principal can tell whether the show will be a big success or not. Under the refinement, each player's initial information partition is

$$(\{Failure,\ Minor\ Success\}, \{Big\ Success\}),$$

instead of the original coarse partition

$$(\{Failure,\ Minor\ Success,\ Big\ Success\}).$$

If the principal would find the investment profitable even if he knew in advance that the show would not be a big success, the refinement does not help him. It changes the behavior of the agent, however, because on observing $\{Failure, Minor\ Success\}$ he is free to embezzle without fear of the oil-boiling output of +100. He would still refrain from embezzling if he observed $\{Big\ Success\}$, but no contract can make the agent not embezzle if he observes $\{Failure, Minor\ Success\}$. If the parameters were such that the principal needed the returns from minor successes to break even, he would not make the investment in the first place were the information sets refined, and the gains from trade would be lost.

6.6   Institutions

Insurance and Incentives in the Law

Usually when agents are risk averse, the first-best cannot be achieved, because some tradeoff must be made between providing the agent with incentives and keeping his compensation from varying too much between states of the world. This is the way economists interpret the everyday idea of fairness and explain why it is so important in our culture: punishing laziness but trying to distinguish between laziness and bad luck is how an optimal contract is devised.

The field of law is well-suited to analysis by principal-agent models. Even in the 19th century, Holmes (1881, p. 31) conjectured in *The Common Law* that the reason why sailors at one time received no wages if their ship was wrecked was to discourage them from taking to the lifeboats too early to save it. The reason why such a legal rule is suboptimal is not that it is unfair—presumably sailors knew the risk before they set out—but because incentive compatibility and insurance work in opposite directions, and insurance is likely to be more important here. If sailors are more risk averse than ship owners, and pecuniary advantage would not add much to their effort during storms, then the owner ought to provide insurance to the sailors by guaranteeing them wages whether the voyage succeeds or not.

Another legal question is who should bear the cost of an accident: the victim (for example, a pedestrian hit by a car) or the person who caused it (the driver). The economist's answer is that it depends on who has the most severe moral hazard. If the pedestrian could have prevented the accident at the lowest cost, he should pay; otherwise, the driver. Insurance or wealth transfer may also enter as considerations. If pedestrians are more risk averse, drivers should bear the cost, and according to some political views, if pedestrians are poorer, drivers should bear the cost. Note that this last consideration—wealth transfer—is not relevant to private contracts. If a principal earning zero profits is required to bear the cost of work accidents, the agent's wage will be lower than if he bore them instead.

Criminal law is also concerned with tradeoffs between incentives and insurance. Holmes (1881, p. 40) also notes, approvingly, that Macaulay's draft of the Indian Penal Code made breach of contract for the carriage of passengers a criminal offense. The reason is that the palanquin-bearers were too poor to pay damages for abandoning their passengers in desolate regions, so the power of the state was needed to provide for heavier punishments than bankruptcy. In general, however, the legal rules actually used seem to diverge more from optimality in criminal law than civil law. If, for example, there is no chance that an innocent man can be convicted of embezzlement, boiling embezzlers in oil might be good policy, but most countries would not allow this. Taking the example a step further, if the evidence for murder is usually less convincing than for embezzling, our analysis could easily indicate that the penalty for murder should be less, but such reasoning offends the common notion of matching the severity of punishment with the crime.

Institutions and Moral Hazard

While agency theory can be used to explain and perhaps improve government policy, it also is useful to explain the development of many curious private institutions. Agency problems are an important hindrance to economic development, and may explain a number of apparently irrational practices. Popkin (1979, pp. 66, 73, 157) notes a variety of these. In Vietnam, for example, absentee landlords were more lenient than local landlords, but improved the land less, as one would expect of principals who suffer from informational disadvantages *vis-à-vis* their agents. Along the pathways in the fields, farmers would plant early-harvesting rice that the farmer's family could harvest by itself in advance of the regular crop, so that hired labor could not grab handfuls as they travelled. In 13th-century England, beans were seldom grown, despite their nutritional advantages, because they were too easy to steal. Some villages tried to solve the problem by prohibiting anyone from entering the beanfields except during certain hours marked by the priest's ringing the church bell, so everyone could tend and watch their beans at the same official time.

In less exotic settings, moral hazard provides another reason besides tax benefits why employees take some of their wages in fringe benefits. Professors are granted some of their wages in university computer time because this induces them to do more research. Having a zero marginal cost of computer time is a way around the moral hazard of slacking on research, despite being a source of moral hazard in wasting computer time. A less typical but more imaginative example is that of the bank in Minnesota which, concerned about its image, gave each employee $100 in credit at certain clothing stores to upgrade their style of dress. By compromising between paying cash and issuing uniforms the bank could hope to raise both its profits and the utility of its employees. ("The $100 Sounds Good, but what do they Wear on the Second Day?" *Wall Street Journal*, 16 October 1987, p. 16.)

Longterm contracts are an important occasion for moral hazard, since so many variables are unforeseen, and hence noncontractible. The term **opportunism** has been used to describe the behavior of agents who take advantage of noncontractibility to increase their payoff at the expense of the principal (see Williamson [1975] and Tirole [1986]). Smith may be able to extract a greater payment from Brown than was agreed in their contract, because when a contract is incomplete, Smith can threaten to harm Brown in some way. This is called **hold-up potential** (Klein, Crawford, & Alchian [1978]). Hold-up potential can even make an agent introduce competing agents into the game, if competition is not so extreme as to drive rents to zero. Michael Granfield tells me that Fairchild once developed a new patent on a component of electronic fuel injection systems that they sought to sell to TRW. TRW offered a much better price if Fairchild was willing to license its patent to other producers, fearing the hold-up potential of buying from just one supplier. TRW could have tried writing a contract to prevent hold-up, but knew that it would be difficult to prespecify all the ways that Fairchild could cause harm, including not only slow delivery, poor service, and low quality, but also sins of omission like failing to sufficiently guard the plant from shutdown due to accidents and strikes.

It should be clear from the variety of these examples that moral hazard is a common problem. Now that the first flurry of research on the principal–agent problem has finished, researchers are beginning to use the new theory to study institutions that were formerly relegated to descriptive "soft" scholarly work.

Recommended Reading

Holmstrom, Bengt & Roger Myerson (1983) "Efficient and Durable Decision Rules with Incomplete Information" *Econometrica*. November 1983. 51, 6: 1799–819.

Milgrom, Paul (1981b) "Good News and Bad News: Representation Theorems and Applications" *Bell Journal of Economics*. Autumn 1981. 12, 2: 380–91.

Problem 6

Worker Effort

A worker can exert two effort levels, *Good* and *Bad*, which cause mistakes with probabilities 0.25 and 0.75. His utility function is $U = 100 - 100/w - G$, where w is his wage and G takes the value 2 if he picks the *Good* effort level and zero otherwise. Whether a mistake is made is contractible, but effort is not. Risk-neutral employers compete for the worker, and his output is worth 0 if a mistake is made, 20 otherwise. (No computation is needed for any of this problem.)

(1)   Will the worker be paid anything if he makes a mistake?
(2)   Will the worker be paid more if he does not make a mistake?
(3)   How would the contract be affected if the employers were also risk averse?
(4)   What would the contract look like if a third category, "slight mistake," with an output of 19, occurs with probability 0.1 after *Bad* effort, and probability 0.0 after *Good* effort?

Notes

N6.1   Categories of Asymmetric Information Models

- The separation of asymmetric information into hidden actions and hidden information is suggested in Arrow (1985) and commented upon in Hart & Holmstrom (1987). The term "hidden knowledge" seems to have become more popular than "hidden information."
- Surveys of agency theory include Baiman (1982) and Hart & Holmstrom (1987). Hess (1983) is a book on organization that covers some of the earlier literature. Sonnenschein (1983) is a nice short essay on hidden information that is relatively nontechnical.
- Empirical work on agency problems includes Joskow (1985, 1987) on coal mining, Masten & Crocker (1985) on natural gas contracts, Monteverde & Teece (1982) on auto components, Murphy (1986) on executive compensation, Rasmusen (1988b) on the mutual organization in banking, Staten & Umbeck (1982) on air traffic controllers and disability payments, and Wolfson (1985) on the reputation of partners in oil-drilling.
- There is a big literature of non-mathematical theoretical papers which look at organizational structure in the light of the agency problem. See Alchian & Demsetz (1972), Fama (1980), Fama & Jensen (1983a), Fama & Jensen (1983b), Jensen & Meckling (1976), and Klein, Crawford, & Alchian (1978).
- For examples of agency problems, see "Many Companies Now Base Workers' Raises on Their Productivity," *Wall Street Journal*, 15 November 1985, pp. 1, 15; "Big Executive Bonuses Now Come with a Catch: Lots of Criticism," *Wall Street Journal*, 15 May 1985, p. 33; "Bribery of Retail Buyers is Called Pervasive," *Wall Street Journal*, 1 April 1985, p. 6; "Some Employers Get Tough on Use of Air-Travel Prizes," *Wall Street Journal*, 22 March 1985, p. 27.
- We have lots of "prinsipuls" in economics. I find this paradigm helpful: "The principal's principal principle was to preserve his principal."
- A common convention in principal–agent models is to make one player male and the other female, so that "his" and "her" can be used to distinguish them. I find this distracting, since gender is irrelevant to most models and adds one more detail for the reader to keep track of. If readers naturally thought "male" when they saw "principal," this would not be a problem—but they do not.

N6.2   A Principal–Agent Model: the Production Game

- In Production Game III, we could make the agent's utility depend on the state of the world as well as on effort and wages. Little would change from the simpler model.

- The model in the text uses "effort" as the action taken by the agent, but effort is used to represent a variety of real world actions. The cost of pilferage by employees is an estimated \$8 billion a year in the USA. Employers have offered rewards for detection, one even offering the option of a yearful of twice-weekly lottery tickets instead of a lump sum. The Chicago department store Marshall Field's, with 14,000 workers, in one year gave out 170 rewards of \$500 each, catching almost 500 dishonest employees. ("Hotlines and Hefty Rewards: Retailers Step Up Efforts to Curb Employee Theft," *Wall Street Journal*, 17 September 1987, p. 37.)

  For an illustration of the variety of kinds of "low effort", see "Hermann Hospital Estate, Founded for the Poor, has Benefited the Wealthy, Investigators Allege," *Wall Street Journal*, 13 March 1985, p. 4, which describes such forms of misbehavior as pleasure trips on company funds, high salaries, contracts for redecorating awarded to girlfriends, phony checks, kicking back real estate commissions, and investing in friendly companies. Nonprofit enterprises, often lacking both principles and principals, are especially vulnerable.

- Linear contracts are unlikely to be optimal (see Laffont & Tirole [1986] for an exception). The reason is that output is just a statistic for effort, and decision rules based on statistics are unlikely to be linear. We could stretch and warp the output function in any way that preserves its ordinality, and output would still contain the same information about effort.

- Suppose that the principal does not observe the variable θ (which might be effort), but he does observe t and x (which might be output and profits). From Holmstrom (1979) and Shavell (1979) we have, restated in my words,

The Sufficient Statistic Condition. *If t is a sufficient statistic for θ relative to x, then the optimal contract needs to be based only on t if both principal and agent have separable utility functions.*

The variable t *is a* **sufficient statistic** *for θ relative to x if, for all t and x,*

$$Prob(\theta|t, x) = Prob(\theta|t). \tag{6.7}$$

This implies, from Bayes's Rule, that $Prob(t, x|\theta) = Prob(x|t)Prob(t|\theta)$; that is, x depends on θ only because x depends on t and t depends on θ.

The Sufficient Statistic Condition is closely related to the Rao-Blackwell Theorem (see Cox & Hinkley [1974] p. 258), which says that the decision rule for non-strategic decisions ought not to be random. Gjesdal (1982) notes that if the utility functions are not separable, the theorem does not apply and randomized contracts may be optimal.

N6.3   The Self-Selection, Participation, and Competition Constraints

- Discussions of the first order condition approach can be found in Grossman & Hart (1983) and Hart & Holmstrom (1987).
- The term "individual rationality constraint" is more common, but "participation constraint" is more sensible and instantly recognizable. The term "competition constraint" is my neologism; "zero profit constraint" is more common.
- **Paying the Agent More than His Reservation Wage.** If agents compete to work for principals, the participation constraint is binding whenever there are only two possible outcomes or the agent's utility function is separable in effort and wages. Otherwise, it might happen that the principal picks a contract giving the agent more expected utility than necessary to keep him from quitting. The reason is that the principal not only wants to keep the agent working, but to choose a high effort.
- If the distribution of output satisfies the Monotone Likelihood Ratio Property, the optimal contract specifies higher pay for higher output. Let $f(q|e)$ be the probability density of output. The MLRP is satisfied if

$$\forall e' > e, \text{ and } q' > q, \ f(q'|e')f(q|e) - f(q'|e)f(q|e') > 0, \tag{6.8}$$

or, in other words, if when $e' > e$, the ratio $f(q|e')/f(q|e)$ is increasing in q. Alternatively, f satisfies the MLRP if $q' > q$ implies that q' is a more favorable message than q in the sense of Milgrom (1981b). Less formally, the MLRP is satisfied if the ratio of the likelihood of a high effort to a low effort rises with observed output. The distribution in the Broadway Game of Section 6.5 violates the MLRP, but the normal, exponential, Poisson, uniform, and chi-square distributions all satisfy it.

- Milgrom has used stochastic dominance to carefully define what we mean by **good news**. Let θ be a parameter about which the news is received in the form of message x or y, and let utility be increasing in θ. The message x is more favorable than y (is "good news") if for every possible nondegenerate prior for $F(\theta)$, the posterior $F(\theta|x)$ first-order dominates $F(\theta|y)$.

We say that probability distribution F **dominates** distribution G in the sense of **first-order stochastic dominance** if the cumulative probability that the variable will take a value less than x is greater for G than for F, i.e. if

$$\text{for any } x, \ F(x) \le G(x), \tag{6.9}$$

and (6.9) is a strong inequality for at least one value of x. The distribution F dominates G in the sense of **second-order stochastic dominance** if the area under the cumulative distribution G up to $G(x)$ is greater than the area under F, i.e. if

$$\text{for any } x, \ \int_{-\infty}^{x} F(y)dy \le \int_{-\infty}^{x} G(y)dy, \tag{6.10}$$

and (6.10) is a strong inequality for some value of x. Equivalently, F dominates G if, limiting U to increasing functions for first-order dominance and increasing concave functions for second-order dominance,

$$\text{for all functions } U, \ \int U(x)dF(x) > \int U(x)dG(x). \tag{6.11}$$

If F is a first-order dominant gamble, it is preferred by all players; if F is a second-order dominant gamble, it is preferred by all risk-averse players. If F is first-order dominant it is second-order dominant, but not vice versa. See Milgrom (1981b) and Copeland & Weston (1983, p. 92) for further details. Rothschild & Stiglitz (1970) shows how two gambles can be related in other ways equivalent to second-order dominance, the most important of which is the **mean preserving spread**. Informally, a mean-preserving spread is a density function which transfers probability mass from the middle of a distribution to its tails.

- Finding general conditions that allow the modeller to characterize optimal contracts is difficult. Much of Grossman & Hart (1983) is devoted to the rather obscure Spanning Condition or Linear Distribution Function Condition, under which the first order condition approach is valid. The survey by Hart & Holmstrom (1987) makes a valiant attempt at explaining the LDFC and the related CDFC.

N6.4   State-Space Diagrams: Insurance Games I and II

- The utility function of the agent in the Insurance Game is **separable** in effort and money: if effort changes from $Careful$ to $Careless$, that does not change the marginal utility of a dollar for a given wealth level. Separability matters to the relative slopes of the dotted and solid indifference curves in the diagrams, because it means the slopes differ at a particular point only because the probabilities of an accident differ because of effort, not because the value of money differs.

- The text uses premiums and payouts to describe insurance. Another way to describe insurance is as the sale of an asset which provides different returns in different states. In Insurance Game I, the insurance company sells a security, C_1, with return $(-6, 6)$

and an expected value of 0, for a price of 0. Although Smith is risk averse, he is eager to buy this risky security because it cancels the risk of his initial asset, the car, which has return (12,0).

In Insurance Game II, which has a higher probability of theft in equilibrium, the insurance company offers a security, C_2, with return $(-9,3)$ and expected value of 0 $(= 0.25[-9] + 0.75[3])$ if Smith chooses *Careless*. The partial insurance contract C_3 is a $(-6, 2)$ security with expected value 0 $(= 0.25[-6] + 0.75[2])$, which when combined with Smith's original asset leaves him with allocation $[0.75, (6, 2)]$, where 0.75 is the chance of a theft and $(6,2)$ is his wealth under the contract.

- When the sale of insurance is monopolized, the market behaves differently. Not only is the quantity of insurance restricted, but the contracts offered may specify both the price and the quantity of insurance an individual can buy.

- Gary Schwartz pointed out to me that one reason why liability insurance rarely has coinsurance is that it covers legal disputes, where the need for the victim's side to have a single voice makes it especially useful to have a single residual claimant. With coinsurance, the benefit from each extra dollar awarded would be split between the insuror and the insured, making litigation less efficient.

- Defining optimality is not always straightforward in models of asymmetric information. Holmstrom & Myerson (1983) distinguish three kinds of efficiency, depending on when expected payoffs are evaluated: **ex ante**, before any player has private information; **interim**, when each player has received private information; and **ex post**, after all information has become common knowledge.

N6.5   Optimal Contracts: the Broadway Game

- Daniel Asquith suggested the idea behind the three-state Broadway Game.

- Franchising is one compromise between selling the store and paying a flat wage. References include Mathewson & Winter (1985), Rubin (1978), and Klein & Saft (1985).

- An interesting line of research looks at how principals can use agents to carry out actions the principals themselves could not credibly perform. In Bernheim & Whinston (1986), competing firms delegate their sales to a single agent who coordinates the cartel. M. Katz (unpub) has investigated observable and unobservable contracts. If only one principal can contract observably with an agent, he can gain an advantage in duopoly output or in bargaining with another principal. If the contract is unobservable, he can gain an advantage only if there is asymmetric information between himself and the agent or the agent is risk averse. In those cases, outsiders can deduce what the contract will be, and react in a way favorable to the principal.

- McAfee & McMillan (1986) is an interesting combination of the principal–agent model with an auction. Agents bid against each other in offering contracts to the principal for some service—say, building a ship. The agents know a portion of their own costs and the principal does not, but since the agents are risk averse, the cost of production ought to be shared between principal and agent.

- Fudenberg & Tirole (unpub) note an interesting problem in the hidden actions model. Optimal contracts impose risk on a risk-averse agent to provide incentives for high effort. If at some point in the game it is common knowledge that the agent has chosen his effort but Nature has not yet moved, the agent bears useless risk. The principal knows the agent has already chosen effort, so the two of them are willing to recontract to put the risk from Nature's move back on the principal. But the expected future recontracting makes a joke of the original contract and reduces the agent's incentives for effort.

- Non-separability of the utility function in effort and wages can make the principal want to randomize the wage. Suppose there are two actions the agent might take. The principal prefers action X, which reduces the agent's risk aversion, to action Y, which increases it. The principal could offer a randomized wage contract, so the agent would choose action X and make himself less risk averse. This randomization is not a mixed strategy. The principal is not indifferent to high and low wages: he prefers to pay a low wage, but we allow him to commit to a random wage earlier in the game.

- Moral hazard is a problem when the principal can neither observe the agent's effort level nor deduce it without assuming equilibrium behavior. Often if the principal could observe the state θ, he could deduce effort from observed output $q(e, \theta)$, or from his own observed utility $V(q - w)$. Moral hazard arises when the functions q and V are not invertible. That $V(q - w)$ is **invertible** means that every point in the space in which $(q - w)$ lies is mapped onto exactly one point in the space in which V lies, and every point in V-space is mapped onto by some point in $(q - w)$-space. Knowing V, you know $(q - w)$, and vice versa.

  In the Broadway Game, moral hazard is a problem even though the state of the world (*Failure* or *Success*) is known and could be conditioned upon in a contract, because even knowing the state, the principal cannot always invert the output to find the effort level. Invertibility, not ignorance of the state, is the key to moral hazard.

7   Moral Hazard: Hidden Information

7.1   Pooling vs. Separating Equilibrium, and the Revelation Principle

In Chapter 7 we will continue with moral hazard by taking up hidden information and a variety of special cases that apply to both kinds of moral hazard. This section introduces hidden information and distinguishes between two kinds of equilibria—pooling and separating—as well as discussing a modelling simplification called the Revelation Principle. Section 7.2 uses diagrams to apply the model to selecting a sales strategy. We then turn to remedies for moral hazard including efficiency wages (7.3), tournaments (7.4), and monitoring (7.5), with a summary in Section 7.6. Section 7.7 finishes with the teams model, the free rider problem, and preference revelation.

Information is complete in moral hazard games, but under hidden information the agent sees some move of Nature that the principal does not. From the principal's point of view, agents come in several types, depending on what they have seen, and his chief concern is to discover the type. The agent may exert effort, but effort's contractibility is unimportant when the principal, ignorant of the state of the world, does not know which effort is appropriate. Just for comparison, let us modify Production Game III from Section 6.2 to become a game of hidden information.

Production Game IV: Hidden Information

Players

The principal and the agent.

Information

Asymmetric, complete, and uncertain.

Actions and Events

(1)   The principal offers the worker a wage contract of the form $w(q, m)$.

(2)   The agent accepts or rejects the principal's offer.

(3)   Nature chooses the state of the world θ, according to probability distribution $F(\theta)$. The agent observes θ, but the principal does not.

(4)   If the agent accepts, he exerts effort e and sends a message m, both observed by the principal.

(5)   Output is $q(e, \theta)$.

Payoffs

If the agent rejects the contract, then $\pi_{agent} = \bar{U}$ and $\pi_{principal} = 0$.

If the agent accepts the contract, then $\pi_{agent} = U(e, w, \theta)$ and $\pi_{principal} = V(q - w)$.

The principal would like to know θ. He would be delighted to employ an honest agent who always chooses $m = \theta$, but in noncooperative games, "tal parola non vale uno zero." Since the agent's words are worthless, the principal must try to design a contract that either provides incentives for truthfulness or takes lying into account.

Pooling and Separating Equilibria

In hidden actions models, the principal tries to construct a contract which will induce the agent to take the single appropriate action. In hidden information models, the principal tries to make different actions attractive under different states of the world, so the agent's choice depends on the hidden state.

If all types of agents pick the same strategy in all states, the equilibrium is **pooling**. *Otherwise, it is* **separating**.

These two terms came up in Section 5.2 in the game PhD Admissions. Neither type of student applied in the pooling equilibrium, but only one type did in the separating equilibrium. The distinction between pooling and separating has nothing to do with the equilibrium concept. A model might have multiple Nash equilibria, some pooling and some separating. Moreover, a single equilibrium—even a pooling one—can include several contracts, but if it is pooling the agent always uses the same strategy, regardless of type. If the agent's equilibrium strategy is mixed, the equilibrium is pooling if the agent always picks the same mixed strategy, even though the messages and efforts would differ across realizations of the game.

A separating contract need not be fully separating. If agents who observe $\theta \leq 4$ accept contract C_1, but other agents accept C_2, then the equilibrium is separating but it does not separate out every type. We say that the equilibrium is **fully revealing** if the agent's choice of contract always conveys his private information to the principal. Between pooling and fully revealing equilibria are the **imperfectly separating** equilibria synonymously called **semi-separating, partially separating, partially revealing**, and **partially pooling**.

The principal's problem, as in Production Game III, is to maximize his profits subject to

(1)   **incentive compatibility** (the agent picks the desired contract and actions);
(2)   **participation** (the agent prefers the contract to his reservation utility).

Under hidden information, the incentive compatibility constraint is sometimes called the **self-selection constraint**, because it induces the different types of agents to pick different contracts. As with hidden actions, if principals compete in offering contracts, a **competition constraint** is added: the equilibrium contract must be as attractive as possible to the agent, since otherwise another principal could profitably lure him away.

An equilibrium must also satisfy a part of the competition constraint not found in hidden actions models: either a **nonpooling constraint** or a **nonseparating constraint** (but not both, of course). If one of several competing principals wishes to construct a pair of separating contracts C_1 and C_2, he must construct them so that not only do agents choose C_1 and C_2 depending on the state of the world (incentive compatibility), but also they prefer (C_1, C_2) to a pooling contract C_3 (nonpooling).

The Revelation Principle

The principal might choose to offer a contract that induces the agent to lie in equilibrium, since he can take lying into account when he designs the contract, but this complicates the analysis. Each state of the world has a single truth, but a continuum of lies: generically speaking, almost everything is false. The Revelation Principle helps us simplify.

> **The Revelation Principle:** *For every contract $w(q, m)$ that leads to lying (that is, to $m \neq \theta$), there is a contract $w^*(q, m)$ with the same outcome for every θ but no inducement for the agent to lie.*

For the Revelation Principle to be useful, we must also make the following assumption, which is usually slipped in unobtrusively.

> **Epsilon Truthfulness:** *If the agent is indifferent between lying and telling the truth, he tells the truth.*

Epsilon truthfulness is an example of a **lexicographic preference**. Just as the words *axis*, *azalea*, and *baa* are ordered letter by letter in the dictionary, with no number of terminal *a*s making up for an initial *b*, so our agent prefers $10 and truth to $10 and lies, but both are preferred to $9.99 and truth.

Suppose that under the contract $w(q, m)$ the agent announces $m = \theta_1$ when the state is really θ_2. The principal can offer the agent another contract in which the agent's wage is the same for all θs, which under epsilon truthfulness leads him to announce the state as θ_2. Although the Revelation Principle does not imply that the first-best contract is possible, and says very little about the outcome, we do learn that the optimal contract can be relatively simple.

Applied to concrete examples, the Revelation Principle may seem obvious. Suppose we are concerned with the effect on the moral climate of cheating on income taxes, but anyone who makes $70,000 a year can claim he makes $50,000 and the government cannot catch him. The Revelation Principle says that we can rewrite the tax code to set the same tax for people earning $70,000 as for those earning $50,000. The same amount is collected, but under epsilon truthfulness nobody lies. Applied to moral education, the principle says that the mother who agrees never to punish her daughter if she tells her everything will never hear any untruths. Clearly, the principle's usefulness is not to improve contracts, only to simplify them. The principal (and the modeller) need only look at contracts which induce truthtelling, so the relevant strategy space is shrunk and we can add a third constraint to help calculate the equilibrium:

(3)   **Truthtelling.** The equilibrium contract induces the agent to choose $m = \theta$.

Unravelling the Truth when Silence is the Only Alternative

The usual hidden information model has no penalty for lying, but let us briefly consider what happens if the agent cannot lie but he can be silent or tell half-truths. Suppose that Nature uses the uniform distribution to assign the variable θ some value in the interval $[0, 10]$ and the agent's payoff is increasing in the principal's estimate of θ. Usually we assume that the agent can lie freely, sending a message m taking any value in $[0, 10]$, but let us assume instead that he cannot lie, although he is free to conceal information. Thus, if $\theta = 2$, he could send the uninformative message $m \geq 0$ (equivalent to no message), or $m \geq 1$, or $m = 2$, but not the lie $m \geq 4$.

Under epsilon truthfulness, when $\theta = 2$ the agent's optimal message is the exact truth: $m = 2$. If he were to choose $m \geq 1$, for example, the principal's first thought might be to estimate θ as the average value of the interval $[1, 10]$, which is 5.5. But the principal would realize that no agent with a value of θ greater than 5.5 would want to send that message. This realization restricts the possible interval to $[1, 5.5]$, which in turn has an average of 3.25. But no agent with $\theta > 3.25$ would send that message. The principal can continue this process of logical **unravelling** to conclude that $\theta = 1$. The message $m \geq 0$ would be even worse, making the principal believe that $\theta = 0$. In this model, "No News is Bad News." The agent would therefore not send the message $m \geq 1$, and under epsilon truthfulness he would prefer $m = 2$ to $m \geq 2$, although the principal would make the same deduction with either message.

Perfect revelation may seem incredible, but that is because the assumptions of the model are rarely satisfied in the real world. In particular, unpunishable lying and genuine ignorance allow information to be concealed. If the seller is free to lie without punishment, then in the absence of other incentives he always pretends that his information is extremely favorable, so nothing he says conveys any information, favorable or unfavorable. If he really is ignorant in some states of the world, then his silence could mean either that he has nothing to say or he has nothing he wants to say. The unravelling argument fails because if he sends an uninformative message the buyers will attach some probability to no news instead of bad news.

7.2   An Example: the Salesman Game

The next game illustrates the differences between pooling and separating equilibria. The manager of a company has told his salesman to investigate a potential customer, who is either a *Pushover* or a *Bonanza*. If he is a *Pushover*, the efficient sales effort is low and sales should be moderate. If he is a *Bonanza*, the effort and sales should be higher.

The Salesman Game

Players

A manager and a salesman.

Information

Asymmetric, complete, and uncertain. The manager is uninformed.

Actions and Events

(1)   The manager offers the salesman a contract of the form $w(q, m)$, where q is sales and m is a message.
(2)   The salesman decides whether or not to accept the contract.
(3)   Nature chooses whether the customer is a *Bonanza* or a *Pushover* with probabilities 0.2 and 0.8. Denote the state variable "customer status" by θ. The salesman observes the state, but the manager does not.
(4)   If the salesman has accepted the contract, he chooses his sales level q, which implicitly measures his effort.

Payoffs

If the salesman rejects the contract, his payoff is $\bar{U} = 8$ and the manager's is zero. If he accepts the contract, then

$$\pi_{manager} = q - w.$$

$$\pi_{salesman} = U(q, w, \theta), \text{ where } \frac{\partial U}{\partial q} < 0, \frac{\partial^2 U}{\partial q^2} < 0, \frac{\partial U}{\partial w} > 0, \frac{\partial^2 U}{\partial w^2} < 0.$$

Figure 7.1 shows the indifference curves of manager and salesman, labelled with numerical values for exposition. The manager's indifference curves are straight lines with slope -1, because he is acting on behalf of a risk-neutral company. If the wage and the quantity both rise by a dollar, profits are unchanged, and the profits do not depend directly on whether θ takes the value *Pushover* or *Bonanza*.

The salesman's indifference curves also slope upwards, because he must receive a higher wage to compensate for the extra effort that makes q greater, and they

Figure 7.1   The Salesman Game

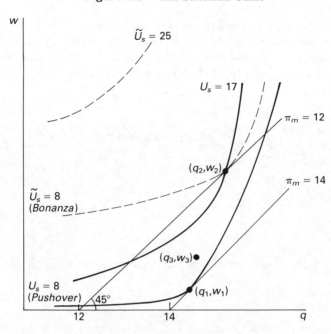

are convex because the marginal utility of dollars is decreasing and the marginal disutility of effort is increasing. As Figure 7.1 shows, the salesman has two sets of indifference curves, solid for pushovers and dotted for bonanzas, since the effort that secures a given level of sales depends on the state.

Because of the participation constraint, the manager must provide the salesman with a contract giving him at least his reservation utility of 8, which is the same in both states. If the true state is that the customer is a bonanza, the manager would like to offer a contract that leaves the salesman on the dotted indifference curve $\tilde{U} = 8$, and the efficient outcome is (q_2, w_2), the point where the salesman's indifference curve is tangent to one of the manager's indifference curves. At that point, if the salesman sells an extra dollar he requires an extra dollar of compensation.

If it were common knowledge that the customer was a bonanza, the principal could choose w_2 so that $U(q_2, w_2, Bonanza) = 8$ and offer the forcing contract

$$w = \begin{cases} 0 & \text{if } q < q_2. \\ w_2 & \text{if } q \geq q_2. \end{cases} \tag{7.1}$$

The salesman would accept the contract and choose $q = q_2$. But if the customer were actually a pushover, the salesman would still choose $q = q_2$, an inefficient outcome that does not maximize profits. High sales would be inefficient because the salesman would be willing to give up more than a dollar of wages to escape having to make his last dollar of sales. Profits would be unmaximized, because the salesman achieves a utility of 17, and he would have been willing to work for less.

The Revelation Principle says that in searching for the optimal contract we need only look at contracts that induce the agent to truthfully reveal what kind of customer he faces. If it required more effort to sell any quantity to the bonanza, as shown in Figure 7.1, then the salesman would always want the manager to believe that he faced a bonanza to extract the extra pay necessary to achieve a utility of 8. The only optimal truthtelling contract is the pooling contract that pays the intermediate wage of w_3 for the intermediate quantity of q_3 and zero for any other quantity, regardless of the message. The pooling contract is a second-best contract, a compromise between the optimum for pushovers and the optimum for bonanzas. The point (q_3, w_3) is closer to (q_1, w_1) than to (q_2, w_2), because the probability of a pushover is higher and the contract must satisfy the participation constraint

$$0.8U(q_3, w_3, Pushover) + 0.2U(q_3, w_3, Bonanza) \geq 8. \tag{7.2}$$

The nature of the equilibrium depends on the shapes of the indifference curves. If they are shaped as in Figure 7.2 then the equilibrium is separating, not pooling, and there does exist a first-best, fully revealing, contract.

$$\text{Separating Contract}: \begin{cases} \text{Agent announces } Pushover: & w = \begin{cases} 0 & \text{if } q < q_1. \\ w_1 & \text{if } q \geq q_1. \end{cases} \\ \text{Agent announces } Bonanza: & w = \begin{cases} 0 & \text{if } q < q_2. \\ w_2 & \text{if } q \geq q_2. \end{cases} \end{cases} \tag{7.3}$$

Figure 7.2   Indifference curves for a separating equilibrium

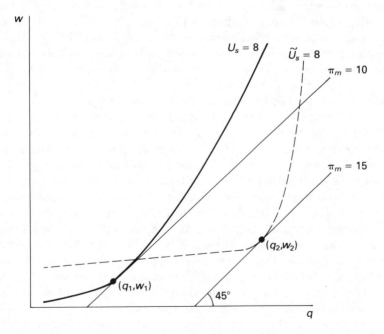

Again, we know from the Revelation Principle that we can narrow attention to contracts that induce the salesman to tell the truth. With the indifference curves of Figure 7.2, contract (7.3) induces the salesman to be truthful and the incentive

compatibility constraint is satisfied. If the customer is a bonanza, but the salesman claims to observe a pushover and chooses q_1, his utility is less than 8 because the point (q_1, w_1) lies below the $\tilde{U} = 8$ indifference curve. If the customer is a pushover and the salesman claims to observe a bonanza, then although (q_2, w_2) does yield the salesman a higher wage than (q_1, w_1), the extra income is not worth the extra effort, because (q_2, w_2) is far below the indifference curve $U = 8$.

The Salesman Game illustrates a number of ideas. It can have either a pooling or a separating equilibrium, depending on the utility function of the salesman. The Revelation Principle can be applied to avoid having to consider contracts in which the manager must interpret the salesman's lies. Also, in contrast to most of our two-valued numerical examples, it shows how to use diagrams when the algebraic functions are intractable or unspecified.

7.3   Efficiency Wages

The next three sections are about remedies for moral hazard that can be applied to either hidden actions or hidden information. Most of the illustrations will be to hidden actions, but usually "low effort" can be replaced by "lying about the hidden information."

Shapiro & Stiglitz (1984) show how involuntary unemployment can be explained by a principal–agent model. When all workers are employed at the market wage, a worker who is caught shirking and fired can immediately find another job just as good. Firing is ineffective, and effective penalties like boiling in oil are excluded from the strategy spaces of legal businesses. Becker & Stigler (1974) have suggested that workers post performance bonds, but if workers are poor this is impractical. Without bonds or boiling in oil, the worker chooses low effort and receives a low wage.

To induce a worker not to shirk, the firm can offer to pay him a premium over the market-clearing wage, which he loses if he is caught shirking and fired. If one firm finds it profitable to raise its wage, however, so do all firms, and one might think that after the wages equalized, the incentive not to shirk would disappear. But when a firm raise its wages, its demand for labor falls, and when all firms raise their wages, the market demand for labor falls, creating unemployment. Even if all firms pay the same wages, a worker has an incentive not to shirk, because if he were fired he would stay unemployed, and even if there is a random chance of leaving the unemployment pool, the unemployment rate rises sufficiently high that workers choose not to risk being caught shirking. The equilibrium is not first-best efficient, because even though the marginal revenue of labor equals the wage, it exceeds the marginal disutility of effort, but it is efficient in a second-best sense. By deterring shirking, the hungry workers hanging around the factory gates are performing a socially valuable function (but they mustn't be paid for it!).

The idea of paying high wages to increase the threat of dismissal is old, and can even be found in *The Wealth of Nations* (Smith [1776] p. 207). What is new in Shapiro & Stiglitz is the observation that unemployment is generated by these "efficiency wages." These firms behave paradoxically. They pay workers more than necessary to attract them, and outsiders who offer to work for less are turned away. Can this explain why "overqualified" jobseekers are unsuccessful and stupid managers are retained? Employers are unwilling to hire someone talented, because he could find another job after being fired for shirking, and trustworthiness matters more than talent in some jobs.

7.4   Tournaments

Games in which relative performance is important are called **tournaments**. Tournaments are similar to auctions, the difference being that the actions of the losers matter directly, unlike in auctions. Like auctions, they are especially useful when the principal wants to elicit information from the agents. A principal-designed tournament is sometimes called **yardstick competition** because the agents provide the measure for their wages.

Farrell (unpub) uses a tournament to explain how "slack" might be the major source of welfare loss from monopoly, an old idea usually prompted by faulty reasoning. The usual claim is that monopolists are inefficient because they, unlike competitive firms, do not have to maximize profits to survive, which presumes that firms care about survival, not profits. Farrell makes a subtler point: although the shareholders of a monopoly maximize profit, the managers maximize their own utility, and moral hazard is severe without the benchmark of other firms' performances.

Let firm Allied have two possible production techniques, *Fast* and *Careful*. Independently for each technique, Nature chooses production cost $c = 1$ with probability θ and $c = 2$ with probability $1 - \theta$. The manager can either choose a technique at random or investigate the costs of both techniques at utility cost α. The shareholders can observe the resulting production cost, but not whether the manager investigates. If they see the manager pick *Fast* and a cost of $c = 2$, they do not know whether he chose it without investigating, or investigated both techniques and found they were both costly. The wage contract is based on what the shareholders can observe, so it takes the form (w_1, w_2), where w_1 is the wage if $c = 1$ and w_2 if $c = 2$. The manager's utility function is $\log w$ if he does not investigate, $\log w - \alpha$ if he does, and a reservation utility of $\log \bar{w}$ if he quits.

If the shareholders want the manager to investigate, the contract must satisfy the self selection constraint

$$U(\text{not investigate}) \leq U(\text{investigate}). \tag{7.4}$$

If the manager investigates, he still fails to find a low cost technique with probability $(1 - \theta)^2$, so (7.4) is equivalent to

$$\theta \log w_1 + (1 - \theta)\log w_2 \leq [1 - (1 - \theta)^2]\log w_1 + (1 - \theta)^2 \log w_2 - \alpha. \tag{7.5}$$

Turning inequality (7.5) into an equality and simplifying, we find

$$\theta(1 - \theta)\log \frac{w_1}{w_2} = \alpha. \tag{7.6}$$

The participation constraint is $U(\bar{w}) = U(\text{investigate})$, or

$$\log \bar{w} = [1 - (1 - \theta)^2]\log w_1 + (1 - \theta)^2\log w_2 - \alpha. \qquad (7.7)$$

Solving equations (7.6) and (7.7) together for w_1 and w_2, we obtain

$$w_1 = \bar{w}e^{\alpha/\theta}.$$
$$w_2 = \bar{w}e^{-\alpha/(1-\theta)}. \qquad (7.8)$$

The expected cost to the firm is

$$[1 - (1 - \theta)^2]\bar{w}e^{\alpha/\theta} + (1 - \theta)^2\bar{w}e^{-\alpha/(1-\theta)}. \qquad (7.9)$$

If the parameters are $\theta = 0.1$, $\alpha = 1$, and $\bar{w} = 1$, the rounded values are $w_1 = 22,026$ and $w_2 = 0.33$, and the expected cost is $4,185$. Quite possibly, the shareholders decide it is not worth making the manager investigate.

But suppose that Allied has a competitor, Brydox, in the same situation. The shareholders of Allied can threaten to boil their manager in oil if Brydox adopts a low cost technology and Allied does not. If Brydox does the same, the two managers are in a Prisoner's Dilemma, both wishing not to investigate, but each investigating from fear of the other. The forcing contract for Allied specifies $w_1 = w_2$ to fully insure the manager, and boiling in oil if Brydox has lower costs than Allied. The contract need satisfy only the participation constraint that $\log w - \alpha = \log \bar{w}$, so $w = 2.72$ and the cost of learning to Allied is only 2.72, not $4,185$. Competition raises efficiency, but not through the threat of firms going bankrupt, but through the threat of managers being fired.

7.5   Monitoring

By **monitoring** we mean the principal's costly acquisition of information about the agent's actions. **Perfect monitoring** tells the principal the exact actions (the information set becomes a singleton). **Imperfect monitoring** refines the principal's information, but not all the way to a singleton.

Monitoring is probably more important in practice than designing contracts carefully, but research has neglected it. In monitoring, as in any sort of learning, a player does something he hopes will refine his information partition. **Auditing** is another sort of learning: the acquisition of information as to whether an agent is lying. The chief difference from monitoring is that the activity of truthtelling, unlike effort, does not generate direct disutility; a contract can induce inefficiently much effort, but not too much truth. A third form of learning is **testing**, which tries to determine the type of an agent under adverse selection. Below are listed four ways to model monitoring of hidden actions that could also be used for auditing or testing.

(1)   Pay X and obtain the true value of e (perfect monitoring).
(2)   Pay X and obtain a noisy statistic for e which has variance α.

(3)   Pay X and obtain the true value of e with probability $(1 - \alpha)$, and no information with probability α.

(4)   Pay X and obtain perfect information on whether e is below α.

Methods (2), (3), and (4) can be adapted so that paying a larger amount produces more information, either by lowering α or, in the case of (4), by increasing the number of αs on which information can be obtained.

The order of actions is important here, as always. It matters whether the agent observes the monitoring action before or after he picks effort. It also matters whether the principal can commit to carry out his monitoring. If he cannot, and monitoring is costly, he might not carry out his threat to monitor: "Quis custodiet ipsos custodes?"

7.6   Alleviating the Agency Problem

A number of ways to alleviate the agency problem are summarized here. Each is illustrated by application to the particular problem of executive compensation, which is empirically important, and interesting both because explicit incentive contracts are used and because they are not used more often (see Baker, Jensen & Murphy [1988]).

(1) **Reputation** (4.6, 4.7, 4.8, 5.4, 5.5)

Managers are promoted on the basis of past effort or truthfulness.

(2) **Second-Best Risk-Sharing Contracts** (6.2, 6.3, 6.4)

The executive receives not only a salary, but call options on the firm's stock. If he reduces the stock value, his options fall in price.

(3) **Boiling in Oil** or **Shifting Support Schemes** (6.5)

If the firm would only become unable to pay dividends if the executive shirked and was unlucky, the threat of firing him when the firm skips a dividend will keep him working hard.

(4) **Selling the Store** (6.5)

The managers buy the firm in a leveraged buyout.

(5) **Efficiency Wages** (7.3)

To make him fear losing his job, an executive is paid a higher salary than his ability warrants.

(6) Tournaments (7.4)

Several vice presidents compete and the winner succeeds the president.

(7) Monitoring (7.5)

The directors hire a consultant to evaluate the executive's performance.

(8) Repetition

Managers are paid less than their marginal products for most of their career, but are rewarded later with higher salaries or generous pensions if their career record has been good.

(9) Changing the Type of the Agent

Older executives encourage the younger by praising ambition and hard work.

We have already talked about all but the last two solutions. Repetition of the game enables the contract to come closer to the first-best if the discount rate is low (Radner [1985]). Production Game III (Section 6.2) failed to attain the first-best because output depended on both the agent's effort and random noise. If the game were repeated 50 times with independent drawings of the noise, the randomness would average out and the principal could form an accurate estimate of the agent's effort. But this is really begging the question by saying that effort can be observed after all.

Changing the agent's type, a solution to both moral hazard and adverse selection, has received little attention in the economics literature, perhaps because it is more psychological than economic. Akerlof (1983), one of the few papers by an economist on the subject of changing type, points out that the moral education of children, as well as their intellectual education, affects their productivity and success. Within the firm, the traditional solution to the agency problems of harem management is particularly striking, but similar if subtler solutions can be found in any modern organization.

7.7   Teams, and the Groves Mechanism

We will close the discussion of moral hazard with a special model focussing on the group of agents rather than the individual.

> A **team** *is a group of agents who independently choose effort levels that result in a single output for the entire group.*

We will look at teams using the following game.

Teams

(Holmstrom [1982])

Players

A principal and n agents.

Information

Asymmetric, complete, and certain.

Actions and Events

(1) The principal offers a contract to each agent i of the form $w_i(q)$, where q is total output.
(2) The agents decide whether or not to accept the contract.
(3) The agents simultaneously pick effort levels e_i, $(i = 1, \ldots, n)$.
(4) Output is $q(e_1, \ldots, e_n)$.

Payoffs

If any agent rejects the contract, all payoffs equal zero. Otherwise,

$$\pi_{principal} = q - \sum_{i=1}^{n} w_i;$$

$$\pi_i = w_i - v_i(e_i), \text{ where } v_i' > 0 \text{ and } v_i'' > 0.$$

Despite the risk neutrality of the agents, "selling the store" fails to work here, because the team of agents still has the same problem as the employer had. The team's problem is cooperation between agents, and the principal is peripheral.

Denote the efficient vector of actions by e^*. An efficient contract is

$$w_i(q) = \begin{cases} b_i & \text{if } q \geq q(e^*). \\ 0 & \text{if } q < q(e^*). \end{cases} \tag{7.10}$$

where $\sum_{i=1}^{n} b_i = q(e^*)$ and $b_i > v_i(e_i^*)$.

Contract (7.10) gives agent i the wage b_i if all agents pick the efficient effort, and nothing if any of them shirks, in which case the principal keeps the output. The teams model gives one reason to have a principal: he is the residual claimant who keeps the forfeited output. Without him, it is questionable whether the agents would carry out the threat to discard the output if, say, it were 99 instead of the efficient 100. There is a perfectness problem: the agents would like to commit in advance to throw away output, but only because they never have to do so in equilibrium. If the modeller wishes to disallow discarding output, he imposes the

budget-balancing constraint that the sum of the wages equals exactly the output, no more and no less. But budget balancing creates a problem for the team that is summarized in Proposition 7.1.

Proposition 7.1

If there is a budget-balancing constraint, no differentiable wage contract $w_i(q)$ generates an efficient Nash equilibrium.

Proof

Agent i's problem is

$$\underset{e_i}{Maximize} \quad w_i(q(e)) - v_i(e_i). \tag{7.11}$$

His first order condition is

$$\left(\frac{dw_i}{dq}\right)\left(\frac{\partial q}{\partial e_i}\right) - \frac{dv_i}{de_i} = 0. \tag{7.12}$$

With budget balancing and a linear utility function, the Pareto optimum maximizes the sum of utilities (something not generally true), so the optimum solves

$$\underset{e_1,\ldots,e_n}{Maximize} \quad q(e) - \sum_{i=1}^{n} v_i(e_i) \tag{7.13}$$

The first order condition is that the marginal dollar contribution to output equal the marginal disutility of effort:

$$\frac{\partial q}{\partial e_i} - \frac{dv_i}{de_i} = 0. \tag{7.14}$$

Equation (7.14) contradicts (7.12), the agent's first order condition, because $\frac{dw_i}{dq}$ is not equal to one. If it were, agent i would be the residual claimant and receive the entire marginal increase in output—but under budget balancing, not every agent can do that. Because each agent bears the entire burden of his marginal effort and only part of the benefit, the contract does not achieve the first-best. Without budget balancing, on the other hand, if the agent shirked a little he would gain the entire leisure benefit from shirking, but he would lose his entire wage under the optimal contract.

Discontinuities in Public Good Payoffs

Ordinarily, there is a free rider problem if several players each pick a level of effort which increases the level of some public good whose benefits they share. Noncooperatively, they choose effort levels lower than if they could make binding promises. Mathematically, let identical risk-neutral players indexed by i choose effort levels e_i to produce amount $q(e_1,\ldots,e_n)$ of the public good, where q is a continuous function. Player i's problem is

$$\underset{e_i}{Maximize} \quad q(e_1, \dots, e_n) - e_i, \tag{7.15}$$

which has first order condition

$$\frac{\partial q}{\partial e_i} - 1 = 0, \tag{7.16}$$

whereas the greater, first-best effort n-vector e^* is characterized by

$$\sum_{i=1}^{n} \frac{\partial q}{\partial e_i} - 1 = 0. \tag{7.17}$$

If the function q is discontinuous at e^* (for example, $q = 0$ if $e_i < e_i^*$ for any i), the strategy combination e^* can be a Nash equilibrium. In the Teams model the same effect is at work. Although the Teams function is not discontinuous, contract (7.10) is constructed to obtain the same incentives as if it were.

The surprising result that the first-best is achieved arises because the discontinuity at e^* makes every player the marginal, decisive player: if he shirks a little, output falls drastically and with certainty. Either of two modifications restores the free rider problem and induces shirking:

(1) Let q be a function not only of effort but of random noise—Nature moves after the players. Uncertainty makes the *expected* output a continuous function of effort.

(2) Let players have incomplete information about the critical value—Nature moves before the players and chooses e^*. Incomplete information makes the *estimated* output a continuous function of effort.

The discontinuity phenomenon is common. References, not all of which note the problem, include:

(1) Effort in teams (Holmstrom [1982], Rasmusen [1987]).
(2) Entry deterrence by an oligopoly (Bernheim [1984b], Waldman [1987]).
(3) Output in oligopolies with trigger strategies (Porter [1983a]).
(4) Patent races (Section 13.4).
(5) Tendering shares in a takeover (Grossman & Hart [1980], Section 13.5).
(6) Preferences for levels of a public good.

Preferences for Public Goods and the Groves Mechanism (Vickrey [1961])

Hidden information is important in public economics, the study of government spending and taxation. The government frequently wants to extract information about the citizen's wealth and preferences over public goods. The government **implements** a **mechanism** to extract the information; as the principal, it chooses a contract which induces the agent to send a true message. In the context of public economics, this is usually adverse selection, not moral hazard, since the agents differ in type, but since both kinds of hidden information can be analyzed this way, I present the topic here.

The example is adapted from Varian (1984, p. 257). The mayor is considering installing a streetlight costing $100. Each of the five houses near the light would be taxed exactly $20, but the mayor will only install it if he decides that the sum of the residents' valuations for it is greater than the cost. The problem is to discover the valuations. If asked, householder Smith says that his valuation is $5,000, and Brown says he would pay $5,000 to keep the street dark, which only indicates that Smith's valuation is more than $20 and Brown's is less. The dominant strategy is to overreport or underreport.

The flawed mechanism just described can be denoted by

$$\left(-20, \sum_{i=1}^{5} m_i \geq 100\right) \tag{7.18}$$

for each resident, which means that he pays 20 and the light is installed if the sum of the messages exceeds 100. If instead, resident j pays his message (or zero if it is negative), the contract is

$$\left(Min[-m_j, 0], \sum_{i=1}^{5} m_i \geq 100\right), \tag{7.19}$$

in which case there is no dominant strategy. Player j would announce $m_j = 0$ if he thought the project would go through without his support, but he would announce up to his valuation if necessary. There is a continuum of Nash equilibria that attain the efficient result. But instead of just ensuring that the correct decision is made in a Nash equilibrium, it might be possible to design a mechanism which makes truthfulness a **dominant strategy mechanism** or **Groves mechanism**. Consider

$$\left(\sum_{i \neq j} m_i - 100, \sum_{i=1}^{5} m_i \geq 100\right). \tag{7.20}$$

Under mechanism (7.20), player j's message does not affect his tax bill except by its effect on whether the streetlight is installed. If player j's valuation is v_j, his full payoff is $v_j + \sum_{i \neq j} m_i - 100$ if $m_j + \sum_{i \neq j} m_i \geq 100$, and zero otherwise. It is not hard to see that he will be truthful in a Nash equilibrium in which the other players are truthful, but we can go further than that. Truthfulness is weakly dominant. Moreover, the players will tell the truth whenever lying would alter the mayor's decision, so epsilon truthfulness is not needed to obtain the efficient result.

Consider a numerical example. Suppose that Smith's valuation is 40 and the sum of valuations is 110, so the project is indeed efficient. If the other players reported their truthful sum of 70, Smith's payoff from truthful reporting is his valuation of 40 minus his tax of 30. Reporting more would not change his payoff, while reporting less than 30 would reduce it to 0.

If we are wondering whether Smith's strategy is dominant, we must also consider his best response when the other players lie. If they underreported 50 instead of the truthful 70, then Smith could make up the difference by overreporting 60, but his payoff would be -10 ($= 40 + 50 - 100$) so he would do better to report the truthful 40, killing the project and leaving him with a payoff of 0. If the other players overreport 80 instead of their truthful 70, then Smith benefits if the project goes through, and he should report at least 40, which he would do under epsilon truthfulness, to obtain his payoff of 40 minus 20.

The problem with a dominant strategy mechanism like the one facing Smith is that it is not budget balancing. The government raises less in taxes than it spends on the project (in fact, the taxes would be negative). Lack of budget balancing is a crucial feature of dominant strategy mechanisms. While the government deficit can be made either positive or negative, it cannot be made zero, unlike in Nash mechanisms.

Recommended Reading

Caillaud, B., Roger Guesnerie, P. Rey, & Jean Tirole (1988) "Government Intervention in Production and Incentives Theory: a Review of Recent Contributions" *Rand Journal of Economics.* Spring 1988. 19, 1: 1–26.

Holmstrom, Bengt (1982) "Moral Hazard in Teams" *Bell Journal of Economics.* Autumn 1982. 13, 2: 324–40.

Lazear, Edward & Sherwin Rosen (1981) "Rank-Order Tournaments as Optimum Labor Contracts" *Journal of Political Economy.* October 1981. 89, 5: 841–64.

Nalebuff, Barry & Joseph Stiglitz (1983) "Prizes and Incentives: Towards a General Theory of Compensation and Competition" *Bell Journal of Economics.* Spring 1983. 14, 1: 21–43.

Shapiro, Carl & Joseph Stiglitz (1984) "Equilibrium Unemployment as a Worker Discipline Device" *American Economic Review.* June 1984. 74, 3: 433–44.

Problem 7

Defense Contractors

The government is considering hiring two defense contractors, Allied and Brydox, to jointly design an airplane, but it does not know how much the design would really cost them. The cost for each firm's part of the project is uniformly distributed between $10 million and $100 million. The social benefit of the design is known to be $50 million. The government wishes to find a mechanism to make the decision of whether to design the airplane, based on social cost and social benefit.

(1) Design a Groves mechanism to make the efficient decision.

(2) Suppose that the cost is actually $20 million for Allied and $10 million for Brydox. How much is each company paid in the mechanism of part (1)?

(3) If the costs were $20 million for Allied and $70 million for Brydox, how much would each be paid?

(4) Does the mechanism reach an efficient result in both cases? Why is the government unlikely to employ it?

Notes

N7.1   Pooling vs. Separating Equilibrium, and the Revelation Principle

- Several surveys of mechanism design exist, including Caillaud et al. (1988), the book by Green & Laffont (1979), several papers in the volume edited by Hurwicz, Schmeidler, & Sonnenschein (1985), and Baron's chapter in the *Handbook of Industrial Organization* edited by Schmalensee and Willig. Using a verbal style, Levmore (1982) discusses hidden information problems in tort damages, corporate freezeouts, and property taxes.
- The Revelation Principle was named by Myerson [1979], but it was already mentioned in Gibbard (1973). A further reference is Dasgupta, Hammond, & Maskin (1979).
- In an important class of problems, a benevolent government tries to assign an efficient contract that allows non-negative profits to a regulated firm that has better information on its costs. Although one might call this adverse selection, I would class such problems as moral hazard with hidden information, since they are very similar to the problem in which the government is trying to induce a firm to accept a contract before either player knows the firm's costs. In this second problem, the firm will not accept the contract unless it meets the participation constraint of allowing non-negative profits. This class of problem is examined in detail by Baron (forth), and has also been studied by, for example, Laffont & Tirole (1986).

  It is often possible to reduce the number of parameters in a moral hazard problem, and the regulation problem illustrates this. It might seem that the government must pick an entire schedule of price-quantity combinations, but given that the government chooses one of these, market demand gives the other. Baron (forth) has a good exposition of this problem.
- Moral hazard frequently occurs in public policy. Should the doctors who prescribe drugs also be allowed to sell them? The question trades off the likelihood of overprescription against the potentially lower cost and greater convenience of doctor-dispensed drugs. See "Doctors as Druggists: Good Rx for Consumers?" *Wall Street Journal*, 25 June 1987, p. 24.
- For a careful discussion of the unravelling argument for information revelation, see Milgrom (1981b).
- A hidden information game requires that the state of the world matter to one of the players' payoffs, but not necessarily in the same way as in Production Game IV. The Salesman Game of Section 7.2 effectively uses the utility function $U(e, w, \theta)$ for the agent and $V(q-w)$ for the principal. The state of the world matters because the agent's disutility of effort varies across states. In other problems, his utility of money might vary across states.

N7.2   An Example: the Salesman Game

- Gonik (1978) describes hidden information contracts used by IBM's subsidiary in Brazil. Salesmen were first assigned quotas. They then announced their own sales forecast as a percentage of quota and chose from among a set of contracts, one for each possible forecast. Inventing some numbers for illustration, if Smith were assigned a quota of 400 and he announced 100 percent, he might get $w = 70$ if he sold 400 and $w = 80$ if he sold 450; but if he had announced 120 percent, he would have gotten $w = 60$ for 400 and $w = 90$ for 450. The contract encourages extra effort when the extra effort is worth the extra sales.
- Sometimes students know more about their class rankings than the professor does. One professor of labor economics is reported to have used a mechanism of the following kind for grading class discussion. Each student i reports a number evaluating other students in the class. Student i's grade is an increasing function of the evaluations given i by other students and of the correlation between i's evaluations and the other students'. There are many Nash equilibria, but telling the truth is a focal point.
- In dynamic games of hidden information with moral hazard, the **ratchet effect** is important: the agent takes into account that his information-revealing choice of contract

this period will affect the principal's offerings next period. A principal might allow high prices to a public utility in the first period to discover that its costs are lower than expected, but in the next period the prices would be lowered: the contract is ratcheted irreversibly to be more severe. Hence, the company might not choose a contract which reveals its costs in the first period. This is modelled in Freixas, Guesnerie & Tirole (1985)

Baron (forth) notes that the principal might purposely design the equilibrium to be pooling in the first period so self selection does not occur. Having learned nothing, he can offer a more effective separating contract in the second period.

- We can only find weak Nash equilibria for most hidden information models, because many contracts can achieve the same outcome by specifying different out-of-equilibrium punishments. In the context of the Salesman Game, the contract could specify either ($w = w_2$ if $q > q_2$) or ($w = 0$ for $q > q_2$). The salesman would choose $q = q_2$ in either case.

N7.3   Efficiency Wages

- For surveys of the efficiency wage literature, see the article by L. Katz (1986) and the book edited by Akerlof & Yellen (1986).
- While the efficiency wage model does explain involuntary unemployment, it does not explain cyclical changes in unemployment.
- The efficiency wage idea is really the same idea as in the Klein & Leffler (1981) model of product quality that we formalized in Section 4.8.

N7.4   Tournaments

- The article which stimulated so much recent interest in tournaments is Lazear & Rosen (1981), which discusses in detail the importance of risk aversion and adverse selection.
- One example of a tournament is the two-year three-man contest in which to choose its new chairman Citicorp named three candidates as vice-chairmen: the head of consumer banking, the head of corporate banking, and the legal counsel. Earnings reports were even split into three components, two of which were the corporate and consumer banking (the third was the "investment" bank, irrelevant to the tournament). See "What Made Reed Wriston's Choice at Citicorp," *Business Week*, 2 July 1984, p. 25.
- General Motors has tried a tournament among its production workers. During a depressed year, management credibly threatened to close down the auto plant with the lowest productivity. Reportedly, this did raise productivity. Such a tournament is interesting because it helps explain why a firm's supply curve could be upward sloping even if all its plants are identical, and why it might hold excess capacity. Should information on a plant's current performance have been released to other plants? See "Unions Say Auto Firms Use Interplant Rivalry to Raise Work Quotas," *Wall Street Journal*, 8 November 1983, p. 1.
- Under adverse selection, tournaments must be used differently than under moral hazard because agents cannot control their effort. Instead, tournaments are used to deter agents from accepting contracts in which they must compete for a prize with other agents of higher ability.
- Interfirm management tournaments run into difficulties when shareholders want managers to cooperate in some arenas. If managers collude in setting prices, for example, they can also collude to make life easier for each other.
- Antle & Smith (1986) is an empirical study of tournaments in managers' compensation. Rosen (1986) is a theoretical model of a labor tournament in which the prize is promotion.
- Suppose a firm conducts a tournament in which the best-performing of its vice-presidents becomes the next president. Should the firm fire the most talented vice-president before it starts the tournament? The answer is not obvious. Maybe in the tournament's

equilibrium, Mr Talent works less hard because of his initial advantage, so that all of the vice-presidents retain the incentive to work hard.

- A tournament can reward the winner, or shoot the loser. Which is better? Nalebuff & Stiglitz (1983) say to shoot the loser, and Rasmusen (1987) finds a similar result for teams, but for a different reason. Nalebuff & Stiglitz's result depends on uncertainty and a large number of agents in the tournament, while Rasmusen's depends on risk aversion. If a utility function is concave because the agent is risk averse, the agent is hurt more by losing a given sum than he would benefit by gaining it. Hence, for incentive purposes the carrot is inferior to the stick, a result unfortunate for efficiency since penalties are often bounded by bankruptcy or legal constraints.

- Using a tournament, the equilibrium effort might be greater in a second-best contract than in the first-best, even though the second-best is contrived to get around the problem of inducing sufficient effort. Also, a pure tournament, in which the prizes are distributed solely according to the ordinal ranking of output by the agents, is often inferior to a tournament in which an agent must achieve a significant margin of superiority over his fellows in order to win (Nalebuff & Stiglitz [1983]). Companies using sales tournaments sometimes have prizes for record yearly sales besides ordinary prizes, and some long distance athletic races have non-ordinal prizes to avoid dull events in which the best racers run "tactical races."

N7.5   Monitoring

- D. Diamond (1984) shows the implications of monitoring costs for the structure of financial markets. A fixed cost to monitoring investments motivates the creation of a financial intermediary to avoid repetitive monitoring by many investors.

- Baron & Besanko (1984) studies auditing in the context of a government agency which can at some cost collect information on the true production costs of a regulated firm.

- Mookherjee & Png (unpub) and Border & Sobel (1987) have examined random auditing in the context of taxation. They find that if a taxpayer is audited he ought to be more than compensated for his trouble, if it turns out he was telling the truth. Under the optimal contract, the truthtelling taxpayer should be delighted to hear that he is being audited. The reason is that a reward for truthfulness widens the differential between the agent's payoff when he tells the truth and when he lies.

- Government action strongly affects what information is available as well as what is contractible. In 1988, for example, the United States passed a law sharply restricting the use of lie detectors for testing or monitoring. Previous to the restriction, about two million workers had been tested each year ("Law Limiting Use of Lie Detectors is Seen Having Widespread Effect" *Wall Street Journal*, 1 July 1988, p. 19).

N7.6   Alleviating the Agency Problem

- Gaver & Zimmerman (1977) describes how a performance bond of 100 percent was required for contractors building the BART subway system in San Francisco. "Surety companies" generally bond a contractor for five to 20 times his net worth, at a charge of 0.6 percent of the bond per year, and absorption of their bonding capacity is a serious concern for contractors in accepting jobs.

- Even if a product's quality need not meet government standards, the seller may wish to bind himself to them voluntarily. Stroh's *Erlanger* beer proudly announces on every bottle that although it is American, "Erlanger is a special beer brewed to meet the stringent requirements of Reinheitsgebot, a German brewing purity law established in 1516." Inspection of household electrical appliances by an independent lab to get the "U_L" listing is a similarly voluntary adherence to standards.

- The stock price is a way of using outside analysts to monitor an executive's performance. When General Motors bought EDS, they created a special class of stock, GM-E, which varied with EDS performance and could be used to monitor it.

N7.7 Teams, and the Groves Mechanism

- **Team theory,** as developed by Marschak & Radner (1972) is an older mathematical approach to organization. In the old usage of "team" (different from the current, Holmstrom [1982] usage), several agents who have different information but cannot communicate it must pick decision rules. The payoff is the same for each agent, and their problem is coordination, not motivation.

- The efficient contract (7.10) supports the efficient Nash equilibrium, but it also supports a continuum of inefficient Nash equilibria. Suppose that in the efficient equilibrium all workers work equally hard. Another Nash equilibrium is for one worker to do no work and the others to work inefficiently hard to make up for him.

- **A Teams Contract with Hidden Information.** In the 1920s, National City Co. assigned 20 percent of profits to compensate management as a group. A management committee decided how to share it, after each officer submitted an unsigned ballot suggesting the share of the fund that Chairman Mitchell should have and a signed ballot giving his estimate of the worth of each of the other eligible officers, himself excluded. (Galbraith [1954] p. 157)

- **A First-Best Budget-Balancing Contract when Agents are Risk Averse.** Proposition 7.1 can be shown to hold for any contract, not just for differentiable sharing rules, but it does depend on risk neutrality and separability of the utility function. Consider the following contract from Rasmusen (1987):

$$w_i = \begin{cases} b_i & \text{if } q \geq q(e^*). \\ 0 & \text{with probability } (n-1)/n \text{ if } q < q(e^*), \\ q & \text{with probability } 1/n \text{ if } q < q(e^*). \end{cases} \tag{7.21}$$

If the worker shirks, he enters a lottery. If his risk aversion is strong enough, he prefers the certain return b_i, so he does not shirk. If agents' wealth is unlimited, then for any positive risk aversion we could construct such a contract, by making the losers in the lottery accept negative pay.

- A teams contract like (7.10) is not a tournament (see Section 7.4). Only absolute performance matters, even though the level of absolute performance depends on what all the players do.

- **The Budget-Balancing Constraint.** The legal doctrine of "consideration" makes it difficult to make binding, Pareto-suboptimal promises. An agreement is not a legal contract unless it is more than a promise: both parties have to receive something valuable for the courts to enforce the agreement.

- Adverse selection can be incorporated into a teams model. A team of workers who may differ in ability produce a joint output, and the principal tries to ensure that only high-ability workers join the team. See Rasmusen & Zenger (unpub).

- Vickrey (1961) first suggested the non-budget-balancing mechanism for revelation of preferences, but it was rediscovered later (Groves [1973]) and became known as the Groves Mechanism.

- Roth (1984) is an interesting analysis of the system by which hospitals and interns are matched after announcing the order of their preferences.

- An article using moral hazard to look at the problems of risk-sharing in optimal taxation, insurance, and price discrimination is Stiglitz (1977). Preference revelation is at the heart of the price discrimination problem, the standard reference for which is Phlips's 1983 book.

8 Adverse Selection

8.1 Introduction: Production Game V

In Chapter 6 we divided games of asymmetric information between moral hazard, in which agents are identical, and adverse selection, in which they are heterogeneous. In both moral hazard with hidden information and adverse selection, the principal tries to sort out agents with different characteristics. But although moral hazard with hidden information is structurally similar to adverse selection, the emphasis is on the agent's action rather than his choice of contract, and agents accept contracts before acquiring information. For comparison with moral hazard, let us consider an adverse selection version of the Production Game described in Sections 6.2, 6.3, and 7.1.

Production Game V: Adverse Selection

Players

The principal and the agent.

Information

Asymmetric, incomplete, and uncertain.

Actions and Events

(0) Nature chooses the agent's ability a, unobserved by the principal, according to distribution $F(a)$.

(1) The principal offers the worker one or more wage contracts $w_1(q), w_2(q), \ldots$

(2) The agent accepts one contract or rejects them all.

(3) Nature chooses a value for the state of the world, θ, according to distribution $G(\theta)$, and $q = q(a, \theta)$.

Payoffs

If the agent rejects all contracts, then $\pi_{agent} = \overline{U}$ and $\pi_{principal} = 0$.

Otherwise, $\pi_{agent} = U(w)$ and $\pi_{principal} = V(q - w)$.

Under adverse selection, it is not the worker's effort, but his ability that is non-contractible. Without uncertainty (move (3)), the principal would provide a single contract specifying high wages for high output and low wages for low output, but unlike under moral hazard, either high or low output might be observed in equilibrium. Another difference is that under adverse selection with uncertainty, multiple contracts may be better than a single contract. The principal might provide a contract with a flat wage to attract the low-ability agents and an incentive contract to attract the high-ability agents.

Production Game V is a fairly complicated game of uncertainty, but although we will return to uncertainty in Section 8.4 using an insurance model, let us start, in Sections 8.2 and 8.3, with a certainty game. The game will model a used car market in which the quality of the car is known to the seller but not the buyer, and the various versions of the game will differ in the types and numbers of the buyers and sellers. Sections 8.4 and 8.5 will go through a model of adverse selection in insurance, where a Nash equilibrium will fail to exist for some parameter values. Section 8.6 will describe a wide variety of other applications of adverse selection.

8.2   Adverse Selection under Certainty: Lemons I and II

Akerlof stimulated an entire field of research with his 1970 model of shoddy used cars ("lemons"), in which adverse selection arises because car quality is better known to the seller than the buyer and they are unlikely ever to repeat the transaction. In agency terms, the principal contracts to buy from the agent a car whose quality, which might be high or low, is noncontractible despite the lack of uncertainty. Such a model may sound like moral hazard with hidden information, but the difference is that in the used car market the seller has private information about his own type before making any kind of agreement. If instead, the seller agreed to resell his car when he first bought it, the model would be moral hazard with hidden information, because there would be no asymmetric information at the time of contracting, just an expectation of future asymmetry.

We will spend considerable time putting twists onto a model of used cars. The game will have one buyer and one seller, but this will simulate competition between buyers, as discussed in Section 6.3, because the seller moves first. If the model had symmetric information there would be no consumer surplus. It will often be convenient to discuss the game as if it had many sellers, interpreting a seller whom Nature randomly assigns a type as a population of sellers of different types, one of whom is drawn by Nature to participate in the game.

Basic Lemons Model

Players

A buyer and a seller.

Information

Asymmetric, incomplete, and certain. The buyer is uninformed.

Actions and Events

(0)   Nature chooses quality type θ for the seller according to the distribution $F(\theta)$. The seller knows θ, but while the buyer knows F, he does not know the θ of the particular seller he faces.

(1)   The seller offers a price P.

(2)   The buyer accepts or rejects.

Payoffs

If the buyer rejects the offer, both players receive payoffs of zero.

Otherwise, $\pi_{buyer} = V(\theta) - P$ and $\pi_{seller} = P - U(\theta)$, where V and U will be defined later.

The payoffs of both players are normalized to equal zero if no transaction takes place. A normalization is part of the notation of the model rather than the substantive assumptions: it just means that the model assigns the players' utility a base value of zero when no transaction takes place. The payoff functions show changes from that base. The seller, for instance, gains P if the sale takes place, but he loses $U(\theta)$ from giving up the car.

There are various ways to specify $F(\theta)$, $U(\theta)$, and $V(\theta)$. We start with identical tastes and two types (Lemons I), and generalize to a continuum of types (Lemons II). In Section 8.3 we specify first that the sellers are identical but value cars more than buyers (Lemons III), and then that the sellers have heterogeneous tastes (Lemons IV). We will look less formally at other modifications involving risk aversion and the relative numbers of buyers and sellers.

Lemons I: Identical Tastes, Two Types of Sellers

Let good cars have quality 6,000 and bad cars (lemons) quality 2,000, so $\theta \in \{2,000, 6,000\}$, and let half the cars in the world be of each type. Assume that both players are risk neutral and value quality at one dollar per unit, so after a trade the payoffs are $\pi_{buyer} = \theta - P$ and $\pi_{seller} = P - \theta$. The extensive form is shown in Figure 8.1.

If he could observe quality at the time of his purchase, the buyer would be willing to accept a contract to pay $6,000 for a good car and $2,000 for a lemon.

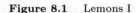

Figure 8.1   Lemons I

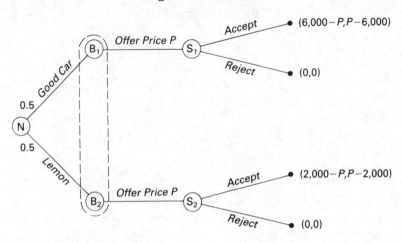

Payoffs to: (Buyer, Seller)

He cannot observe quality, and we assume that he cannot enforce a contract based on his discoveries once the purchase is made. Given these restrictions, if the seller offers $4,000, a price equal to the average quality, the buyer will deduce that the seller does not have a good car. The very fact that the car is for sale demonstrates its low quality. Knowing that for $4,000 he would be sold only lemons, the buyer would refuse to pay more than $2,000. Let us assume that an indifferent seller sells his car, in which case half of the cars are traded in equilibrium, all of them lemons.

You might suggest to the owner of a good car that he wait until all the lemons have been sold and then sell his own car, since everyone knows that only good cars have remained unsold. But allowing such behavior changes the model by adding a new action. If it were anticipated, the owners of lemons would also hold back and wait for the price to rise. Such a game could be formally analyzed as a war of attrition (Section 3.3).

The outcome that half the cars are held off the market is interesting, though not startling. It is a formalization of Groucho Marx's wisecrack that he would refuse to join a club that would accept him as a member. Lemons II will have a more dramatic outcome.

Lemons II: Identical Tastes, a Continuum of Types of Sellers

One might wonder whether the outcome of Lemons I was an artifact of the assumption of just two types. Lemons II generalizes the game by allowing the seller to be any of a continuum of types. We will assume that the quality types are uniformly distributed between 2,000 and 6,000. The average quality is $\bar{\theta} = 4,000$, which is therefore the price the buyer would be willing to pay for a car of unknown quality if all cars were on the market. The probability density is zero except on the support

[2,000, 6,000], where it is $f(\theta) = 1/(6,000 - 2,000)$, and the cumulative density is

$$F(\theta) = \int_{2,000}^{\theta} f(x)dx. \tag{8.1}$$

After substituting the uniform density for $f(\theta)$ and integrating (8.1) we obtain

$$F(\theta) = \frac{\theta}{4,000} - 0.5. \tag{8.2}$$

The payoff functions are the same as in Lemons I.

The equilibrium price must be less than \$4,000 in Lemons II because, as in Lemons I, not all cars are put on the market at that price. Owners are willing to sell only if the quality of their cars is less than 4,000, so while the average quality of all used cars is 4,000, the average quality offered for sale is 3,000. The price cannot be \$4,000 when the average quality is 3,000, so the price must drop at least to \$3,000. If that happens, the owners of cars with values from 3,000 to 4,000 pull their cars off the market and the average of those remaining is 2,500. The acceptable price falls to \$2,500, and the unravelling continues, just as in Section 7.1, until the price reaches its equilibrium level of \$2,000. But at $P =2,000$ the number of cars on the market is infinitesimal. The market has completely collapsed!

Figure 8.2   Lemons II: identical tastes

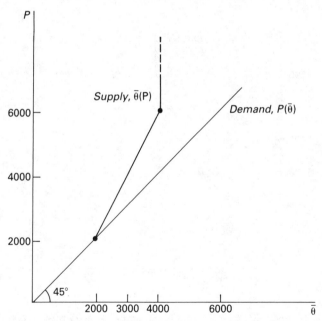

Figure 8.2 puts the price of used cars on one axis and the average quality of cars offered for sale on the other. Each price leads to a different average quality, $\bar{\theta}(P)$, and the slope of $\bar{\theta}(P)$ is greater than one because average quality does not rise proportionately with price. If the price rises, the quality of the *marginal* car

offered for sale equals the new price, but the quality of the *average* car offered for sale is much lower. In equilibrium, the average quality must equal the price, so the equilibrium lies on the 45° line through the origin. That line is a demand schedule of sorts, just as $\bar{\theta}(P)$ is a supply schedule. The only intersection is the point ($2,000, 2,000).

8.3   Heterogeneous Tastes: Lemons III and IV

The outcome that no cars are traded is extreme, but there is no efficiency loss in either Lemons I or Lemons II. Since all the players have identical tastes, it does not matter who ends up owning the cars. But the players of this section, whose tastes differ, have real need of a market.

Lemons III: Buyers Value Cars More than Sellers

Assume that sellers value their cars at exactly their qualities θ, but buyers have valuations 20 percent greater and they outnumber the sellers. The payoffs if a trade occurs are $\pi_{buyer} = 1.2\theta - P$ and $\pi_{seller} = P - \theta$. In equilibrium, the sellers will capture the gains from trade.

Figure 8.3   Lemons III: adverse selection

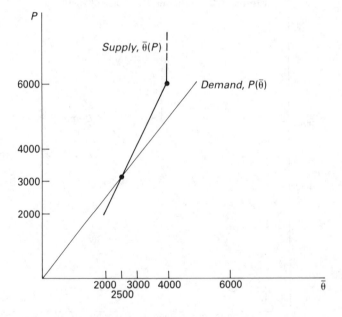

In Figure 8.3, the curve $\bar{\theta}(P)$ is much the same as in Lemons II, but the equilibrium condition is no longer that price and average quality lie on the 45° line, but that they lie on the demand schedule $P(\bar{\theta})$, which has a slope of 1.2 instead of 1.0. The demand and supply schedules intersect only at $(P = \$3,000, \bar{\theta}(P) = 2,500)$. Because buyers are willing to pay a premium, we only see **partial adverse selection**; the equilibrium is partially pooling in the terminology of Section 7.1. The

Figure 8.4   Lemons IV: sellers' valuations differ

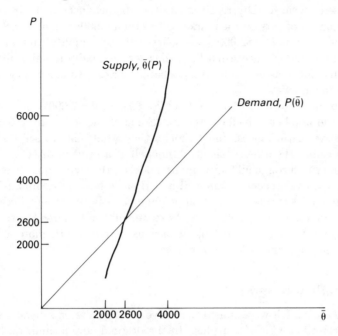

outcome is inefficient, because in a world of perfect information all the cars would be owned by the "buyers," who value them more, but under adverse selection they only end up owning the low-quality cars.

Lemons IV: Sellers' Valuations Differ

In Lemons IV, we dig a little deeper to explain why trade occurs, and we model sellers as consumers whose valuations of quality have changed since they bought their cars. For a particular seller, the valuation of one unit of quality is $1+\varepsilon$, where the random disturbance ε can be either positive or negative and has an expected value of zero. The disturbance could arise because of the seller's mistake—he did not realize how much he would enjoy driving when he bought the car—or because conditions changed—he switched to a job closer to home. Payoffs if a trade occurs are $\pi_{buyer} = \theta - P$ and $\pi_{seller} = P - (1+\varepsilon)\theta$.

If $\varepsilon = -0.15$ and $\theta = 2{,}000$, then \$1,700 is the lowest price at which player i would resell his car. The average quality of cars offered for sale at price P is the expected quality of cars valued by their owners at less than P, i.e.,

$$\bar{\theta}(P) = E\left(\theta \mid (1+\varepsilon)\theta \leq P\right). \tag{8.3}$$

Suppose that a large number of new buyers, greater in number than the sellers, appear in the market, and their valuation of one unit of quality is \$1. The demand schedule, shown in Figure 8.4, is the 45° line through the origin. Figure 8.4 shows one possible shape for the supply schedule $\bar{\theta}(P)$, although to specify it precisely we would have to specify the distribution of the disturbances.

In contrast to Lemons I, II, and III, if $P \geq \$6,000$ some car owners would be reluctant to sell, because they received positive disturbances to their valuations. The average quality of cars on the market is less than 4,000 even at $P = \$6,000$. On the other hand, even if $P = \$2,000$ some sellers with low quality cars *and* negative realizations of the disturbance still sell, so the average quality remains above 2,000. Under some distributions of ε, a few sellers hate their cars so much they would pay to have them taken away.

The equilibrium drawn in Figure 8.4 is ($P = \$2,600$, $\bar{\theta} = 2,600$). Some used cars are sold, but the number is inefficiently low. Some of the sellers have high quality cars but negative disturbances, and although they would like to sell their cars to someone who values them more, they will not sell at a price of \$2,600.

A theme running through all four Lemons models is that when quality is unknown to the buyer, less trade occurs. Lemons I and II show how trade diminishes, while Lemons III and IV show that the disappearance can be inefficient because some sellers value cars less than some buyers. Next we will use Lemons III, the simplest model with gains from trade, to look at various markets with more sellers than buyers, excess supply, and risk-averse buyers.

More Sellers than Buyers

In analyzing Lemons III, we assumed that buyers outnumbered sellers. As a result, the sellers received producer's surplus. In the original equilibrium, all the sellers with quality less than 3,000 offered a price of \$3,000 and earned surpluses of up to \$1,000. There were more buyers than sellers, so every seller who wished to sell was able, but the price equalled the buyers' expected utility, so no buyer who failed to purchase was dissatisfied. The market cleared.

If, instead, sellers outnumber buyers, what price should a seller offer? At \$3,000, not all would-be sellers can find buyers. A seller who proposed a lower price would find willing buyers despite the somewhat lower expected quality. The buyer's trade-off between lower price and lower quality is shown in Figure 8.3, in which the expected consumer surplus is the vertical distance between the price (the height of the supply schedule) and the demand schedule. When the price is \$3,000 and the average quality is 2,500, the buyer expects a consumer surplus of zero, which is $\$3,000 - \$1.2 \cdot 2,500$. The combination of price and quality that buyers like best is (\$2,000, 2,000), because if there were sufficiently many sellers with quality $\theta = 2,000$ each of the buyers would pay $P = \$2,000$ for a car worth \$2,400 to him, acquiring a surplus of \$400. If there were fewer sellers, the equilibrium price would be higher and some sellers would receive producer surplus.

Heterogeneous Buyers: Excess Supply

If buyers have different valuations for quality, the market might not clear, as C. Wilson (1980) points out. Assume that the number of buyers willing to pay \$1.2 per unit of quality exceeds the number of sellers, but buyer Smith is an eccentric whose demand for high quality is unusually strong. He would pay \$100,000 for a car of quality 5,000 or greater, and \$0 for any lower quality.

In Lemons III without Smith, the outcome is a price of \$3,000, an average market quality of 2,500, and a market quality range between 2,000 and 3,000. Smith would be unhappy with this, since he has probability zero of finding a car he likes. In fact, he would be willing to accept a price of $6,000, so that all the cars, from quality 2,000 to 6,000, would be offered for sale and the probability that he buys a satisfactory car would rise from 0 to 0.25. But Smith would not want to buy all the cars offered to him, so the equilibrium has two prices, $3,000 and $6,000, with excess supply at the higher price.

Strangely enough, Smith's demand function is upward sloping. At a price of $3,000, he is unwilling to buy; at a price of $6,000, he is willing, because expected quality rises with price. This does not contradict basic price theory, for the standard assumption of *ceteris paribus* is violated. As the price increases, the quantity demanded would fall if all else stayed the same, but all else does not—quality rises.

Risk Aversion

We have implicitly assumed, by the choice of payoff functions, that the buyers and sellers are both risk neutral. What happens if they are risk averse—that is, the marginal utilities of wealth and car quality are diminishing? Again we will use Lemons III and the assumption of many buyers.

On the seller's side, risk aversion changes nothing. The seller runs no risk because he knows exactly the price he receives and the quality he surrenders. But the buyer does bear risk, because he buys a car of uncertain quality. Although he would pay $3,600 for a car he knows has quality 3,000, if he is risk averse he will not pay that much for a car with expected quality 3,000 but actual quality of possibly 2,500 or 3,500: he would obtain less utility from adding 500 quality units than from subtracting 500. The buyer would pay perhaps $2,900 for a car whose expected quality is 3,000, and the demand schedule is nonlinear, lying everywhere below the demand schedule of the risk-neutral buyer. As a result, the equilibrium has a lower price and average quality.

8.4   Adverse Selection under Uncertainty: Insurance Game III

The term "adverse selection," like "moral hazard," comes from insurance. Insurance pays more if there is an accident than otherwise, so it benefits accident-prone customers more than safe ones and a firm's customers are "adversely selected" to be accident-prone. The classic article on adverse selection in insurance markets is Rothschild & Stiglitz (1976), which begins, "Economic theorists traditionally banish discussions of information to footnotes." How things have changed! Within ten years, information problems came to dominate research in both microeconomics and macroeconomics.

We will follow Rothschild & Stiglitz in using state-space diagrams, and we will use a version of Section 6.4's Insurance Game. Under moral hazard, Smith chose whether to be *Careful* or *Careless*. Under adverse selection, Smith cannot affect the probability of a theft, which is chosen by Nature. Rather, Smith is either *Safe* or *Unsafe*, and while he cannot affect the probability that his car will be stolen, he does know what the probability is.

Insurance Game III

Players

Smith and two insurance companies.

Information

Asymmetric, incomplete, and uncertain. The insurance companies are uninformed.

Actions and Events

(0)   Nature chooses Smith to be either *Safe*, with probability 0.6, or *Unsafe*, with probability 0.4. Smith knows his type, but the insurance companies do not.

(1)   Each insurance company offers its own contract (x, y) under which Smith pays premium x unconditionally and receives the compensation y if there is a theft.

(2)   Smith picks a contract.

(3)   Nature chooses whether there is a theft, using probability 0.5 if Smith is *Safe* and 0.75 if he is *Unsafe*.

Payoffs

Smith's payoff depends on his type and the contract (x, y) that he accepts. Let $U' > 0$ and $U'' < 0$.

$$\pi_{Smith}(Safe) \quad = \quad 0.5U(12 - x) + 0.5U(0 + y - x).$$
$$\pi_{Smith}(Unsafe) \quad = \quad 0.25U(12 - x) + 0.75U(0 + y - x).$$

The companies' payoffs depend on what types accept their contracts, as shown in Table 8.1.

Table 8.1 Insurance Game III: payoffs

Company payoff	Types of customers
0	no customers
$0.5x + 0.5(x - y)$	just *Safe*
$0.25x + 0.75(x - y)$	just *Unsafe*
$0.6[0.5x + 0.5(x - y)] + 0.4[0.25x + 0.75(x - y)]$	*Unsafe* and *Safe*

Smith is *Safe* with probability 0.6 and *Unsafe* with probability 0.4. Without insurance, Smith's dollar wealth is 12 if there is no theft and 0 if there is, depicted in Figure 8.5 as his endowment in state-space, $\omega = (12, 0)$. If Smith is *Safe*, a theft

Figure 8.5   Insurance Game III: nonexistence of pooling equilibrium

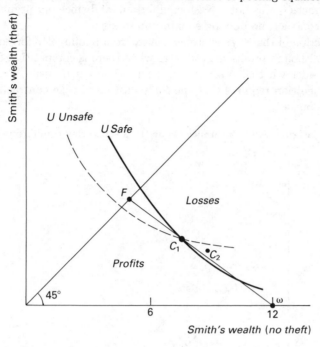

occurs with probability 0.5, but if he is *Unsafe* the probability is 0.75. Smith is risk averse (because $U'' < 0$) and the insurance companies are risk neutral.

If an insurance company knew that Smith was *Safe*, it could offer him insurance at a premium of 6 with a payout of 12 after a theft, leaving Smith with an allocation of $(6, 6)$. This is the most attractive contract that is not unprofitable, because it fully insures Smith. Whatever the state, his allocation is 6.

Figure 8.5 shows the indifference curves of Smith and an insurance company. The insurance company is risk neutral, so its indifference curve is the straight line ωF if Smith is a customer regardless of his type. The insurance company is indifferent between ω and C_1, at both of which its expected profits are zero. Smith is risk averse, so his indifference curves are closest to the origin along the 45 degree line where his wealth in the two states is equal. He has two sets of indifference curves, solid if he is *Safe* and dotted if he is *Unsafe*.

Figure 8.5 shows why no Nash pooling equilibrium exists. To make zero profits, the equilibrium must lie on the line ωF. It is easiest to think about these problems by imagining an entire population of Smiths, whom we will call "customers." Pick a contract C_1 anywhere on ωF and think about drawing the indifference curves for the *Unsafe* and *Safe* customers that pass through C_1. *Safe* customers are always willing to trade *Theft* wealth for *No Theft* wealth at a higher rate than *Unsafe* customers. At any point, therefore, the slope of the solid (*Safe*) indifference curve is steeper than that of the dotted (*Unsafe*) curve. Since the slopes of the dotted and the solid indifference curves differ, we can insert another contract, C_2, between them and just barely to the right of ωF. The *Safe* customers prefer contract C_2 to C_1, but the *Unsafe* customers stay with C_1, so C_2 is profitable—since C_2 only attracts *Safes*, it need not be to the left of ωF to avoid losses. But then the original contract C_1 was not a Nash equilibrium, and since our argument holds for any pooling contract, no pooling equilibrium exists.

The attraction of the *Safe* customers away from pooling is referred to as **cream skimming**, although profits are still zero when there is competition for the cream. We next consider whether a separating equilibrium exists, using Figure 8.6. The zero profit condition requires that the *Safe* customers take contracts on ωC_4 and the *Unsafes* on ωC_3.

Figure 8.6   Insurance Game III: a separating equilibrium

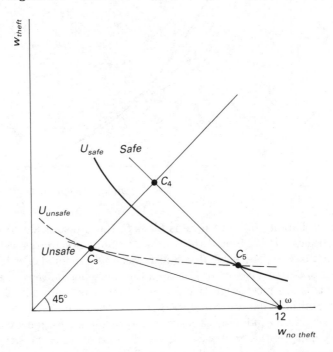

The *Unsafes* will be completely insured in any equilibrium, although at a high price. On the zero-profit line ωC_3, the contract they like best is C_3, which the *Safes* are not tempted to take. The *Safes* would prefer contract C_4, but C_4 uniformly dominates C_3, so it would attract *Unsafes* too, and generate losses. To avoid attracting *Unsafes*, the *Safe* contract must be below the *Unsafe* indifference curve. Contract C_5 is the fullest insurance the *Safes* can get without attracting *Unsafes*: it satisfies the self selection and competition constraints.

Contract C_5, however, might not be an equilibrium either. Figure 8.7 is the same as Figure 8.6 with a few additional points marked. If one firm offered C_6, it would attract both types, *Unsafe* and *Safe*, away from C_3 and C_5, because it is to the right of the indifference curves passing through those points. Would C_6 be profitable? That depends on the proportions of the different types. The assumption on which the equilibrium of Figure 8.6 is based is that the proportion of *Safes* is 0.6, so that the zero-profit line for pooling contracts is ωF and C_6 would be unprofitable. In Figure 8.7 it is assumed that the proportion of *Safes* is higher, so the zero-profit line for pooling contracts would be $\omega F'$ and C_6, lying to its left, is profitable. But we already showed that no pooling contract is Nash, so C_6 cannot be an equilibrium. Since neither a separating pair like (C_3, C_5) nor a pooling contract like C_6 is an equilibrium, no equilibrium whatsoever exists.

Figure 8.7   Insurance Game III: no equilibrium exists

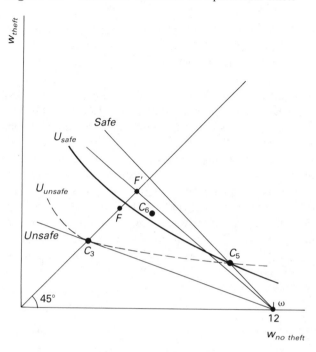

The essence of nonexistence here is that if separating contracts are offered, some company is willing to offer a superior pooling contract; but if a pooling contract is offered, some company is willing to offer a separating contract that makes it unprofitable. A monopoly would have a pure strategy equilibrium, but in a competitive market only a mixed strategy Nash equilibrium exists (see Dasgupta & Maskin [1986b]).

8.5   Other Equilibrium Concepts: Wilson and Reactive Equilibrium

In Insurance Game III, any pooling contract is vulnerable to a cream-skimming contract that draws away the *Safes*, but this is a little strange, because it seems that after that happens the now-unprofitable old pooling contract (which was soaking up the *Unsafes*) would be withdrawn. The game tree does not reflect this, nor does the Nash equilibrium concept.

One way to obtain a pure strategy equilibrium is to redefine the equilibrium concept. C. Wilson (1980) suggests that the pooling equilibrium is legitimate because a principal (uninformed player) thinking about introducing the new contract would realize that it would be unprofitable once the old contract was withdrawn.

A **Wilson equilibrium** *is a set of contracts such that when the agents (informed players) choose among them so as to maximize profits,*

(1) All contracts make nonnegative profits; and

(2) No new contract (or set of contracts) could be offered that would make positive profits even after all contracts that would make negative profits as a result of its entry were withdrawn.

The Wilson equilibrium is the same as the Nash separating equilibrium if that exists, and otherwise it is the pooling contract most preferred by the *Safes*. In Figure 8.6, the Wilson equilibrium is the same as the Nash equilibrium, the separating pair (C_3, C_5). In Figure 8.7, where no Nash equilibrium exists, the Wilson equilibrium is the zero-profit pooling contract, F'. It is on the line $\omega F'$, so it satisfies part (1) of the definition. It provides the fullest insurance of any zero-profit pooling contract, so no new pooling contract would be more attractive, and while some new separating contract might be profitable if the *Unsafes* stayed with F', any such contract would cause F' to be withdrawn and would henceforth be unprofitable.

The idea of Wilson equilibrium can be also incorporated into the game by modifying the game tree instead of redefining the equilibrium concept, as suggested by Fernandez & Rasmusen (unpub), to obtain the Wilson outcome as the perfect equilibrium of the modified game.

Wilson Equilibrium

(1)   Principals simultaneously offer contracts, called "old contracts."
(2)   Principals may simultaneously offer other contracts, called "new contracts."
(3)   Principals may simultaneously withdraw any old contracts.
(4)   Agents choose from among the remaining old and new contracts, and trading occurs.

In the perfect Bayesian equilibrium of this game, the principals offer the contracts that form a Wilson equilibrium in move (1). The approach of changing the game tree may seem more complicated than changing the equilibrium concept, but that is because it clearly delineates the somewhat vague intuition behind the equilibrium concept. Making use of the Wilson concept is not just a technical assumption: it is assuming that the market has a particular structure.

Riley (1979b) uses reasoning similar to Wilson's to justify his concept of "reactive equilibrium." Under this concept, an equilibrium is a set of contracts such that though some new contract might be profitable, that new contract would itself become unprofitable if a second new contract were introduced. More formally, following Engers & Fernandez (1987),

A **reactive equilibrium** *is a set of contracts S yielding nonnegative profits*

*such that for any nonempty set of contracts S' (the defection), where $S \cup S'$
is closed, there exists a closed set of contracts S'' (the reaction) such that:*

(1) S' incurs losses when only these three sets are tendered, and;

*(2) S'' does not incur losses when these three sets are tendered, whether or
not other contracts are also offered.*

In both Figure 8.6 and Figure 8.7, the reactive equilibrium is the separating pair
(C_3, C_5). That pair yields zero profits, and while in Figure 8.7 there is a profitable
deviation (C_6), that deviation would become unprofitable if a cream-skimming
contract were added as a reaction. Moreover, condition (2) of the definition is met,
because if the reactive cream-skimming contract is chosen carefully, no additional
contracts can be added which make it unprofitable, given that C_6 continues to be
offered.

A separating reactive equilibrium always exists because any pooling contract
disrupting it could be reacted against: reactive equilibrium makes constructive
use of the nonexistence of a pooling equilibrium. The Wilson concept is based
on withdrawing contracts in response to deviation, whereas the reactive concept
is based on adding them. As a result, when Nash equilibrium does not exist, the
Wilson concept favors a pooling equilibrium, while the reactive concept favors a
separating equilibrium.

8.6   Applications

Price Dispersion

Usually the best model for explaining price dispersion is a search model—Salop &
Stiglitz (1977), for example, which is based on buyers whose search costs differ. But
although we passed over it quickly in Section 8.3, the lemons model with Smith, the
quality-conscious consumer, generated not only excess supply, but price dispersion.
Cars of the same average quality were sold for $3,000 and $6,000.

Similarly, while the most obvious explanation for why brands of stereo amplifiers
sell at different prices is that customers are willing to pay more for higher quality,
adverse selection contributes another explanation. Consumers might be willing
to pay high prices because they know that high-priced brands could include both
high-quality and low-quality amplifiers, whereas low-priced brands are invariably
low quality. The low-quality amplifier ends up selling at two prices: a high price in
competition with high-quality amplifiers, and, in different stores or under a different
name, a low price aimed at customers less willing to trade dollars for quality.

This explanation does depend on sellers of amplifiers incurring a large enough
fixed set-up or operating cost. Otherwise, too many low-quality brands would crowd
into the market, and the proportion of high-quality brands would be too small for
consumers to be willing to pay the high price. The low-quality brands would benefit
as a group from entry restrictions: too many of them spoil the market, not through
price competition but through degrading the average quality.

Health Insurance and Medicare

Medical insurance is subject to adverse selection because some people are healthier than others. The variance in health is particularly large among old people, who have had difficulty in obtaining insurance at all. Under basic economic theory this is a puzzle: the price should rise until supply equals demand. The problem is pooling: when the price of insurance is appropriate for the average old person, healthier ones stop buying. The price must rise to keep profits nonnegative, and the market disappears, just as in Lemons II.

If the facts indeed fit this story, adverse selection is an argument for government-enforced pooling. If all old people are required to purchase government insurance (Medicare in the United States), then while the healthier of them may be worse off, the vast majority could be helped.

Using adverse selection to justify medicare, however, points out how dangerous many of the models in this book can be. For policy questions, the best default opinion is that markets are efficient. On closer examination, we have found that many markets are inefficient because of strategic behavior or information asymmetry. It is dangerous, however, to immediately conclude that the government should intervene, because the same arguments applied to government show that the cure might be worse than the disease. The analyst of health care needs to take seriously the moral hazard and rent-seeking that arise from government insurance. Doctors and hospitals will increase the cost and amount of treatment if the government pays for it, and the transfer of wealth from young people to old, which is likely to swamp the gains in efficiency, might distort the shape of the government program from the economist's ideal.

Henry Ford's Five-Dollar Day

In 1914 Henry Ford made a much-publicized decision to raise the wage of his auto workers to $5 a day, considerably above the level needed to attract them. This pay hike occurred without pressure from the workers, who were non-unionized. Why did Ford do it?

The pay hike could be explained by either moral hazard or adverse selection. In accordance with the idea of efficiency wages (Section 7.3), Ford might have wanted workers who worried about losing their premium job at his factory, because they would work harder and refrain from shirking. Adverse selection could also explain the pay hike: by raising his wage Ford attracted a mixture of low and high quality workers, rather than low quality alone (see Raff & Summers [1987]).

Bank Loans

Suppose that two people come to you for an unsecured loan of $10,000. One offers to pay an interest rate of 10 percent and the other offers 200 percent. Who do you accept? Like the car buyer who chooses to buy at a high price, you may choose to lend at a low interest rate.

If a lender raises his interest rate, both his pool of loan applicants and their behavior changes because adverse selection and moral hazard contribute to a rise in default rates. Borrowers who expect to default are less concerned about the high interest rate than dependable borrowers, so the number of loans shrinks and the default rate rises (see Stiglitz & Weiss [1981]). In addition, some borrowers shift to higher-risk projects with greater chance of default but higher yields when they are successful. In Section 13.2 we will go through the model of D. Diamond (unpub) which looks at this problem.

Whether because of moral hazard or adverse selection, asymmetric information can also result in excess demand for bank loans. The savers who own the bank do not save enough at the equilibrium interest rate to provide loans to all the borrowers who want loans. Thus, the bank makes a loan to John Smith, while denying one to Joe, his observationally equivalent twin. Policymakers should carefully consider any laws that rule out arbitrary loan criteria or require banks to treat all customers equally. Banks might wish to restrict their loans to blue-eyed people neither from prejudice nor because they really are better credit risks, but because it is useful to ration loans according to some criterion arbitrary enough to avoid the moral hazard of favoritism by loan officers.

Bernanke (1983) suggests adverse selection in bank loans as an explanation for the Great Depression in the United States. The difficulty in explaining the Depression is not so much the initial stock market crash as the persistence of the unemployment that followed. Bernanke notes that the crash wiped out local banks and dispersed the expertise of the loan officers. After the loss of this expertise, the remaining banks were less willing to lend because of adverse selection, and it was difficult for the economy to recover.

The Bid–Ask Spread

A common feature of asset markets is an individual called the "marketmaker" (the "specialist" on the New York Stock Exchange), who stands ready to buy or sell at any time so that the market can always be relied upon to trade the asset. One might think that the bid and the ask prices would be very close in a competitive market because the transactions cost of providing the marketmaking service is low. In actuality, however, the spread is large enough to be a major cost of trading, even in markets like NASDAQ that have competing marketmakers.

Adverse selection can explain the bid–ask spread. The price of a particular stock, which is an estimate, based on public information, of the firm's value, can either rise or fall. If marketmakers compete in offering liquidity services to traders no better informed than themselves, the bid–ask spread should be narrowed to the transactions cost—nearly zero. But if there also exist some informed traders, who have better information than the marketmakers, then the bid–ask spread must be greater. If the marketmakers only break even trading with the uninformed, they consistently lose money trading with the informed. Hence, the bid–ask spread must be large enough that the profits from the uninformed balance the losses from the informed. The marketmaker's problem is adverse selection because he would like to trade with the uninformed at a small bid–ask spread, and not trade with the informed at all, but he must pool the two types.

Wouldn't the uninformed traders bypass the marketmaker and trade with each other to avoid the spread? Their problem is that they, like the marketmaker, cannot tell the difference between informed and uninformed traders. Trading with the marketmaker, they pay the spread, but they have the assurance (in a reputable market) that they are not trading with an insider. The uninformed trader, in fact, faces a worse adverse selection problem than the marketmaker, because the marketmaker can use information from the market volume and the peculiarities of individual purchases to adjust the prices and spreads. Being the marketmaker has some features of a natural monopoly, because whichever marketmaker engages in the most trading receives the most information and can best adjust his prices.

Solutions to Adverse Selection

Even in markets where it apparently does not occur, the threat of adverse selection, like the threat of moral hazard, can be an important influence on market institutions. Adverse selection can be circumvented in a number of ways besides the contractual solutions we have been analyzing. I will mention some of them in the context of the used car market.

One set of solutions consists of ways to make car quality contractible. Buyers who find that their car is defective may have recourse to the legal system if the sellers were fraudulent, although in the United States the courts are too slow and costly to be fully effective. Other government bodies such as the Federal Trade Commission may do better by issuing regulations particular to the industry. Even without regulation, private warranties—promises to repair the car if it breaks down—may be easier to enforce than oral claims, by disspelling ambiguity about what quality was guaranteed.

Testing (the equivalent of monitoring in moral hazard) is always used to some extent. The prospective driver tries the car on the road, inspects the body, and otherwise tries to reduce information asymmetry. At a cost, he could even reverse the asymmetry by hiring mechanics to learn more about the car than the owner knows. The rule is not always *caveat emptor*; what should one's response be to an antique dealer who offers to pay $500 for an apparently worthless old chair?

Reputation can solve adverse selection, just as it can solve moral hazard, but only if the transaction is repeated and the other conditions of the models in Chapters 4 and 5 are met. An almost opposite solution is to show that there are innocent motives for a sale; that the owner of the car has gone bankrupt, for example, and his creditor is selling the car cheaply to avoid the holding cost.

Penalties not strictly economic are also important. One example is the social ostracism inflicted by the friend to whom a lemon has been sold; the seller is no longer invited to dinner. Or, the seller might have moral principles that prevent him from defrauding buyers. Such principles, provided they are common knowledge, would help him obtain a higher price in the used car market. Akerlof himself has worked on the interaction between social custom and markets in his 1980 and 1983 articles. The second of these looks directly at the value of inculcating moral principles, using theoretical examples to show that parents might wish to teach their children principles, and that society might wish to give hiring preference to students from elite schools.

It is by violating the assumptions needed for perfect competition that asymmetric information enables government and social institutions to raise efficiency. This points to a major reason for studying asymmetric information: where it is important, non-economic interference can be helpful instead of harmful. I find the social solutions particularly interesting since, as mentioned earlier in connection with health care, government solutions introduce agency problems as severe as the information problems they solve. Non-economic behavior is important under adverse selection, in contrast to under perfect competition, which permits the Invisible Hand to guide the market to efficiency regardless of the moral beliefs of the traders. If everyone were honest, the lemons problem would disappear because the sellers would truthfully disclose quality. If some fraction of the sellers were honest, but buyers could not distinguish them from the dishonest sellers, the outcome would presumably be somewhere between the outcomes of complete honesty and complete dishonesty. The subject of market ethics is important, and would profit from investigation by scholars trained in economic analysis.

Recommended Reading

Akerlof, George (1970) "The Market for Lemons: Quality Uncertainty and the Market Mechanism" *Quarterly Journal of Economics.* August 1970. 84, 3: 488–500.

Nalebuff, Barry & David Scharfstein (1987) "Testing in Models of Asymmetric Information" *Review of Economic Studies.* April 1987. 54(2), 178: 265–78.

Rothschild, Michael & Joseph Stiglitz (1976) "Equilibrium in Competitive Insurance Markets: An Essay on the Economics of Imperfect Information" *Quarterly Journal of Economics.* November 1976. 90, 4: 629–49.

Stiglitz, Joseph & Andrew Weiss (1981) "Credit Rationing in Markets with Imperfect Information" *American Economic Review.* June 1981. 71, 3: 393–410.

Problem 8

Worker Ability

Workers come in two types, *Good* and *Bad*, who make mistakes with probabilities 0.25 and 0.75. Both are risk averse. Whether a mistake is made is contractible. Employers are risk neutral.

(1)   Draw a state-space diagram with a separating equilibrium.
(2)   Show why no pooling equilibrium exists.
(3)   Draw a state-space diagram so there is no separating equilibrium.
(4)   Find the Wilson and Reactive equilibria in the diagram of part (3).

Notes

N8.1   Introduction: Production Game V

- For an example of an adverse selection model in which workers also choose effort level, see Akerlof (1976) on the "rat race." The model is not moral hazard, because while the employer observes effort, the worker's types—their utility costs of hard work—are known only to themselves.
- Gresham's Law ("Bad money drives out good") is a statement of adverse selection.

Only debased money will be circulated if the payer knows the quality of his money better than the receiver. The same result occurs if quality is common knowledge, but for legal reasons the receiver is obligated to take the money, whatever its quality. An example of the first is Roman coins with low silver content; and of the second, Zambian currency with an overvalued exchange rate.

- Most adverse selection models have types that could be called "good" and "bad," because one type of agent would like to pool with the other, who would rather be separate. It is also possible to have a model in which both types would rather separate—types of workers who prefer night shifts and day shifts, for example—or two types who both prefer pooling—male and female college students.

N8.2   Adverse Selection under Certainty: Lemons I and II

- Dealers in new cars and other durables have begun offering "extended-service contracts" in recent years. These contracts, offered either by the manufacturers or by independent companies, pay for repairs after the initial warranty expires. For reasons of moral hazard or adverse selection, the contracts usually do not cover damage from accidents. Oddly enough, they also do not cover items like oil changes despite their usefulness in prolonging engine life. Such contracts have their own problems, as shown by the fact that several of the independent companies went bankrupt in the late 1970s and early 80s, making their contracts worthless. See "Extended-Service Contracts for New Cars Shed Bad Reputation as Repair Bills Grow," *Wall Street Journal*, 10 June 1985, p. 25.
- Suppose that the cars of Lemons II lasted two periods and did not physically depreciate. A naive economist looking at the market would see new cars selling for $6,000 (twice $3,000) and old cars selling for $2,000 and conclude that the service stream had depreciated by 33 percent. Depreciation and adverse selection are hard to untangle using market data.
- Bond (1982) is an empirical paper on prices of used trucks.
- Lemons II uses a uniform distribution. For a general distribution F, the average quality $\bar{\theta}(P)$ of cars with quality P or less is

$$\bar{\theta}(P) = E(\theta | \theta \leq P) = \frac{\int_{-\infty}^{P} xF'(x)dx}{F(P)}. \tag{8.4}$$

Equation (8.4) also arises in physics (the center of gravity) and nonlinear econometrics (computing likelihoods). Think of $\bar{\theta}(P)$ as a weighted average of the values of θ up to P, the weights being the densities. Having multiplied by all these weights in the numerator, we have to divide by their "sum," $F(P) = \int_{-\infty}^{P} F'(x)dx$, in the denominator, giving rise to equation (8.4).

N8.3   Heterogeneous Tastes: Lemons III and IV

- You might object to a model in which the buyers of used cars value quality more than the sellers, since the sellers are often richer people. Remember that quality here is "quality of used cars," which is different from "quality of cars." The utility functions could be made more complicated without abandoning the basic model. We could specify something like $\pi_{buyer} = \theta + k/\theta - P$, where $\theta^2 > k$. Such a specification implies that the lower is the quality of the car, the greater is the difference between the valuations of buyer and seller.
- In Akerlof (1970) the quality of new cars is uniformly distributed between zero and two, and the model is set up differently than here, with the demand and supply curves offered by different types of traders and net supply and gross supply presented rather confusingly. Usually the best way to model a situation in which traders sell some of their endowment and consume the rest is to use only gross supplies and demands. Each old owner supplies his car to the market, but in equilibrium he might buy it back, having

better information about his car than the other consumers. Otherwise, it is easy to count a given unit of demand twice, once in the demand curve and once in the net supply curve.

- **Lemons III′: Minimum Quality of Zero** If the minimum quality of car in Lemons III were 0, not 2,000, the resulting game (Lemons III′) is close to the original Akerlof (1970) specification. As Figure 8.8 shows, the supply schedule and the demand schedule intersect at the origin, so that the equilibrium price is 0 and no cars are traded. The market has shut down entirely, because of the unravelling effect described in Lemons II. Even though the buyers are willing to accept a quality lower than the dollar price, the price buyers are willing to pay does not rise with quality as fast as the price needed to extract that average quality from the sellers, and a car of the minimum quality is valued exactly the same by buyers and sellers. A 20 percent premium on zero is still zero. The efficiency implications are even stronger than before, because at the optimum all the old cars are sold to new buyers, but in equilibrium, none are.

Figure 8.8   Lemons III′: buyers value cars more; minimum quality is zero

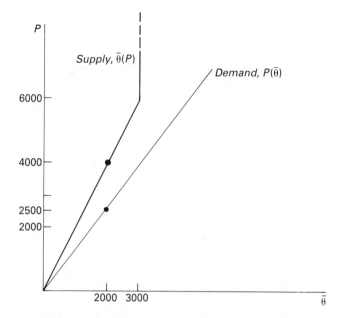

- See Stiglitz (1987) for a good survey of the relation between price and quality. Leibenstein (1950) uses diagrams to analyze the implications of individual demand being linked to the market price of quantity in markets for "bandwagon," "snob," and "Veblen" goods. See also "Pricing of Products is Still an Art, Often Having Little Link to Costs," *Wall Street Journal*, 25 November 1981, p. 29.
- Risk aversion is concerned only with variability of outcomes, not their level. If the quality of used cars ranges from 2,000 to 6,000, buying a used car is risky. If all used cars are of quality 2,000, buying a used car is riskless, because the buyer knows exactly what he is getting.

  In the insurance game of Section 8.4, the separating contract for the *Unsafe* consumer fully insures him: he bears no risk. But in constructing the equilibrium, we had to be very careful to keep the *Unsafes* from being tempted by the risky contract designed for the *Safes*. Risk is a bad thing, but as with old age, the alternative is worse. If Smith were sure to have his car stolen, he would bear no risk, because he would be sure to have low utility.
- To the buyers in Lemons IV, the average quality of cars for a given price is stochastic

because they do not know which values of ε were realized. To them, the curve $\overline{\theta}(P)$ is only the *expectation* of the average quality.

N8.4   Adverse Selection under Uncertainty: Insurance Game III

- Markets with two types of customers are very common in insurance, because it is easy to distinguish male from female, both those types are numerous, and the difference between them is important. Males under age 25 pay almost twice the auto insurance premiums that females do, and females pay 10 to 30 percent less for life insurance. The difference goes both ways, however: Aetna charges a 35-year-old woman 30 to 50 percent more than a man for medical insurance. One market in which rates do not differ much is disability insurance. Women do make more claims, but the rates are the same because relatively few women buy the product (*Wall Street Journal*, 27 August 1987, p. 25).

N8.5   Other Equilibrium Concepts: Wilson and Reactive Equilibrium

- Engers & Fernandez (1987) show how to transform a simultaneous move game into a sequential move game such that the reactive equilibrium of the original game is one of the perfect Bayesian equilibria of the transformed game.
- In adverse selection games it often matters whether the informed player or the uninformed player offers the contract. Wilson and reactive equilibrium are important when the uninformed player offers the contract, since it is only he who runs the risk of receiving something unexpected in the transaction and might want to withdraw an offer. The issues involved are the same as in the difference between screening and signalling, which will be discussed at length in Chapter 9.

N8.6   Applications

- Bagehot (1971) is the earliest reference on the adverse selection explanation for the bid–ask spread. Copeland & Galai (1983) is an empirical study of the bid–ask spread in options markets.
- Economics professors sometimes make use of self selection for student exams. One of my colleagues put the following instructions on an MBA exam, after stating that either Question 5 or 6 must be answered.

  > The value of Question 5 is less than that of Question 6. Question 5, however, is straightforward and the average student may expect to answer it correctly. Question 6 is more tricky: only those who have understood and absorbed the content of the course well will be able to answer it correctly... For a candidate to earn a final course grade of A or higher, it will be *necessary* for him to answer Question 6 successfully.

  Making the question even more self-referential, he asked the students for an explanation of its purpose.

  Another of my colleagues tried asking who in his class would be willing to skip the exam and settle for an A−. Those students who were willing received an A−. The others got As. But nobody had to take the exam (this method did upset a few people). More formally, Guasch & Weiss (1980) have looked at adverse selection and the willingess of workers with different abilities to take tests.
- Nalebuff & Scharfstein (1987) have written on testing, generalizing Mirrlees (1974), who showed how a forcing contract in which output is costlessly observed might attain efficiency by punishing only for very low output. In Nalebuff & Scharfstein, testing is costly and agents are risk averse. They develop an equilibrium in which the employer tests workers with small probability, using high-quality tests and heavy punishments to attain almost the first-best. Under a condition which implies that large expenditures on each test can eliminate false accusations, they show that the principal will test workers

with small probability, but use expensive, accurate tests when he does test a worker, and impose a heavy punishment for lying.

9 Signalling

9.1 Introduction

Signalling is a way for an agent to communicate his type under adverse selection. The signalling contract specifies a wage that depends on an observable characteristic—the signal—which the agent chooses for himself after Nature chooses his type. The extensive forms of two kinds of models with signals are shown in Figures 6.1d and 6.1e. If the agent chooses his signal before the contract is offered, he is signalling to the principal. If he chooses the signal afterwards, the principal is screening him. What we will see is that the order of moves is important, signalling costs must differ between agent types for signalling to be useful, and the outcome is often inefficient.

Section 9.2 begins with signalling models in which a worker chooses one of two education levels to signal his ability. Section 9.3 steps back from the technical detail to more practical considerations in applying the model to education. Section 9.4 turns the game into a screening model and generalizes to a continuum of education levels. Section 9.5 switches to diagrams and applies signalling to new stock issues to show how two signals are used when the agent has two unobserved characteristics.

9.2 The Informed Player Moves First: Signalling

Spence (1973) introduced the idea of signalling in the context of education. We will construct a series of models which formalize the notion that education is useless to increase a worker's ability, but useful to demonstrate that ability to employers. Let half of the workers have the type "high ability" and half "low ability," where ability is a number denoting the dollar value of a worker's output. Output is assumed to be a noncontractible variable. This is a simplifying assumption, imposed so that the contracts are functions only of the signals, rather than a combination of the signal and the output. If output is contractible, it should be in the contract, as we have seen in Chapter 6.

Employers do not observe worker ability, but they do know the distribution of abilities, and they observe the worker's education, which also takes two levels. Workers choose their education levels before employers choose compensation schemes to attract them. To simplify, we will specify that the players are one worker and two employers. The employers compete profits down to zero and the worker receives the gains from trade. The worker's strategy is his education level and his choice of contract. The employers' strategies are the sets of contracts they offer giving wages as functions of the education level.

Education I

Players

A worker and two employers.

Information

Asymmetric, incomplete, and certain.

Actions and Events

(0) Nature chooses the worker's ability $a \in \{2, 5.5\}$, the *Low* and *High* ability each having probability 0.5. The variable a is observed by the worker, but not by the employers.
(1) The worker chooses education level $y \in \{0, 1\}$.
(2) The employers each offer a wage contract $w(y)$.
(3) The worker accepts a contract.
(4) Output equals a.

Payoffs

The worker's payoff is his wage minus his cost of education, and the employer's is his profit.

$$\pi_{worker} = w - 8y/a \text{ if the worker accepts contract } w.$$

$$\pi_{employer} = \begin{cases} a - w & \text{for the employer whose contract is accepted.} \\ 0 & \text{for the other employer.} \end{cases}$$

The payoffs assume both that education is costly for the worker and that it is more costly if his ability takes a low value. The cost's dependence on ability is what permits separation to occur. As in any hidden information game, we must think about both pooling and separating equilibria. Education I has both. In the pooling equilibrium, which we will call PE 1.1, both types of workers pick zero education and the employers pay the zero-profit wage of 3.75 (= [2+5.5]/2) regardless of the education level. PE 1.1 needs to be specified as a perfect Bayesian equilibrium rather than simply Nash, because the interpretation that the uninformed player (the employer) puts on out-of-equilibrium behavior is important. Here, the equilibrium needs to specify the employer's beliefs when he observes $y = 1$, since that is never observed in equilibrium. In PE 1.1, the beliefs are "passive conjectures" (see Section 5.2): employers believe that a worker who chooses $y = 1$ is *Low* with probability 0.5. Given this belief, both types of workers realize that education is useless, and the model reaches the unsurprising outcome that workers do not bother to acquire unproductive education.

Pooling Equilibrium 1.1
(PE 1.1)
$$\left\{ \begin{array}{l} y(Low) = y(High) = 0 \\ w(0) = w(1) = 3.75 \\ Prob(a = Low|y = 1) = 0.5 \end{array} \right\}$$

Under certain other beliefs, the pooling equilibrium breaks down. Under the belief $Prob(a = Low|y = 1) = 0$, for example, employers believe that any worker who acquired education is a *High*, so pooling is not Nash: the *High* workers are tempted to deviate and acquire education. This leads to the separating equilibrium for which signalling is best known, in which the high-ability worker acquires education to prove to employers that he really has high ability.

Separating Equilibrium 1.2
(SE 1.2)
$$\left\{ \begin{array}{l} y(Low) = 0, y(High) = 1 \\ w(0) = 2, w(1) = 5.5 \end{array} \right\}$$

Following the method used in Sections 6.3 and 7.1, we will show that SE1.2 is an equilibrium by using constraints. A pair of separating contracts must maximize the utility of the *Highs* and the *Lows* subject to the participation constraints that the firms can offer the contracts without making losses and the incentive compatibility constraints that the *Lows* are not attracted to the *High* contract, nor the *Highs* to the *Low* contract. The participation constraints are that

$$w(0) \leq a_L = 2 \text{ and } w(1) \leq a_H = 5.5. \tag{9.1}$$

Competition between the employers makes the expressions in (9.1) hold as equalities. The *Lows*' incentive compatibility constraint is

$$U_L(y = 0) \geq U_L(y = 1), \tag{9.2}$$

which in Education I is

$$w(0) - 0 \geq w(1) - 8/2. \tag{9.3}$$

Since in SE 1.2 the separating wage of the *Lows* is 2, and the separating wage of the *Highs* is 5.5 from (9.1), this incentive compatibility constraint is satisfied.
The incentive compatibility constraint for the *Highs* is

$$U_H(y = 1) \geq U_H(y = 0), \tag{9.4}$$

which in Education I is

$$w(1) - 8/5.5 \geq w(0) - 0. \tag{9.5}$$

Constraint (9.5) is satisfied by SE 1.2.

We can test the equilibrium by looking at the best responses. Given the worker's strategy and the other employer's strategy, each employer must pay the worker his full output or lose him to the other employer. Given the employers' contracts, the *Low* has a choice between the payoff 2 ($=2-0$) for ignorance and 1.5 ($= 5.5 - 8/2$) for education, so he picks ignorance. The *High* has a choice between the payoff 2 ($= 2 - 0$) for ignorance and 4.05 ($= 5.5 - 8/5.5$, rounded) for education, so he picks education.

Unlike the pooling equilibrium, the separating equilibrium does not need to specify beliefs. Either of the two education levels might be observed in equilibrium, so Bayes's Rule always tells the employers how to interpret what they see: education means the agent is *High* and lack of it that he is *Low*. This does not mean that a worker cannot deviate, but deviation will not change the employer's beliefs. If a *High* worker deviates by choosing $y = 0$ and tells the employers he is a *High* who would rather pool than separate, the employers disbelieve him and offer him the *Low* wage of 2, not the pooling wage of 3.75 or the *High* wage of 5.5.

A strong case can be made that the beliefs in the separating equilibrium are more sensible. Harking back to the equilibrium refinements of Section 5.2, recall that one suggestion (from Cho & Kreps [1987]) is to discover whether one type of player could not possibly benefit from deviating, no matter how the uninformed player changed his beliefs as a result. Here, the *Low* worker could never benefit from deviating from PE1.1. Under the passive conjectures specified, the *Low* gets a payoff of 3.75 in equilibrium versus -0.25 ($= 3.75 - 8/2$) if he deviates and becomes educated. Under the most favorable belief possible, that a worker who deviates is *High* with probability one, the *Low* would get a wage of 5.5 if he deviated, but his payoff from deviating would only be 1.5 ($= 5.5 - 8/2$). So the more reasonable belief seems to be that a worker who acquires education is a *High*, which does not support the pooling equilibrium.

The nature of the separating equilibrium lends support to the claim that education *per se* is useless, or even pernicious. While we may be reassured by the fact that Professor Spence himself thought it worthwhile to become Dean of Harvard College, the implications are disturbing, and suggest that we should think seriously about how well the model applies to the real world. We will do that in Section 9.3.

Education II: Modelling Trembles so Nothing is Out of Equilibrium

The pooling equilibrium of Education I required the modeller to specify the employers' out-of-equilibrium beliefs. An equivalent model constructs the game tree to support the beliefs, instead of introducing them via the equilibrium concept. This approach was briefly mentioned in connection with PhD Admissions in Section 5.2. The advantage is that the assumptions on beliefs are put in the rules of the game along with the other assumptions. So let us replace Nature's move with

Actions and Events

(0)   Nature chooses worker ability $a \in \{2, 5.5\}$, each ability having probability
      0.5. (a is observed by the worker, but not by the employer.) With probability
      0.001, Nature endows a worker with free education.

. . .

Payoffs

$$\pi_{worker} = \begin{cases} w - 8y/a & \text{if the worker accepts contract } w \text{ (ordinarily).} \\ w & \text{if the worker accepts contract } w \text{ (with free education).} \end{cases}$$

. . .

With probability 0.001 the worker receives free education regardless of his ability. If
the employer sees a worker with education, he knows that the worker might be one
of this rare type, in which case the probability that the worker is *Low* is 0.5. Both
$y = 0$ and $y = 1$ can be observed in any equilibrium and Education II has almost
the same two equilibria as Education I without the need to specify beliefs. The separating equilibrium did not depend on beliefs, and remains an equilibrium. What
was Pooling Equilibrium 1.1 becomes "almost" a pooling equilibrium—almost all
workers behave the same, but the small number with free education behave differently. The two types of greatest interest—the *High* and *Low*—are not separated,
but the ordinary workers are separated from the workers whose education is free.
Even that small amount of separation allows the employers to use Bayes's Rule and
eliminates the need for exogenous beliefs.

Education III: No Separating Equilibrium, Two Pooling Equilibria

Let us next modify Education I by changing the possible worker abilities from
$\{2, 5.5\}$ to $\{2, 12\}$. The separating equilibrium vanishes, but a new pooling equilibrium emerges. In equilibria PE 3.1 and PE 3.2, both pooling contracts pay the
same zero-profit wage of 7 ($= [2+12]/2$), and both types of agents acquire the same
amount of education, but the amount depends on the equilibrium.

Pooling Equilibrium 3.1
(PE 3.1)
$$\left\{ \begin{array}{l} y(Low) = y(High) = 0 \\ w(0) = w(1) = 7 \\ Prob(a = Low|y = 1) = 0.5 \text{ (passive conjectures)} \end{array} \right\}$$

Pooling Equilibrium 3.2
(PE 3.2)
$$\left\{ \begin{array}{l} y(Low) = y(High) = 1 \\ w(0) = 2, w(1) = 7 \\ Prob(a = Low|y = 0) = 1 \end{array} \right\}$$

PE 3.1 is similar to the pooling equilibrium in Education I and II, but PE 3.2 is inefficient. Both types of workers receive the same wage, but they incur the education costs anyway. Each type is frightened to do without education because the employer would pay him not the average, pooling wage, but the wage appropriate to the *Low*.

PE 3.2 also illustrates a fine point of the definition of pooling, because although the two types of workers adopt the same strategies, the equilibrium contract offers different wages for different education. The implied threat to pay a low wage to an uneducated worker never needs to be carried out, so the equilibrium is still pooling. Notice that perfectness does not rule out threats based on beliefs. The model imposes these beliefs on the employer, and he would carry out his threats because he believes they are best responses. The employer receives a higher payoff under some beliefs than under others, but he is not free to choose his beliefs.

Following the approach of Education II, we could eliminate PE 3.2 by adding the exogenous probability 0.001 that either type is completely unable to buy education. Then no behavior is out-of-equilibrium, and we end up with PE 3.1 because the only rational belief is that if $y = 0$ is observed the worker has equal probability of being *High* or *Low*. To eliminate PE 3.1 requires less reasonable beliefs; for example, a probability 0.001 that a *Low* gets free education together with probability zero that a *High* does.

9.3   General Comments on Signalling in Education

Signalling and Similar Phenomena

The distinguishing feature of signalling is that the agent's action, although not directly related to output, is useful because it is related to ability. For the signal to work, it must be less costly for an agent with higher ability. Separation can occur in Education I because when the principal pays a greater wage to educated workers, only the *High*s, whose utility costs of education are lower, are willing to acquire it. That is why a signal works where a simple message would not: actions speak louder than words.

Signalling is outwardly similar to other solutions to adverse selection. The high-ability agent finds it cheaper than the low-ability one to build a reputation, but the reputation-building actions are based directly on his high ability: in a typical reputation model he shows ability by producing high output period after period. Also, the nature of reputation is to require several periods of play, which signalling does not.

Another form of communication is possible when some observable variable not under the control of the worker is correlated with ability. Age, for example, is correlated with reliability, so an employer pays older workers more, but the correlation does not arise because it is easier for reliable workers to acquire the attribute of age. Because age is not an action chosen by the worker, we would not need game theory to model it.

Problems in Applying Signalling to Education

On the empirical level, the first question to ask of a signalling model of education is, "What is education?", which for operational purposes means, "In what units is education measured?" Two possible answers are "years of education" and "grade point average." If the sacrifice of a year of earnings is greater for a low-ability worker, years of education can serve as a signal. If less intelligent students must work harder to get straight As, then grade point average can also be a signal.

Layard & Psacharopoulos (1974) reject signalling as an important motive for education for three reasons. First, dropouts get as high a rate of return on education as those who complete degrees, so the signal is not the diploma, although it might be the years of education. Second, wage differentials between different education levels rise with age, although one would expect the signal to be less important after the employer has acquired more observations on the worker's output. Third, testing is not widely used for hiring, despite its low cost relative to education. In fact, tests are available but unused: students commonly take tests like the American SAT whose results they could credibly communicate to employers, and their scores correlate highly with later grade point average. One would also expect an employer to prefer to pay an 18-year-old low wages for four years to determine his ability, rather than waiting for his grades as a history major. No good answer has been found to this objection.

Productive Signalling

Even if education were largely signalling, we might not want to close the schools. Signalling might be wasteful in a pooling equilibrium like PE3.2, but in a separating equilibrium it can be second-best efficient, for at least three reasons. First, it allows the employer to match workers with jobs suited to their talents. If the only jobs available were "professor" and "secretary," then in a pooling equilibrium both *High* and *Low* workers would be employed, but they would be randomly allocated to the two jobs. Pridefully, I think that research would suffer. Typing might too.

Second, signalling keeps talented workers from moving to jobs where their productivity is lower, but their talent is known. Without signalling, a talented worker might withdraw from a corporation and start his own company, where he would be less productive but better paid. The naive observer would see that corporations hire only one type of worker (*Low*), and imagine there was no welfare loss.

Third, if ability is endogenous—moral hazard rather than adverse selection—signalling encourages workers to acquire ability and all workers might acquire a positive amount of the signal. One of my teachers said that you always understand your next-to-last econometrics class. Suppose that solidly learning econometrics increases the student's ability, but a grade of A is not enough to show that he solidly learned the material. To signal his newly acquired ability, the student must also take "Time Series," which he cannot pass without solid econometrics. "Time Series" might be useless in itself, but if it did not exist, the students would not learn basic econometrics either.

9.4   The Informed Player Moves Second: Screening

Education IV: Screening with a Discrete Signal

Players

A worker and two employers.

Information

Asymmetric, incomplete, and certain.

Actions and Events

(0)   Nature chooses worker ability $a \in \{2, 5.5\}$, each ability having probability 0.5. Employers do not observe ability, but the worker does.
(1)   Each employer offers a wage contract $w(y)$.
(2)   The worker chooses education level $y \in \{0, 1\}$.
(3)   The worker accepts a contract.
(4)   Output equals a.

Payoffs

$$\pi_{worker} \quad = \quad w - 8y/a \text{ if the worker accepts contract } w.$$

$$\pi_{employer} \quad = \quad \begin{cases} a - w & \text{for the employer whose contract is accepted.} \\ 0 & \text{for the other employer.} \end{cases}$$

Education IV has no Nash pooling equilibrium, because if one employer tried to offer the zero profit pooling contract, $w(0) = 3.75$, the other employer would offer $w(1) = 5.5$ and draw away all the *Highs*. The unique equilibrium is

Separating Equilibrium 4.1 $\left\{ \begin{array}{l} y(Low) = 0, y(High) = 1 \\ w(0) = 2, w(1) = 5.5 \end{array} \right\}$
(SE 4.1)

Beliefs do not need to be specified in a screening model. The uninformed player moves first, so his beliefs after seeing the move of the informed player are irrelevant. The informed player is fully informed, so his beliefs are not affected by what he observes. This is much like simple adverse selection, in which the uninformed player moves first, offering a set of contracts, after which the informed player chooses one of them. The modeller does not need to refine perfectness in a screening model, although he might be tempted to abandon it altogether in favor of reactive or Wilson equilibrium. The similarity between adverse selection and screening is strong enough that Education IV would not have been out of place in Chapter 8, but since the context is so similar to the signalling models of education, I have waited to present it.

The restriction of education to just two levels can be misleading, so we will now allow a continuum, in a game otherwise the same as Education IV.

Education V: Screening with a Continuous Signal

Players

A worker and two employers.

Information

Asymmetric, incomplete, and certain.

Actions and Events

(0) Nature chooses worker ability $a \in \{2, 5.5\}$, each ability having probability 0.5. Employers do not observe ability, but the worker does.
(1) Each employer offers a wage contract $w(y)$.
(2) The worker chooses education level $y \in [0, 1]$.
(3) The worker chooses a contract.
(4) Output equals a.

Payoffs

$$\pi_{worker} = w - 8y/a \text{ if the worker accepts contract } w.$$
$$\pi_{employer} = \begin{cases} a - w & \text{for the firm whose contract is accepted.} \\ 0 & \text{for the other firm.} \end{cases}$$

Pooling equilibria generally do not exist in screening games with continuous signals, and separating equilibria are also sometimes lacking. Education V, however, does have a separating Nash equilibrium, with a unique equilibrium path.

Separating Equilibrium 5.1
(SE 5.1)
$$\left\{ \begin{array}{l} y(Low) = 0, y(High) = 0.875 \\ w = \begin{cases} 2 & \text{if } y < 0.875 \\ 5.5 & \text{if } y \geq 0.875 \end{cases} \end{array} \right\}$$

In any separating contract, the *Lows* must be paid a wage of 2 for an education of 0, because this is the most attractive contract that breaks even. The separating contract for the *Highs* must maximize their utility subject to the constraints discussed in Education I. When the signal is continuous, the constraints are especially useful to the modeller for calculating the equilibrium. The participation constraints are that

$$w(0) \leq a_L = 2 \text{ and } w(y^*) \leq a_H = 5.5, \tag{9.6}$$

where y^* is the separating value of education that we are trying to find. Competition turns (9.6) into equalities. The incentive compatibility constraint is

$$U_L(y = 0) \geq U_L(y = y^*),$$ (9.7)

which in Education V is

$$w(0) - 0 \geq w(y^*) - 8y^*/2.$$ (9.8)

Since the separating wage is 2 for the *Lows* and 5.5 for the *Highs*, constraint (9.8) is satisfied as an equality if $y^* = 0.875$, which is the crucial education level in SE 5.1. The nonpooling constraint is

$$U_H(y = y^*) \geq U_H(pooling),$$ (9.9)

or, for Education V,

$$w(y^*) - 8 \cdot y^*/5.5 \geq 3.75.$$ (9.10)

Since the payoff of *Highs* in the separating contract is 4.23 ($= 5.5 - 8 \cdot 0.875/5.5$, rounded), the nonpooling constraint is satisfied.

No Pooling Equilibrium in Education V

Education V lacks a pooling equilibrium, which would require the outcome $\{y = 0, w(0) = 3.75\}$, shown as C_1 in Figure 9.1. If one employer offered a pooling contract requiring more than zero education (such as in PE 3.2), the other employer could make the more attractive offer of the same wage for zero education. The wage

Figure 9.1   Education V: no pooling Nash equilibrium

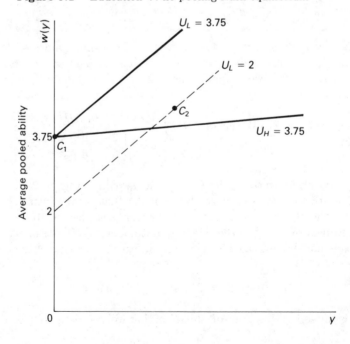

is 3.75 to ensure zero profits. The rest of the wage function—the wages for positive education levels—can take a variety of shapes, so long as the wage does not rise so fast with education that the *Highs* are tempted to become educated.

But no equilibrium has these characteristics. In a Nash equilibrium, no employer can offer a pooling contract, because the other employer could always profit by offering a separating contract paying more to the educated. One such separating contract is C_2 in Figure 9.1, which pays 4.9 to workers with an education of $y = 0.5$ and yields a payoff of 4.17 $(= 4.9 - [8 \cdot 0.5]/5.5$, rounded) to the *Highs* and 2.9 $(= 4.9 - 8 \cdot 0.5/2)$ to the *Lows*. Only *Highs* prefer C_2 to the pooling contract C_1, which yields payoffs of 3.75 to both *High* and *Low*, and if only *Highs* accept C_2, it yields positive profits to the employer.

Nonexistence of a pooling equilibrium in screening models is a general result. In any screening model, the *Lows* have greater costs of education, which is equivalent to steeper indifference curves. (The linearity of the curves in Education V is unimportant.) Any pooling equilibrium must, like C_1, lie on the vertical axis where education is zero and the wage equals the average ability. A separating contract like C_2 can always be found to the northeast of the pooling contract, between the indifference curves of the two types, and it will yield positive profits by attracting only the *Highs*.

Education VI: No Nash Equilibrium

In Education V, we showed that screening models have no pooling equilibria. In Education VI, the parameters are changed a little to eliminate even the separating equilibrium. Let the proportion of *Highs* be 0.9 instead of 0.5, so that the zero-profit pooling wage is 5.15 $(= 0.9[5.5] + 0.1[2])$ instead of 3.75. Consider the separating contracts C_3 and C_4, shown in Figure 9.2 (see p.216), calculated in the same way as SE 5.1. (C_3, C_4) is the most attractive pair of contracts that separates *Highs* from *Lows* by satisfying constraint (9.7). *Low* workers accept contract C_3, obtain $y = 0$, and receive a wage of 2, their ability. *Highs* accept contract C_4, obtain $y = 0.875$, and receive a wage of 5.5, their ability. Education is not attractive to *Lows* because the *Low* payoff from pretending to be *High* is 2 $(= 5.5 - 8 \cdot 0.875/2)$, no better than the *Low* payoff of 2 from C_3 $(= 2 - 8 \cdot 0/2)$.

The wage of the pooling contract C_5 is 5.15, so that even the *Highs* strictly prefer C_5 to (C_3, C_4). But our reasoning that no pooling equilibrium exists is still valid; some contract C_6 would attract all the *Highs* from C_5. No Nash equilibrium in pure strategies exists, either separating or pooling.

Screening and Adverse Selection

We also encountered a nonexistence problem in the adverse selection game of Section 8.4, where, just as in screening, the informed player takes no action until after the uninformed player has offered a set of contracts. Screening models behave much the same way as simple adverse selection models, in contrast to signalling models.

Compare Education V, the screening model of education, with the Rothschild & Stiglitz (1976) Insurance Game III of Section 8.4. Both have contracts which give something of value in exchange for costs incurred to communicate the informed players' private information. In Education V, the contracts specified a higher wage for higher costs of education that communicated high ability. In Insurance Game III, the contracts specified a lower premium for higher coinsurance rates (less complete insurance). In both models, the two types of informed player separate, picking two different levels of the costly variable. The outcomes are only second-best efficient because in Insurance Game III the *Safe* type is not fully insured and in Education V the *High* type incurs the cost of education. Insurance Game III specifies two distinct contracts, as opposed to the $w(y)$ function of Education V, but since the worker only picks one of two education levels in equilibrium, the difference is more apparent than real. The fact that Education V is a game of certainty and Insurance Game III is a game of uncertainty is also unimportant.

What then is the difference between the two games? The main difference is that in Education V, it is possible to conceive of the education apart from the wage contract, whereas in Insurance Game III the signal is communicated by choice of the insurance contract. The inefficiency is more striking in the screening game, where the cost of communication is distinct from the act of accepting a particular contract.

Wilson Equilibrium and Reactive Equilibrium

As with Insurance Game III, it is possible to go beyond Nash equilibrium in Education VI. Is it reasonable to say that a pooling equilibrium could always be broken by a contract which draws away the *Highs*? After the *Highs* departed, the old pooling contract, which would still soak up all the *Lows*, would be unprofitable. In Figure 9.2, if C_5 is withdrawn after C_6 is offered, the *Lows* prefer C_6 to the

Figure 9.2   Education VI: no Nash equilibrium

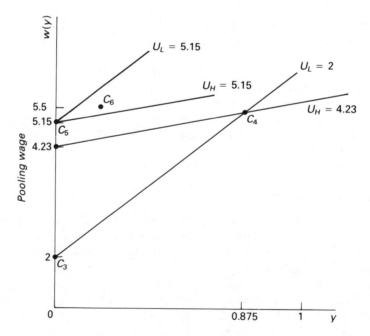

zero they obtain from unemployment, and C_6 becomes a pooling contract. This is irrelevant to the question of whether C_5 is a Nash equilibrium, but it might lead one to doubt the wisdom of the equilibrium concept.

Under the concept of the **Wilson equilibrium** that we discussed in Section 8.5, the pooling equilibrium is legitimate, because an employer thinking about introducing the new equilibrium-breaking contract would realize that the new contract would be unprofitable once the old contract was withdrawn. **Reactive equilibrium** can also be applied, and generates a separating equilibrium. Under its reasoning, the separating equilibrium cannot be broken by a pooling contract, because the pooling contract would in turn be broken by a second separating contract. The Wilson C_5 and the reactive (C_3, C_4) are the two clear candidates for equilibrium in Figure 9.2. Alternately, we could restructure the model so that the worker moves first—our assumption in Section 9.2. While avoiding the existence problem, that introduces the need to think about out-of-equilibrium beliefs, and it really models a different situation, in which workers cannot change their education in response to employers' contracts.

Summary of Education Models

Because of signalling's complexity, most of this chapter has been devoted to elaboration of the education model. We began with Education I, which showed how with two types and two signal levels the perfect Bayesian equilibrium could be either separating or pooling. Education II took the same model and replaced the specification of out-of-equilibrium beliefs with an additional move by Nature, while Education III changed Education I's parameters to increase the difference between types and show how signalling could continue with pooling. After a purely verbal discussion of how to apply signalling models, we looked at screening, in which the employer moves first. Education IV was a screening version of Education I, while Education V broadened the model to allow a continuous signal, which eliminates pooling equilibria. Education VI modified the parameters of Education V to show that sometimes no pure strategy Nash equilibrium exists at all.

Throughout, the implicit assumption was maintained that all the players were risk neutral. Risk neutrality is unimportant, because there is no uncertainty in the model and the agents bear no risk. Only if the worker types differed in their degrees of risk aversion could contracts attempt to use the difference to support a separating equilibrium. If the principal were risk averse he might offer a wage less than the average productivity in the pooling equilibrium, but he is under no risk at all in the separating equilibrium, because it is fully revealing. The models are also games of certainty, and this too is unimportant. If output were uncertain, agents would just look at the expected payoffs rather than the actual payoffs, and very little would change.

We could extend the education models further—allowing more than two levels of ability would be a high priority—but instead, let us turn to the financial markets and look graphically at a model with two continuous characteristics of type and two continuous signals.

9.5   Two Signals: Underpricing of Stock

One signal might not be enough to communicate two of an agent's characteristics. In the game Underpricing, the two characteristics are the mean and variance of the value of a stock "going public." Empirically, it has been found that companies consistently issue stock at a price so low that it rises sharply in the days after the issue, an abnormal return estimated to average 11.4 percent (Copeland & Weston [1983], p. 334). The model tries to explain this using the percentage of the stock retained by the original owner and the amount of underpricing as two signals.

Underpricing

(Hwang [unpub])

Players

The entrepreneur and many investors.

Information

Asymmetric, incomplete, and uncertain. The investors are uninformed.

Actions and Events

(See Figure 2.3a for a time line.)

(0)   Nature chooses the expected value (μ) and variance (σ^2) of a share of the firm using some distribution F.
(1)   The entrepreneur retains fraction α of the stock and offers to sell the rest at a price per share of P_0.
(2)   The investors decide whether to accept or reject the offer.
(3)   The market price becomes P_1, the investors' estimate of μ.
(4)   Nature chooses the value V of a share using some distribution $G(\mu, \sigma^2)$. With probability θ, V is revealed to the investors and becomes the market price.
(5)   The entrepreneur sells his remaining shares at the market price.

Payoffs

$$\pi_{entrepreneur} = U([1-\alpha]P_0 + \alpha[\theta V + (1-\theta)P_1]), \text{ where } U' > 0 \text{ and } U'' < 0.$$
$$\pi_{investors} = (1-\alpha)(V - P_0) + \alpha(1-\theta)(V - P_1).$$

The entrepreneur's payoff is the utility of the value of the shares he issues at P_0 plus the value of those he sells later at the price P_1 or V. The investors' payoff is the true value of the shares they buy minus the prices they pay.

Underpricing subsumes the simpler model of Leland & Pyle (1977), in which σ^2

is common knowledge and if the entrepreneur chooses to retain a large fraction of the shares, the investors deduce that the stock value is high. The one signal in that model is fully revealing because holding a larger fraction exposes the undiversified entrepreneur to a larger amount of risk, which he is unwilling to accept unless the stock value is greater than investors would guess without the signal.

If the variance of the project is high, that also increases the risk to the undiversified entrepreneur, which is important even though the investors are risk neutral and do not care directly about the value of σ^2. Since the risk is greater when variance is high, the signal α is more effective and retaining a smaller amount allows the entrepreneur to sell the remainder at the same price as a larger amount for a lower-variance firm. Figure 9.3 (see p. 220) shows the signalling schedules for two variance levels.

In Underpricing, σ^2 is not known to the investors, so the signal is no longer fully revealing. An α equal to 0.1 could mean either that the firm has a low value with low variance, or a high value with high variance. But the entrepreneur can use a second signal, the price at which the stock is issued, and by observing α and P_0, the investors can deduce μ and σ^2.

Using particular numbers for concreteness, consider the entrepreneur's options if he decides to retain $\alpha = 0.1$. He could offer the stock without a discount, at $P_0 = 90$, in which case the price would not rise afterwards, or he could offer a discount, setting $P_0 = 80$, after which it would rise to $P_1 = \mu = 120$. In discounting, he would not do so well on the fraction $1 - \alpha$ that he sold initially, but when in move (5) he sold the fraction he retained, he would do better on average. If $\mu = 90$, the entrepreneur would not underprice, because he would only get 80 in the initial sale, and with probability θ the true value would be revealed by Nature and the retained shares could only be sold at an expected price of 90. If the variance of the stock were lower, the entrepreneur would pick a smaller discount and be willing to hold a larger fraction. Figure 9.4 (see p. 220) shows the different combinations of initial price and fraction retained that might be used.

This model explains why new stock is issued at a low price. The entrepreneur knows that the price will rise, but only if he issues it at a low initial price to show that the variance is high. The price discount shows that signalling by holding a large fraction of stock is unusually costly, but he is nonetheless willing to signal. The discount is costly because he is selling stock at less than its true value, and retaining stock is costly because he bears extra risk, but both are necessary to signal that the stock is valuable.

Recommended Reading

Engers, Maxim & Luis Fernandez (1987) "Market Equilibrium with Hidden Knowledge and Self-Selection" *Econometrica*. March 1987. 55, 2: 425–39.

Leland, Hayne & David Pyle (1977) "Informational Asymmetries, Financial Structure, and Financial Intermediation" *Journal of Finance*. May 1977. 32, 2: 371–87.

Figure 9.3   Underpricing: how the signal changes with the variance

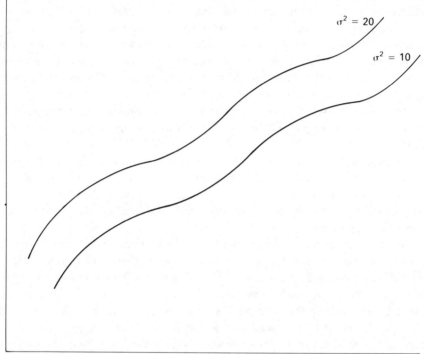

μ (= expected value) unobserved

σ^2 is unobserved

$\sigma^2 = 20$

$\sigma^2 = 10$

α (= fraction retained) observed

Figure 9.4   Underpricing: different ways to signal a given μ

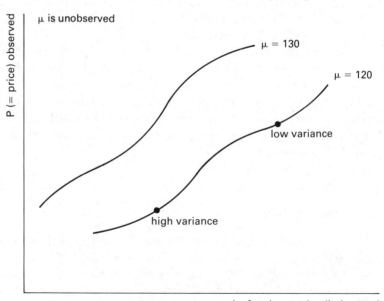

P (= price) observed

μ is unobserved

$\mu = 130$

$\mu = 120$

low variance

high variance

α (= fraction retained) observed

Problem 9

Warranties

Twenty telephone manufacturers each have the same capacity and the same marginal cost of 1.0. The telephones of Allied break down with probability 0.1, while the others break down with probability 0.8. Each company decides whether to offer a warranty that pays out 20 if the telephone breaks. A large number of competing consumers are offered prices by the firms. A consumer gets utility of 10 from a working telephone, but he does not know which company has higher quality.

(1) Should this be modelled as screening or signalling?
(2) Find a separating equilibrium for a signalling model.
(3) Show that there is no pooling equilibrium under *any* beliefs in this signalling model.
(4) Find the equilibrium if Brydox enters with phones that break down with probability 0.4.

Notes

N9.1   Introduction

- The term "signalling" was introduced by Spence (1973). The games in this book take advantage of hindsight to build simpler and more rational models of education than in his original article, which used a rather strange equilibrium concept: a strategy combination from which no worker has incentive to deviate and under which the employer's profits are zero. Under that concept, the firm's incentives to deviate are irrelevant.

  The distinction between signalling and screening is relatively new, and has been attributed to Stiglitz & Weiss (unpub). The literature has shown wide variation in the use of both terms, and "signal" is such a useful word that it is often used in models that have no signalling of the kind discussed in this chapter.

- Empirical signalling papers include Layard & Psacharopoulos (1974) on education and Staten & Umbeck (1986) on occupational diseases.

- Applications of signalling include prices in C. Wilson (1980) and Stiglitz (1987), payment of dividends in Ross (1977), bargaining (Section 10.5) and greenmail (Section 13.5).

- Legal bargaining is one area of application for signalling. See Grossman & Katz (1983). Reinganum (1988) has a nice example of the value of precommitment in legal signalling. In her model, a prosecutor who wishes to punish the guilty and release the innocent wishes, if parameters are such that most defendants are guilty, to commit to a pooling strategy in which his plea bargaining offer is the same whatever the probability that a particular defendant will be found guilty.

- The peacock's tail may be a signal. Zahavi (1975) suggests that a large tail may benefit the peacock because by hampering him it demonstrates to potential mates that he is fit enough to survive even with a handicap.

- **Advertising.** Advertising is a natural application for signalling. The literature includes Nelson (1974), written before signalling was well-known, Kihlstrom & Riordan (1984) and Milgrom & Roberts (1986). I will briefly describe a model based on Nelson's. Firms take one of two types, low quality or high quality. Consumers do not know that a firm exists until they receive an ad from it, and they do not know its quality until they buy its product. They are unwilling to pay more than zero for low quality, but any product is costly to produce. This is not a reputation model, because it is finite in length and quality is exogenous.

  If the cost of an ad is greater than the profit from one sale, but less than the profit from repeat sales, then high rates of advertising are associated with high product quality. A firm with low quality would not advertise, but a firm with high quality would.

The model can work even if consumers do not understand the market, so it does not have to be a signalling model. But if consumers do understand, signalling reinforces the result. Consumers know that firms which advertise must have high quality, so they are willing to try them. This understanding is important, because if consumers knew that 90 percent of firms were low quality but did not understand that only high quality firms advertise, they would not respond to the ads they received. This should bring to mind Section 5.2's game PhD Admissions.

N9.2   The Informed Player Moves First: Signalling

- It is important to signalling that the employers see only the worker's action, and not his strategy. If employers believe that the actual equilibrium in Education I is the separating equilibrium, then when they see a worker getting education of zero, they deduce that he is low ability, and the worker can do nothing to persuade them otherwise. If the worker could commit to a strategy of the form $(y|a = 2, y|a = 5.5)$, instead of just picking an action observed by the employer, then the only equilibrium would be the pooling equilibrium.

- Signalling games (but not screening games) have at least one perfect Bayesian equilibrium in pure strategies. The games Education I, II, and III have either one or two pure strategy equilibria because they only allow two levels of education, but if education could take any value in the continuum $[0,1]$, instead of being either 0 or 1, then a continuum of equilibria would exist.

- If there are just two workers in the population, then the model is different depending on whether:

  (1)   Each is *High* ability with objective probability 0.5, so possibly both are *High* ability; or

  (2)   One of them is *High* and the other is *Low*, so only the subjective probability is 0.5.

  The outcomes are different because in case (2) if one worker credibly signals he is *High* ability, the employer knows the other one must be *Low* ability.

- Table 9.1 shows the relation between the order of play and the appropriate equilibrium concept.

Table 9.1: A summary of adverse selection models

Name	Who moves first	Equilibrium concepts
simple adverse selection	uninformed (principal, employer)	Nash, Wilson, Reactive
screening	uninformed (principal, employer)	Nash, Wilson, Reactive
signalling	informed (agent, worker)	perfect Bayesian, sequential, proper, etc.

N9.5   Two Signals: Underpricing of Stock

- Matthews & Moore (1987) use multiple signals to model warranty issue, and Hughes (1986) to model the role of the investment banker in new stock issues. See also the more general Engers (1987).

Part III

Applications

10 Bargaining

10.1 The Basic Bargaining Problem: Splitting a Pie

Part III of this book is designed to stretch your muscles by providing a large number of applications of the techniques from Parts I and II. The next four chapters lack the unity of the earlier chapters in two ways: they may be read in any order, and they hit only the highlights of each topic. Moreover, while bargaining (Chapter 10) and auctions (Chapter 11) are reasonably well-defined areas of research, Chapters 12 and 13 are selections of models from the sprawling literature of industrial organization. All that the chapters have in common is that they use new theory to answer old questions.

Bargaining theory attacks a kind of price determination ill-described by standard economic theory. In markets with many participants on one side or the other, standard theory does a good job of explaining prices. In competitive markets we find the intersection of the supply and demand curves, while in markets monopolized on one side we find the monopoly or monopsony output. The problem is when there are few players on each side. Early in one's study of economics, one learns that under bilateral monopoly (one buyer and one seller), standard economic theory is inapplicable because the traders must bargain. In the chapters on asymmetric information we would have come across this repeatedly except for our assumption that either the principal or the agent faced competition.

Sections 10.1 and 10.2 introduce the archetypal bargaining problem, Splitting a Pie, ever more complicated versions of which make up the rest of the chapter. Section 10.2, where we take the original rules of the game and apply the Nash bargaining solution, is our one dip into cooperative game theory in this book. Section 10.3 looks at bargaining as a finitely repeated process of offers and counteroffers, and Section 10.4 as an infinitely repeated process. Section 10.5 returns to a finite number of repetitions (two, in fact), but with incomplete information.

Splitting a Pie

Players

Smith and Brown.

Information

Imperfect, symmetric, complete, and certain.

Strategies

The players choose shares θ_s and θ_b of the pie simultaneously.

Payoffs

If $\theta_s + \theta_b \leq 1$, each player gets the fraction he chose: $\begin{cases} \pi_s = & \theta_s. \\ \pi_b = & \theta_b. \end{cases}$

If $\theta_s + \theta_b > 1$, then $\pi_s = \pi_b = 0$.

Splitting a Pie resembles Chicken from Section 3.3, except that it has a continuum of Nash equilibria: any strategy combination (θ_s, θ_b) such that $\theta_s + \theta_b = 1$ is Nash. The Nash concept is at its worst here, because the assumption that the equilibrium being played is common knowledge is very strong when there is a continuum of equilibria. The idea of the focal point (Section 1.5) might help to choose a single Nash equilibrium. The strategy space of Chicken is discrete and it has no symmetric pure strategy equilibrium, but the strategy space of Splitting a Pie is continuous, which permits a symmetric pure strategy equilibrium to exist. That equilibrium is the even split, (0.5,0.5), which is a focal point.

If the players moved in sequence, the game would have a tremendous first-mover advantage. If Brown moved first, the unique Nash outcome would be (0,1), although only weakly because Smith would be indifferent as to his action. Smith accepts the offer if he chooses θ_s to make $\theta_s + \theta_b = 1$, but if we added only an epsilonsworth of ill will to the model he would pick $\theta_s > 0$ and reject the offer.

For most of this chapter we will make an assumption which looks like the opposite of ill will: that a player accepts an offer if he is indifferent between accepting and rejecting. This is just a technical assumption, because all it does is prevent behavior from changing discontinuously when Brown offers zero instead of a small finite share. The problem arises only because of the continuous strategy space. If the pie could only be split into fractions using the discrete unit 0.01, an offer of 0.02 or 0.01 from Brown would unquestionably be accepted by Smith, so it seems reasonable that an offer of 0.00 would also be accepted. Assuming that an indifferent player accepts the offer solves a technical problem that could be solved in other, less elegant, ways.

The acceptance assumption must be used with care. It rules out mixed strategies, because, as was noted in Section 3.2, a player using a mixed strategy is always indifferent between two pure strategies. Moreover, the discreteness argument given above is not always valid, as will be seen in Section 10.5.

Another objection to the acceptance assumption, that the players should derive utility from hurting "unfair" players, is easier to answer. If Brown offers Smith a mere epsilon, everyday experience tells us that Smith will indignantly refuse the offer. Guth et al. (1982) have also found in experiments that people turn down bargaining offers they perceive as unfair. But if Smith's grudge is important, it can be explicitly incorporated into the payoffs, and if that is done, the acceptance assumption is still useful. There is still some offer that Smith is indifferent about accepting, but because of his grudge it is the particular share, greater than zero, where hurting Brown is no longer worth sacrificing Smith's own small share of the pie.

In many applications this version of Splitting a Pie is unacceptably simple, because if the two players find their fractions add to more than one they have a chance to change their minds. In labor negotiations, for example, if manager Brown makes an offer which union Smith rejects, they do not immediately forfeit the gains from combining capital and labor. They lose a week's production and make new offers. The recent trend in research has been to model such a sequence of offers, but before we do that, let us see how cooperative game theory deals with the original game.

10.2   The Nash Bargaining Solution

When game theory was young and games were static, a favorite approach was to decide upon some characteristics an equilibrium should have based on notions of fairness or efficiency, mathematicize the characteristics, and maybe add a few other axioms to make the equilibrium turn out neatly. Nash (1950a) did this for the bargaining problem in what is perhaps the best known application of cooperative game theory. Nash's objective was to pick axioms that would characterize the agreement the two players would anticipate making with each other. He used a game only a little more complicated than Splitting a Pie. In the Nash model, the two players can have different utilities if they do not come to an agreement, and the utility functions can be nonlinear in shares of the pie. Figures 10.1a and 10.1b compare the two games.

In Figure 10.1, the shaded region denoted by X is the set of feasible payoffs, which we will assume to be convex. The disagreement point is $\bar{U} = (\bar{U}_s, \bar{U}_b)$. The Nash bargaining solution, $U^* = (U_s^*, U_b^*)$, is a function of $\bar{U}$ and X. The axioms that generate the concept are

(1) Invariance

For any strictly increasing linear function F,

$$U^*[F(\bar{U}), F(X)] = F[U^*(\bar{U}, X)]. \tag{10.1}$$

This says that the solution is independent of the units in which we measure utility.

Figure 10.1   (a) Splitting a Pie; (b) the Nash Bargaining Game

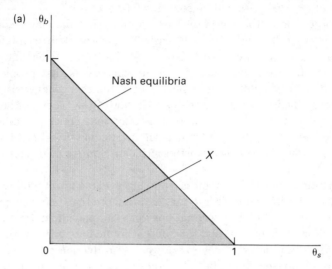

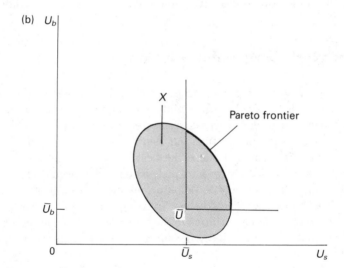

(2) Efficiency

The solution is Pareto-optimal, so that not both players can be made better off. In mathematical terms,

$$(U_s, U_b) > U^* \Rightarrow (U_s, U_b) \notin X. \tag{10.2}$$

(3) Independence of Irrelevant Alternatives.

If we drop some possible utility combinations from X, leaving the smaller set Y, then if U^* was not one of the dropped points, U^* does not change.

$$U^*(\bar{U}, X) \in Y \subseteq X \Rightarrow U^*(\bar{U}, Y) = U^*(\bar{U}, X). \tag{10.3}$$

(4) Anonymity (or Symmetry)

Switching the labels on players Smith and Brown does not affect the solution.

The axiom of Independence of Irrelevant Alternatives is the most debated of the four, but if I were to complain, it would be about the axiomatic approach, which depends heavily on the intuition behind the axioms. Everyday intuition says that the outcome should be efficient and symmetric, so that other outcomes can be ruled out *a priori*. But most of the games in the earlier chapters of this book turn out to have reasonable but inefficient outcomes, and games like Chicken have reasonable asymmetric outcomes.

Whatever their drawbacks, these axioms fully characterize the Nash solution. It can be proven that if U^* satisfies the four axioms above, then it is the unique strategy combination such that

$$U^* = \mathop{Argmax}_{U \in X, U \geq \bar{U}} \ (U_s - \bar{U}_s)(U_b - \bar{U}_b). \tag{10.4}$$

Splitting a Pie is a simple enough game that not all the axioms are needed to generate a solution. If we put the game in this context, however, problem (10.4) becomes

$$\mathop{Maximize}_{\theta_s, \theta_b \mid \theta_s + \theta_b \leq 1} \ (\theta_s - 0)(\theta_b - 0), \tag{10.5}$$

which generates the first order conditions

$$\theta_s - \lambda = 0, \ \ \text{and} \ \ \theta_b - \lambda = 0, \tag{10.6}$$

where λ is the Lagrange multiplier on the constraint. From (10.6) and the constraint, we obtain $\theta_s = \theta_b = 1/2$, the even split that we found as a focal point of the noncooperative game.

Although Nash's objective was simply to characterize the anticipations of the players, I perceive a heavier note of morality in cooperative game theory than in noncooperative game theory. Cooperative outcomes are neat, fair, beautiful, and efficient. In the next few sections we will look at noncooperative bargaining models that, while plausible, lack every one of those features. Cooperative game theory may be useful for ethical decisions, but its attractive features are inappropriate for most economic situations, and the spirit of the axiomatic approach is very different from the utility maximization of current economic theory.

10.3   Alternating Offers over Finite Time

In the past few years, use of the perfectness concept has made bargaining a hot research area which has provided a fertile testing range for the exotic refinements of perfect Bayesian equilibrium. The two-player, symmetric-information problem of the next two sections, however, is relatively simple. The actions are the same as in Splitting a Pie, but with many periods of offers and counteroffers. This means that strategies are no longer just actions, but rather rules for choosing actions based on the actions chosen in earlier periods.

Alternating Offers

Players

Smith and Brown.

Information

Perfect and certain.

Actions and Events

(1)   Smith makes an offer θ_1.
(1*) Brown accepts or rejects.
(2)   Brown makes an offer θ_2.
(2*) Smith accepts or rejects.

. . .

(T)   Smith offers θ_T.
(T*) Brown accepts or rejects.

Payoffs

The discount factor is $\delta \leq 1$.

If Smith's offer is accepted by Brown in round m,

$$\pi_s = \delta^m \theta_m,$$
$$\pi_b = \delta^m (1 - \theta_m).$$

(If Brown's offer is accepted, reverse the subscripts.)

If no offer is ever accepted, both payoffs equal 0.

When a game has many rounds we need to decide whether discounting is appropriate (see Section 4.5). Recall that if the discount rate is r then the discount factor is $\delta = 1/(1 + r)$, so without discounting, $r = 0$ and $\delta = 1$. Whether to discount depends on the particular model. Whether discounting is appropriate to the situation being modelled depends on whether delay should matter to the payoffs because the bargaining occurs over real time (Section 2.2) or the game might suddenly end

(Section 4.5). The game Alternating Offers can be interpreted in two ways, depending on whether it occurs over real time or not. If the players made all the offers and counteroffers between dawn and dusk of a single day, discounting would be inconsequential because essentially no time has passed. If each offer consumed a week of time, on the other hand, the delay before the pie was finally consumed would be important to the players and their payoffs should be discounted.

Even if discounting is appropriate to the situation, the modeller must ask himself whether the added complexity of discounting is accompanied by a significant change in the outcome of the game or a surprising demonstration that the outcome is unchanged. If discounting makes no difference, it should usually be dropped so that it does not obscure more important features of the model.

Consider first the game without discounting. There is a unique subgame perfect outcome—Smith gets the entire pie—which is supported by a number of different equilibria. In each equilibrium, Smith offers $\theta_s = 1$ each period, but they differ in when Brown accepts the offer. All of them are weak equilibria because Brown is indifferent between accepting and rejecting, but if we make the acceptance assumption, the only equilibrium is for Brown to accept in the first period.

Smith owes his success to his ability to make the last offer. When Smith claims the entire pie in the last period, Brown gains nothing by refusing to accept. What we have here is not really a first-mover advantage, but a last-mover advantage, a difference not apparent in the one-period model.

In the game with discounting, the total value of the pie is 1 in the first period, δ in the second, and so forth. In period T, if it is reached, Smith would offer 0 to Brown, keeping 1 for himself, and Brown would accept under our assumption on indifferent players. In period $T - 1$, Brown could offer Smith δ, keeping $(1 - \delta)$ for himself, and Smith would accept, although he could receive a greater share by refusing, because that greater share would arrive later and be discounted.

By the same token, in period $T - 2$, Smith would offer Brown $\delta(1 - \delta)$, keeping $1 - \delta(1 - \delta)$ for himself, and Brown would accept, since with a positive share Brown also prefers the game to end soon. In period $T - 3$, Brown would offer Smith $\delta[1 - \delta(1 - \delta)]$, keeping $1 - \delta[1 - \delta(1 - \delta)]$ for himself, and Smith would accept, again to prevent delay. Table 10.1 shows the progression of Smith's shares when $\delta = 0.9$.

Table 10.1 Alternating Offers over Finite Time

Round	Smith's share	Brown's share	Total value	Who offers
$T-3$	0.819	0.181	0.9^{T-4}	Brown
$T-2$	0.91	0.09	0.9^{T-3}	Smith
$T-1$	0.9	0.1	0.9^{T-2}	Brown
T	1	0.0	0.9^{T-1}	Smith

As we work back from the end, Smith always does a little better when he makes the offer than when Brown does, but if we consider just the class of periods in which Smith makes the offer, Smith's share falls. If we were to continue to work back for a large number of periods, Smith's offer in a period in which he makes the offer would approach $\frac{1}{1+\delta}$, which equals about 0.53 if $\delta = 0.9$. The reasoning behind that precise expression is given in the next section. In equilibrium, the very first offer would be accepted, since it is chosen precisely so that the other player can do no better by waiting.

10.4   Alternating Offers over Infinite Time

The Folk Theorem of Section 4.6 says that when discounting is small and a game is repeated an infinite number of times, there are many equilibrium outcomes. That does not apply to the bargaining game, however, because it is not a repeated game. It ends when one player accepts an offer, and only the accepted offer is relevant to the payoffs, not the earlier proposals. In particular, there are no out-of-equilibrium punishments such as enforce the Folk Theorem's outcomes.

Let players Smith and Brown have discount factors of δ_s and δ_b which are not necessarily equal, but are strictly positive. In the unique subgame perfect outcome for the infinite-period bargaining game, Smith's share is

$$\theta_s = \frac{1 - \delta_b}{1 - \delta_s \delta_b},\tag{10.7}$$

which if $\delta_s = \delta_b = \delta$ is equivalent to

$$\theta_s = \frac{1}{1 + \delta}.\tag{10.8}$$

If the discount rate is large, Smith gets most of the pie: a 1,000 percent discount rate ($r = 10$) makes $\delta = 0.091$ and $\theta_s = 0.92$ (rounded), which makes sense, since under such extreme discounting the second period hardly matters and we are almost back to the simple game of Section 10.1. At the other extreme, if r is small then the pie is almost evenly split: if $r = 0.01$, then $\delta \approx 0.99$ and $\theta_s \approx 0.503$.

It is crucial that the discount rate be strictly greater than zero, even if only by a little. Otherwise, the game has the same continuum of perfect equilibria as in Section 10.1. Since nothing changes over time, there is no incentive to come to an early agreement. When discount rates are equal, the intuition behind the result is that since a player's cost of delay is proportional to his share of the pie, if Smith were to offer a grossly unequal split such as (0.7, 0.3), then Brown, with less to lose by delay, would reject the offer. Only if the split were almost even would Brown accept, as we will now prove.

Proposition 10.1 (Rubinstein [1982])

In the discounted infinite game, the unique perfect equilibrium outcome is

$$\theta_s = \frac{1 - \delta_b}{1 - \delta_s \delta_b},$$

where Smith is the first mover.

Proof

We found that in the T-period game Smith gets a larger share in a period in which he makes the offer. Denote by M the maximum nondiscounted share, taken over all the perfect equilibria that might exist, that Smith can obtain in a period in which he makes the offer. Consider the game starting at t. Smith is sure to get no more than M, as noted in Table 10.2. (Brown would thus get $1 - M$, but that is not relevant to the proof.)

Table 10.2 Alternating Offers over Infinite Time

Round	Smith's share	Brown's share	Who offers
$t-2$	$1 - \delta_b(1 - \delta_s M)$		Smith
$t-1$		$1 - \delta_s M$	Brown
t	M		Smith

The trick is to find a way besides M to represent the maximum Smith can obtain. Consider the offer made by Brown at $t-1$. Smith will accept any offer which gives him more than the discounted value of M received one period later, so Brown can make an offer of $\delta_s M$ to Smith, retaining $1-\delta_s M$ for himself. At $t-2$, Smith knows that Brown will turn down any offer less than the discounted value of the minimum Brown can look forward to receiving at $t-1$. Smith therefore cannot offer any less than $\delta_b (1 - \delta_s M)$, or retain for himself any more than $1 - \delta_b (1 - \delta_s M)$ at $t-2$.

Now we have two expressions for "the maximum Smith can receive," which we can set equal to each other.

$$M = 1 - \delta_b (1 - \delta_s M). \tag{10.9}$$

Solving equation (10.9) for M, we obtain

$$M = \frac{1 - \delta_b}{1 - \delta_s \delta_b}. \tag{10.10}$$

We can repeat the argument using m, the minimum of Smith's share. If Smith can expect at least m at t, then Brown cannot receive more than $1 - \delta_s m$ at $t-1$. At $t-2$ Smith knows that if he offers Brown the discounted value of that amount, Brown will accept, so Smith can guarantee himself $1 - \delta_b (1 - \delta_s m)$, which is the same as the expression we found for M. The smallest perfect equilibrium share that Smith can receive is the same as the greatest, so the equilibrium outcome must be unique.

No Discounting, but a Fixed Bargaining Cost

There are two ways to model bargaining costs per period: proportional to the remaining value of the pie as we did above, or as fixed costs each period, which is analyzed next (again following Rubinstein [1982]). To understand the difference, think of labor negotiations during a construction project. If a strike slows down completion, there are two kinds of losses. One is the loss from delay in renting or selling the new building, a loss proportional to its value. The other is the loss from late-completion penalties in the contract that might take the form of a fixed penalty each week. The two kinds of costs have very different effects on the bargaining process.

To represent this second kind of cost, assume that there is no discounting, but whenever Smith or Brown makes an offer, he incurs the cost c_s or c_b. In every subgame perfect equilibrium, Smith makes an offer and Brown accepts, but there are three possible cases.

(1) Delay Costs are Equal: $c_s = c_b = c$

The Nash indeterminacy of Section 10.1 remains almost as bad; any fraction such that each player gets at least c is supported by some perfect equilibrium.

(2) Delay Hurts Brown More: $c_s < c_b$

Smith gets the entire pie. Brown has more to lose than Smith by delaying, and delay does not change the situation except by diminishing the wealth of the players. The game is stationary, because it looks the same to both players no matter how many periods have already elapsed. If in any period t Brown offered Smith x, then in period $(t-1)$ Smith could offer Brown $(1-x-c_b)$, keeping $(x+c_b)$ for himself. In period $(t-2)$, Brown would offer Smith $(x+c_b-c_s)$, keeping $(1-x-c_b+c_s)$ for himself, and in periods $(t-4)$ and $(t-6)$ Brown would offer $(1-x-2c_b+2c_s)$ and $(1-x-3c_b+3c_s)$. As we work backwards, Smith's advantage rises to $\beta(c_b-c_s)$ for an arbitrarily large integer β. Looking ahead from the start of the game, Brown is willing to give up and accept zero.

(3) Delay Hurts Smith More: $c_s > c_b$

Smith gets a share worth c_b and Brown gets $(1-c_b)$. The cost c_b is a lower bound for the share of Smith, the first mover, because if Smith knows Brown will offer $(0,1)$ in the second period, Smith can offer $(c_b, 1-c_b)$ in the first period and Brown will accept.

10.5   Incomplete Information

Instant agreement has characterized even the multiperiod games of complete information discussed so far. Under incomplete information, knowledge can change over the course of the game and bargaining can last more than one period in equilibrium, a result that might be called inefficient, but is certainly realistic. Models with complete information have difficulty explaining such things as strikes or wars, but if over time an uninformed player can learn the type of the informed player by observing what offers are made or rejected, such unfortunate outcomes can arise. The literature on bargaining under incomplete information is recent, but vast. For this section, I have chosen to use a model based on the first part of Fudenberg &

Tirole (1983), but it is only a particular example of how one could construct such a model, and not a good indicator of what results are to be expected from bargaining.

Bargaining with Incomplete Information

Players

A seller, and a buyer called Buyer(100) or Buyer(150) depending on his type.

Information

Asymmetric, incomplete, and certain.

Actions and Events

(0)   Nature picks the buyer's type, his valuation of the object being sold, which is $b = 100$ with probability γ and $b = 150$ with probability $(1 - \gamma)$.
(1)   The seller offers price p_1.
(2)   The buyer accepts or rejects p_1 (acceptance ends the game).
(3)   The seller offers a second price p_2.
(4)   The buyer accepts or rejects p_2.

Payoffs

$$\pi_{seller} = \begin{cases} p_1 & \text{if } p_1 \text{ is accepted;} \\ \delta p_2 & \text{if } p_2 \text{ is accepted;} \\ 0 & \text{if no offer is accepted.} \end{cases}$$

$$\pi_{buyer} = \begin{cases} b - p_1 & \text{if } p_1 \text{ is accepted;} \\ \delta(b - p_2) & \text{if } p_2 \text{ is accepted;} \\ 0 & \text{if no offer is accepted.} \end{cases}$$

If p_2 is accepted, the buyer's payoff is $\delta(b - p_2)$, rather than $\delta b - p_2$, because the present value of cash paid in the second period is less than that of cash paid in the first period. Consumption in the second period provides less pleasure, but payment provides less pain. Let us set $\delta = 0.9$ for the numerical computations.

As described, the game describes an encounter between one buyer and one seller. Another interpretation is that the game is between a continuum of buyers and one seller, in which case the analysis works out the same way, but the verbal translation is different. As noted in Section 3.2, models with a continuum of buyers are sometimes easier to understand, because the two types of buyers can be interpeted as different fractions of the population, and mixed strategies can be interpreted as fractions of the total population using different pure strategies.

If this were a game of complete information, then in equilibrium the seller, who has the last-mover advantage, would choose $p_1 = 100$ or $p_1 = 150$ and the buyer would accept. If the buyer did not accept, the seller would offer the same price again, and the buyer would accept that second offer. With incomplete information, however, the probability that the buyer is a Buyer(100) determines whether the equilibrium is pooling at a low price, or separating at a high price with some offers rejected.

The incomplete information game is a screening model, like those we saw in Chapter 9, because the uninformed player moves first. Screening models with a continuum of types do not have pooling equilibria, but the assumption that there are only two types will permit one in this model. If the buyer made the offers, instead of the seller, this would be a signalling game, in which out-of-equilibrium beliefs would be more important. This game, however, is not a pure screening model, because it lasts two periods and the uninformed player does have a chance to make his second move after the informed player has moved and possibly thereby revealed information.

A Pooling Equilibrium with Many Low-Valuation Buyers: $\gamma = 0.5$

We will start by assuming that $\gamma = 0.5$, so the probability is fairly high that the buyer has the valuation 100. We will weaken the assumption that an indifferent player always accepts an offer, assuming not that the buyer always accepts if he is indifferent but that he accepts if he is indifferent in the second period. This modified assumption is harmless because the seller could always lower his second-period price by ε to make acceptance a dominant strategy, but it will be important to allow mixed strategies in the first period.

Pooling Equilibrium

$p_1 = 100$.
Buyer(100) accepts $p_1 \leq 100$.
Buyer(150) accepts $p_1 \leq 105$.
$p_2 = 100$.
Buyer(100) accepts $p_2 \leq 100$.
Buyer(150) accepts $p_2 \leq 150$.
Seller's beliefs out of equilibrium: If a buyer rejects $p_1 = 100$, then he is
     Buyer(100) with probability γ (passive conjectures).
Outcome: $p_1 = 100$, accepted by the buyer.

Let us test that this is a perfect Bayesian equilibrium. As always, we work back from the end. Both types of buyers have a dominant strategy for the last move: accept any offer below b. Given the parameters, the seller should not raise p_2 above 100, because with probability $\gamma = 0.5$ he would lose a profit of 100 and his potential revenue is no greater than 150.

Buyer(150), although looking ahead to $p_2 = 100$, is willing to pay more than 100 in period one because of discounting. His payoff is the same from accepting $p_1 = 105$ as from accepting $p_2 = 100$, because his nominal surplus of 50 from accepting the lower price is discounted to a utility value of 45. Buyer(100), however, is never willing to pay more than 100, and discounting is irrelevant because he has no surplus to look forward to anyway.

The seller knows that even if $p_1 = 105$ he will still sell to Buyer(150), but if he tries that and finds no buyer in the first period, he has delayed receipt of his payment, which is discounted. Since $100 > 97.5(= [1 - \gamma] \cdot [105] + \gamma \cdot \delta \cdot [100])$, the seller prefers the safe present price of 100 to the alternative of a gamble between a present 105 and a future 100.

Out-of-equilibrium beliefs are specified for the equilibrium, but they do not really matter. Whatever inference the seller may draw if the buyer refuses $p_1 = 100$, the inference never induces the buyer to change his actions. It might be that the seller believes a refusal indicates that the buyer's value is 150, so $p_2 = 150$, but that does not change the buyer's incentive to accept $p_1 = 100$.

A Separating Equilibrium with Few Low-Valuation Buyers: $\gamma = 0.05$

If the proportion of low-valuation buyers is as small as $\gamma = 0.05$, the equilibrium is separating and in mixed strategies.

Mixed Strategy Separating Equilibrium

$p_1 = 150$.
Buyer(100) accepts $p_1 \leq 100$.
Buyer (150) accepts p_1 with probability $m(p_1)$, where

$$
\begin{cases}
m = & 1 \text{ if } p_1 \leq 105. \\
m = & \alpha \text{ if } 105 < p_1 \leq 150, \\
& (\text{where } 0 \leq \alpha \leq 0.89) \\
m = & 0 \text{ if } p_1 > 150.
\end{cases}
$$

$p_2 = 150$ if the seller believes that he faces a Buyer(100) with probability less than $1/3$.
Otherwise, $p_2 = 100$.
Buyer(100) accepts $p_2 \leq 100$.
Buyer(150) accepts $p_2 \leq 150$.
Outcome: $p_1 = 150$ and is sometimes accepted by Buyer(150). $p_2 = 150$ and is accepted by Buyer(150). Buyer(100) never accepts an offer.

The observed outcome is simple—the price always stays at 150, some buyers accept in each period, and some buyers never accept—but this is quite a complicated equilibrium. As we will see, it is not even fully determined, because the mixing probability α can take any of a continuum of values.

The strategies in the second-period game are simple enough. In the second period, the buyer accepts if the price is less than his valuation and the seller trades off a safe 100 against a gamble between 0 and 150. He is indifferent between them if

$$100 = 0 \cdot Prob(\text{Buyer}(100)) + 150 \cdot [1 - Prob(\text{Buyer}(100))], \qquad (10.11)$$

which yields a critical value of $Prob(\text{Buyer}(100)) = 1/3$. If neither type of buyer accepted first-period offers, the second-period belief would be $Prob(\text{Buyer}(100)) = \gamma$, which we assumed to be 0.05, so the second-period price would be 150.

The first-period strategies are more complicated. The Buyer(150)'s first-period strategy is not the pure strategy of accepting the offer $p_1 = 150$, because if he always accepted in the first period the seller would lower the price in the second period, knowing that a buyer who rejected the first-period offer must be a Buyer(100). Anticipating the price drop, the Buyer(150) would refuse $p_1 = 150$, which contradicts the reason for the drop.

In equilibrium, it must be that after a refusal in the first period the seller puts a high enough probability on Buyer(150) that he decides to keep the price high in the second period. For the seller to want to keep $p_2 = 150$, the probability that a buyer who refused the first-period offer is a Buyer(150) must be at least 2/3 from equation (10.11). If both p_1 and p_2 equal 150, the buyer is indifferent as to when he accepts, so he is willing to follow a mixed strategy. We can calculate the mixing probability $m(150)$ by finding the value that makes the seller willing to keep the price at 150 in the second period. The probability that a buyer is a Buyer(150) is equal to $1 - \gamma$ in the first period, but a Buyer(150) only rejects the first-period offer with probability $1 - m(150)$. Therefore, in the second period

$$Prob(\text{Buyer}(100)) = \frac{\gamma}{\gamma + [1 - m(150)][1 - \gamma]}. \tag{10.12}$$

Plugging in $\gamma = 0.05$ and $Prob(\text{Buyer}(100)) = 1/3$, equation (10.12) can be solved to yield $m(150) = 0.89$ (rounded). The calculation has ensured that if the Buyer(150) accepts the first offer with probability 0.89, the probability that a second-period buyer has valuation 100 is 1/3. The value $\alpha = 0.89$ is the maximum equilibrium probability that a Buyer(150) refuses to buy in the first period, but a smaller value for α would support an equilibrium *a fortiori* since the probability that a refuser had valuation 150 would be even greater than 2/3.

There is a continuum of equilibria, which differ in their values for α. Two values are focal points, 0 and 0.89. The value of 0 is a pure strategy, and has the attraction of simplicity. The value of 0.89 is Pareto-efficient, because it is the largest equilibrium probability of the buyer accepting immediately, which avoids the lost utility from delay. The qualitative difference between the two equilibria is that in the pure strategy equilibrium no buyers accept the offer in the first period.

Out-of-Equilibrium Behavior

Although the description above was titled "Mixed Strategy Separating Equilibrium," it describes only part of the equilibrium. A full description would specify each player's actions at each node of the game tree given the past history of the game, including out-of-equilibrium paths that start with deviations such as $p_1 = 139$ or $p_1 = 140$. As we will see in describing the path starting with the seller deviation $p_1 = 140$, there would be an overwhelming amount of possible out-of-equilibrium behavior.

Consider what happens after the seller offers a price of 140 in the first period. For the same reasons as we described for $p_1 = 150$, the equilibrium cannot be in pure strategies. The equilibrium strategies in the out-of-equilibrium subgame are for the Buyer(150) to mix between accepting and rejecting, and for the seller to mix between $p_2 = 100$ and $p_2 = 150$. Here, unlike on the equilibrium path, the seller must also mix, because otherwise the buyer would strongly prefer to accept 140 rather than wait for 150. The seller is willing to mix only if he believes that there is exactly a 1/3 probability that the buyer is a Buyer(100), so the buyer's strategy is $m(150) = 0.89$, as before. Denote the seller's mixing probability of $p_2 = 100$ by μ. It must take a value that makes the buyer indifferent between accepting and rejecting, so

$$150 - p_1 = \mu(150 - 100) + (1 - \mu) \cdot 0, \tag{10.13}$$

which solves to $\mu = 3 - p_1/50$, or $\mu = 0.2$ for $p_1 = 140$.

These calculations have been intricate and hard, but they are good examples of the kind of care that must be taken with out-of-equilibrium behavior. The most important other lesson of this model is that bargaining can lead to inefficiency. In the separating equilibrium, some of the Buyer(150)s delay their transactions until the second period, which is inefficient since the payoffs are discounted. Moreover, the Buyer(100)s never buy at all, and the potential gains from trade are lost.

Recommended Reading

Nash, John (1950a) "The Bargaining Problem" *Econometrica*. January 1950. 18, 2: 155–62.

Rubinstein, Ariel (1982) "Perfect Equilibrium in a Bargaining Model" *Econometrica*. January 1982. 50, 1: 97–109.

Sutton, John (1986) "Non-Cooperative Bargaining Theory: An Introduction" *Review of Economic Studies*. October 1986. 53(5), 176: 709–24.

Problem 10

Bilateral Monopoly

One seller of steel faces one buyer. The marginal cost of steel production is $MC = 10$, and the marginal benefit to the buyer of an extra unit of steel is $MB = 210 - 2q$. The two players must agree on a price p. Once they have decided on that price, the quantity traded is the minimum of what the Buyer is willing to buy and the Seller is willing to sell.

(1) What are the payoffs as functions of p and q?

(2) Show the set of Pareto-optimal prices, and the resulting quantities, in a diagram in (p, q)-space (only one of these points maximizes the sum of the surpluses).

(3) Taking the Pareto-optimal prices from part (2), show the payoffs for the quantities 50, 75, and 100, in a diagram in (π_s, π_b)-space.

(4) If the Seller offers a price, and the Buyer can do no more than pick a quantity to buy at that price, what price is offered and what payoffs result? What is the price and the payoffs if the Buyer gets to make the price offer?

(5) Considering just the Pareto-optimal (p, q) combinations, find π_s as a function of π_b and solve the problem $\{Max \ \pi_s\pi_b\}$ to find the Nash solution for (π_b, π_s) and (p, q).

Notes

N10.2   The Nash Bargaining Solution

- See Binmore, Rubinstein, & Wolinsky (1986) for a comparison of the cooperative and noncooperative approaches to bargaining. For overviews of cooperative game theory see Luce & Raiffa (1957) and Shubik (1982).
- While the Nash bargaining solution can be generalized to n players (see Harsanyi [1977], p. 196), the possibility of interaction between coalitions of players introduces new complexities. Solutions such as the Shapley value try to account for these complexities.

   The **Shapley value** satisfies the properties of Invariance, Anonymity, Efficiency, and Linearity in the variables from which it is calculated. Let S_i denote a **coalition** containing player i; that is, a group of players including i that makes a sharing agreement. Let $v(S_i)$ denote the sum of the values of the utilities of the players in coalition S_i, and $v(S_i - \{i\})$ denote the sum of the utilities in the coalition created by removing i from S_i. Finally, let $c(s)$ be the number of coalitions of size s containing player i. The Shapley value for player i is then

$$\phi_i = \frac{1}{n} \sum_{s=1}^{n} \frac{1}{c(s)} \sum_{S_i \ of \ size \ s} [v(S_i) - v(S_i - \{i\})]. \tag{10.14}$$

   The motivation for the Shapley value is that player i receives the average of his marginal contributions to different coalitions that might form. Gul (forth) has provided a noncooperative interpretation.

N10.4   Alternating Offers over Infinite Time

- The proof of Proposition 10.1 is not from the original Rubinstein (1982), but is adapted from Shaked & Sutton (1984). The maximum rather than the supremum can be used because of the assumption that indifferent players always accept offers.
- In extending Alternating Offers to three players, there is no obviously best way to specify how players make and accept offers. Haller (1986) shows that for at least one specification, the outcome is not similar to the Rubinstein (1982) outcome, but rather is a return to the indeterminacy of the game without discounting.

N10.5   Incomplete Information

- There are quite a few models of bargaining under asymmetric information. Fudenberg & Tirole (1983) uses a two-period model with two types of buyers and two types of sellers. Sobel & Takahashi (1983) builds a model with either T or infinite periods, a continuum of types of buyers, and one type of seller. Cramton (1984) uses an infinite number of periods, a continuum of types of buyers, and a continuum of types of sellers. Rubinstein (1985a) uses an infinite number of periods, two types of buyers, and one type of seller, but the types of buyers differ not in their valuations, but in their discount rates. Rubinstein (1985b) puts emphasis on the choice of out-of-equilibrium conjectures. W. Samuelson (1984) looks at the case where one bargainer knows the size of the pie better than the other bargainer. Perry (1986) uses a model with fixed bargaining costs and asymmetric information in which each bargainer makes an offer in turn, rather than one offering and the other accepting or rejecting.
- The asymmetric information model in Section 10.5 has **one-sided** asymmetry in the information: only the buyer's type is private information. Fudenberg & Tirole (1983) and others have also built models with **two-sided** asymmetry, in which buyers' and sellers' types are both private information. In such models a multiplicity of perfect Bayesian equilibria can be supported for a given set of parameter values. Out-of-equilibrium beliefs become quite important, and provided much of the motivation for the exotic refinements mentioned in Section 5.2.

- There is no separating equilibrium if instead of discounting, the asymmetric information model has fixed-size per-period bargaining costs, unless the bargaining cost is higher for the high-valuation buyer than for the low-valuation. If, for example, there is no discounting, but a cost of c is incurred each period that bargaining continues, no separating equilibrium is possible. That is the typical signalling result. In a separating equilibrium the buyer tries to signal a low valuation by holding out, which fails unless it really is less costly for a low-valuation buyer to hold out. See Perry (1986) for a model with fixed bargaining costs which ends after one round of bargaining.

11 Auctions

11.1 Introduction

Because auctions are stylized markets with well-defined rules, modelling them with game theory is particularly appropriate. Moreover, several of the motivations behind auctions are similar to the motivations behind the asymmetric information contracts of Part II of this book. Besides the mundane reasons such as speed of sale that make auctions important, auctions are useful for a variety of informational purposes. Often the buyers know more than the seller about the value of what is being sold, and the seller, not wanting to suggest a price first, uses an auction as a way to extract information. Art auctions are a good example, because the value of a painting depends on the buyer's tastes, which are known only to himself.

Auctions are also useful for agency reasons, because they hinder dishonest dealing between the seller's agent and the buyer. If the mayor were free to offer a price for building the new city hall and accept the first contractor who showed up, the lucky contractor would probably be the one who made the biggest political contribution. If the contract is put up for auction, cheating the public is more costly, and the difficulty of rigging the bids may outweigh the political gain.

We will spend most of this chapter on the effectiveness of different kinds of auction rules in extracting surplus from buyers, which requires considering the strategies with which they respond to the rules. Section 11.2 classifies auctions based on the relationships between different buyers' valuations of what is being auctioned, and explains the possible auction rules and the bidding stategies optimal for each rule. Section 11.3 compares the outcomes under the various rules. Section 11.4 discusses optimal strategies under common-value information, which can lead bidders into the "winner's curse" if they are not careful. Section 11.5 is about information asymmetry in common-value auctions.

11.2 Auction Classification and Private-Value Strategies

Private, Common, and Correlated Values

Auctions differ enough for an intricate classification to be useful. One way to

245

classify auctions is based on differences in the values buyers put on what is being auctioned. We will call the dollar value of the utility that player i receives from an object its **value** to him, V_i, and we will call his *estimate* of its value his **valuation**, $\hat{V}_i$.

In a **private-value** auction, each player knows his value with certainty, although he may still have to estimate the values of the other players. An example is the sale of antique chairs to people who will not resell them. The values need not be independent. If it were common knowledge that the values were either all high or all low, for example, that could affect the choice of auction rules by the seller and the estimates made by buyers of each others' values. A player's value equals his valuation in a private-value auction.

If an auction is to be private-value, it cannot be followed by costless resale of the object. If there were resale, a bidder's valuation would depend on the price at which he could resell, which would depend on the other players' valuations.

What is special about a private-value auction is that a player cannot extract any information about his own value from the valuations of the other players. Knowing all the other bids in advance would not change his valuation, although it might well change his strategy. The outcomes would be similar even if he had to estimate his own value, so long as the behavior of other players did not help him to estimate it, so this kind of auction could just as well be called "private-valuation."

In a **common-value** auction, the players have identical values, but each player forms his own valuation by estimating with his private information. An example is bidding for US Treasury bills. A player's valuation in that auction would change if he could sneak a look at the other players' valuations, because they are all trying to estimate the same true value.

The **correlated-value** auction is a general category which includes the common-value auction as an extreme case. In this auction, the valuations of the different players are correlated, but their values may differ. Practically every auction we see is correlated-value, but, as always in modelling, we must trade off descriptive accuracy against simplicity, and private-value vs. common-value is an appropriate simplification.

Auction Rules and Private-Value Strategies

Auctions have as many different sets of rules as poker games do. Cassady's 1967 book describes a myriad of rules, but here I will just list the main varieties and describe the equilibrium private-value strategies. In teaching this material, I ask each student to pick a valuation between 80 and 100, after which we conduct auctions of the various kinds. I advise the reader to try this. Pick two valuations and try out sample strategy combinations for the different auctions as they are described. Even though the values are private, it will immediately become clear that the best-response bids still depend on the strategies the bidder thinks other players have adopted.

The types of auctions to be described are:

(1)   English.
(2)   First-price sealed-bid.

(3)   Second-price sealed-bid.
(4)   Dutch.

(1) English (first-price open-cry)

Rules.   Each bidder is free to revise his bid upwards. When no bidder wishes to revise his bid further, the highest bidder wins the object and pays his bid.

Strategies.   A player's strategy is his series of bids as a function of (1) his value, (2) his prior estimate of other players' valuations, and (3) the past bids of all the players. His bid can therefore be updated as his information set changes.

Payoffs.   The winner's payoff is his value minus his highest bid.

A player's dominant strategy in a private-value English auction is to keep bidding some small amount epsilon more than the previous high bid until he reaches his valuation, and then to stop. This is optimal because he always wants to buy the object if the price is less than its value to him, but he wants to pay the lowest price possible. All bidding ends when the price reaches the valuation of the player with the second-highest valuation. The optimal strategy is independent of risk neutrality if players know their own values with certainty rather than estimating them, although risk-averse players who must estimate their values should be more conservative in bidding.

In correlated-value open-cry auctions, the bidding procedure is important. The most common procedures are (1) for the auctioneer to raise prices at a constant rate, (2) for him to raise prices at whatever rate he thinks appropriate, and (3) for the bidders to raise prices as I specified in the rules above. A fourth procedure is often the easiest to model: the **open exit** auction, in which the price rises continuously and players must publicly announce that they are dropping out (and cannot reenter) when the price becomes unacceptably high. In an open exit auction the players have more evidence available about each others' valuations than if they could drop out secretly.

(2) First-price sealed-bid

Rules.   Each bidder submits one bid, in ignorance of the other bids. The highest bidder pays his bid and wins the object.

Strategies.   A player's strategy is his bid as a function of his value and his prior beliefs about other players' valuations.

Payoffs.   The winner's payoff is his value minus his bid.

Suppose Smith's value is 100. If he bid 100 and won when the second bid was 80, he would wish that he had bid only $80 + \varepsilon$. If it is common knowledge that the second-highest value is 80, Smith's bid should be $80 + \epsilon$. If he is not sure about the second-highest value, the problem is difficult and no general solution has been discovered. The tradeoff is between bidding high—thus winning more often—and bidding low—thus benefiting more if the bid wins. The optimal strategy, whatever it may be, depends on risk neutrality and beliefs about the other bidders, so the equilibrium is less robust than the equilibria of English and second-price auctions.

Nash equilibria can be found for more specific first-price auctions. Suppose that there are N risk-neutral bidders, and that Nature assigns them values independently using a uniform density from 0 to some amount $\bar{v}$. Denote player i's value by v_i, and let us consider the strategy for player 1. If some other player has a higher value, then in a symmetric equilibrium player 1 is going to lose the auction anyway, so we can ignore that possibility in finding his optimal bid. Player 1's equilibrium strategy is to bid epsilon above his expectation of the second-highest value, conditional on his bid being the highest (i.e., assuming that no other bidder has a value over v_1).

If we assume that v_1 is the highest value, the probability density for player 2's value, which is uniformly distributed between 0 and v_1, equals v is $1/v_1$, and the probability that v_2 is less than or equal to v is v/v_1. The probability that v_2 lies in $[v, v + dv]$ and is the second-highest value is

$$Prob(v_2 = v)Prob(v_3 \leq v)Prob(v_4 \leq v) \cdots Prob(v_N \leq v)dv, \tag{11.1}$$

which equals

$$\left(\frac{1}{v_1}\right)\left(\frac{v}{v_1}\right)^{N-2} dv. \tag{11.2}$$

Since there are $N - 1$ players besides player 1, the probability that one of them has the value v, and v is the second-highest is $N - 1$ times expression (11.2). The expectation of v is the integral of v over the range 0 to v_1,

$$\begin{aligned} E(v) &= \int_0^{v_1} v(N-1)(1/v_1)[v/v_1]^{N-2}dv \\ &= (N-1)\frac{1}{v_1^{N-1}} \int_0^{v_1} v^{N-1}dv \\ &= \frac{(N-1)v_1}{N}. \end{aligned} \tag{11.3}$$

Thus we find that player 1 ought to bid a fraction $\frac{N-1}{N}$ of his own value, plus epsilon.

The previous example is an elegant result, but it is not a general rule. Suppose Smith knows that Brown's value is 0 or 100 with equal probability, and Smith's value of 400 is known by both players. Brown bids either 0 or 100 in equilibrium, and Smith always bids $(100 + \epsilon)$, because his value is so high that winning is more important than paying a low price.

If Smith's value were 102 instead of 400, the equilibrium would be much different. Smith would use a mixed strategy, and while Brown would still offer 0 if his value were 0, if his value were 100 he would use a mixed strategy too. No pure strategy could be part of a Nash equilibrium, because if Smith always bid a value $x < 100$,

Brown would always bid $x + \varepsilon$, in which case Smith would deviate to $x + 2\varepsilon$, and if Smith bid $x \geq 100$ he would be paying 100 more than necessary half the time.

(3) Second-price sealed-bid

Rules. Each bidder submits one bid, in ignorance of the other bids. The bids are opened, and the highest bidder pays the amount of the second-highest bid and wins the object.

Strategies. A player's strategy is his bid as a function of his value and his prior belief about other players' valuations.

Payoffs. The winner's payoff is his value minus the second-highest bid that was made.

Second-price auctions are similar to English auctions. I have never heard of them actually being carried out, but they are useful for modelling. Bidding one's valuation is the dominant strategy: a player who bids less is more likely to lose the auction, but pays the same price if he does win. The structure of the payoffs is reminiscent of the Groves mechanism of Section 7.7, because in both games a player's strategy affects some major event (who wins the auction or whether the project is undertaken), but his strategy affects his own payoff only via that event. In the auction's equilibrium, each player bids his value and the winner ends up paying the second-highest value. If players know their own values, the outcome does not depend on risk neutrality.

(4) Dutch (Descending)

Rules. The seller announces a bid, which he continuously lowers until some buyer stops him and takes the object at that price.

Strategies. A player's strategy is when to stop the bidding as a function of his valuation and his prior beliefs as to other players' valuations.

Payoffs. The winner's payoff is his value minus his bid.

The Dutch auction is **strategically equivalent** to the first-price sealed-bid auction, which means that there is a one-to-one mapping between the strategy sets and the equilibria of the two games. The reason for the strategic equivalence is that no relevant information is disclosed in the course of the auction, only at the end, when it is too late to change anybody's behavior. In the first-price auction a player's bid is irrelevant unless it is the highest, and in the Dutch auction a player's stopping price is also irrelevant unless it is the highest. The equilibrium price is calculated the same way for both auctions.

Dutch auctions are actually used. One example is the Ontario tobacco auction, which uses a clock four feet in diameter marked with quarter cent gradations. Each of six or so buyers has a stop button. The clock hand drops a quarter cent at a time, and the stop buttons are registered so that ties cannot occur (tobacco buyers need reflexes like race-car drivers). The farmers who are selling their tobacco watch from an adjoining room and can later reject the bids if they feel they are too low (a form of reserve price). 2,500,000 lb. a day can be sold using the clock (Cassady [1967] p. 200).

Dutch auctions are common in less obvious forms. Filene's is one of the biggest stores in Boston, and Filene's Basement is its most famous department. In the basement are a variety of marked-down items formerly in the regular store, each with a price and date attached. The price customers pay at the register is the price on the tag minus a discount which depends on how long ago the item was dated. As time passes and the item remains unsold, the discount rises from 10 to 50 to 70 percent. The idea of predictable time discounting has also recently been used by bookstores ("Waldenbooks to Cut Some Book Prices in Stages in Test of New Selling Tactic," *Wall Street Journal*, 29 March 1988, p. 34).

11.3   Comparing Auction Rules

Equivalence Theorems

When one mentions auction theory to an economic theorist, the first thing that springs to his mind is the idea that various kinds of auctions are the same in some sense. Milgrom & Weber (1982) give a good summary of how and why this is true. Regardless of the information structure, the Dutch and first-price sealed-bid auctions are the same in the sense that the strategies and the payoffs associated with the strategies are the same. That equivalence does not depend on risk neutrality, but let us assume that all players are risk neutral for the next few paragraphs.

In private independent-value auctions, the second-price sealed-bid and English auctions are the same in the sense that the bidder who values the object most highly wins and pays the valuation of the second-highest valuer, but the strategies are different in the two auctions. In all four kinds of private independent-value auctions discussed, the seller's expected price is the same. This fact is the biggest result in auction theory: the **revenue equivalence theorem** (Vickrey [1961]).

The revenue equivalence theorem does not imply that in every realization of the game all four auction rules yield the same price, only that the expected price is the same. The difference arises because in the Dutch and first-price sealed-bid auctions, the winning bidder has estimated the value of the second-highest bidder, and that estimate, while correct on average, is above or below the true value in particular realizations. The variance of the price is higher in those auctions because of the additional estimation, which means that a risk-averse seller should use the English or second-price auction.

Whether the auction is private-value or not, the Dutch and first-price sealed-bid auctions are strategically equivalent. If the auction is correlated-value and there are three or more bidders, the open exit English auction leads to greater revenue than the second-price sealed-bid auction, and both yield greater revenue than the first-price sealed-bid auction (Milgrom & Weber [1982]). If there are just two bidders, however, the open exit English auction is no better than the second-price sealed-bid auction, because the open exit feature—knowing when non-bidding players drop out—is irrelevant.

A question of less practical interest is whether an auction form is Pareto-optimal; that is, does the auctioned object end up in the hands of whoever values it most? In a common-value auction this is not an interesting question, because all bidders value the object equally. In a private-value auction, all of the auction forms—first-price, second-price, Dutch, and English—are Pareto-optimal. They are also optimal in a correlated-value auction if all players draw their information from the same distribution and the equilibrium is in symmetric strategies.

Auctions with risk-averse bidders are difficult to analyze. One known fact is that in a private-value auction the first-price sealed-bid auction yields a greater expected revenue than the English or second-price auctions. That is because by increasing his bid from the level optimal for a risk-neutral bidder, the risk-averse bidder insures himself. If he wins, his surplus is slightly less because of the higher price, but he is more likely to win and avoid a surplus of zero. Thus, the buyers' risk aversion helps the seller.

Hindering Buyer Collusion

As I mentioned at the start of this chapter, one motivation for auctions is to discourage collusion between players. Some auctions are more vulnerable to this than others. Robinson (1985) has pointed out that whether the auction is private-value or common-value, the first-price sealed-bid auction is superior to the second-price sealed-bid or English auctions for deterring collusion among bidders.

Consider a buyer's cartel in which buyer Smith has a private value of 20, the other buyers' values are each 18, and they agree that everybody will bid 5 except Smith, who will bid 6 (we will not consider the rationality of this choice of bids, which might be based on avoiding legal penalties). In an English auction this is self-enforcing, because if somebody cheats and bids 7, Smith is willing to go all the way up to 20 and the cheater will end up with no gain from his deviation. Enforcement is also easy in a second-price sealed-bid auction, because the cartel agreement can be that Smith bids 20 and everyone else bids 6.

In a first-price sealed-bid auction, however, it is hard to prevent buyers from cheating on their agreement in a one-shot game. Smith does not want to bid 20, because he would have to pay 20, but if he bids anything less than the other players' value of 18 he risks them overbidding him. The buyer will end up with a price of 18, rather than the 6 he would receive in an English auction with collusion.

11.4   Common-Value Auctions and the Winner's Curse

In Section 11.2 we distinguished private-value auctions from common-value auctions, in which the values of the players are identical but their valuations may differ. All four sets of rules discussed in Section 11.2 can be used for common-value auctions, but the optimal strategies are different. In common-value auctions, the players can extract useful information about the object's value to themselves from the bids of the other players. Surprisingly enough, a buyer can use the information from other buyers' bids even in a sealed-bid auction, as will be explained below.

When I teach this material I bring a jar of pennies to class and ask the students to bid for it in an English auction. All but two of the students get to look at the jar before the bidding starts, and everybody is told that the jar contains more than 5 and less than 100 pennies. Before the bidding starts, I ask each student to write down his best guess of the number of pennies. The two students who do not get to see the jar are like "technical analysts," those peculiar people who try to forecast stock prices using charts showing the past movements of the stock while remaining ignorant of the stock's "fundamentals."

A common-value auction in which all the bidders knew the value would not be very interesting, but more commonly, as in the penny jar example, the bidders must estimate the common value. The obvious strategy, especially following our discussion of private-value auctions, is for a player to bid up to his unbiased estimate of the number of pennies in the jar. But in fact this strategy makes the winner's payoff negative, because the winner is the bidder who has made the largest positive error in his valuation. The bidders who underestimated the number of pennies, on the other hand, lose the auction, but their payoff is limited to a downside value of zero, which they would receive even if the true value were common knowledge. Only the winner suffers from overbidding: he has stumbled into **the winner's curse**. When other players are better informed, it is even worse for an uninformed player to win. Anyone, for example, who wins an auction against 50 experts should worry about why they all bid less.

To avoid the winner's curse, players should scale down their estimates to form their bids. The mental process is a little like deciding how much to bid in a private-value first-price sealed-bid auction, in which bidder Smith estimates the second-highest value conditional upon himself having the highest value and winning. In the common-value auction, Smith estimates his own value, not the second-highest, conditional upon himself winning the auction. He knows that if he wins using his unbiased estimate, he probably bid too high, so after winning with such a bid he would like to retract it. Ideally, he would submit a bid of $[X$ *if I lose, but* $(X - Y)$ *if I win]*, where X is his valuation conditional upon losing and $(X - Y)$ is his lower valuation conditional upon winning. If he still won with a bid of $(X - Y)$ he would be happy; if he lost, he would be relieved. But Smith can achieve the same effect by simply submitting the bid $(X - Y)$ in the first place, since the size of losing bids is irrelevant.

Another explanation of the winner's curse can be devised from the Milgrom definition of "bad news" (Milgrom [1981b], note N6.5). Suppose that the government is auctioning off the mineral rights to a plot of land with common value V, and bidder i has valuation $\hat{V}_i$. Suppose also that the bidders are identical in everything but their valuations, which are based on the various information sets Nature has assigned them, and that the equilibrium is symmetric, so the equilibrium bid function $b(\hat{V}_i)$ is the same for each player. If Bidder 1 wins with a bid $b(\hat{V}_1)$ that is based on his prior valuation $\hat{V}_1$, his posterior valuation $\tilde{V}_1$ is

$$\tilde{V}_1 = E(V|\hat{V}_1, b(\hat{V}_2) < b(\hat{V}_1), \ldots, b(\hat{V}_n) < b(\hat{V}_1)). \tag{11.4}$$

The news that $b(\hat{V}_2) < \infty$ would be neither good nor bad, since it conveys no information, but the information that $b(\hat{V}_2) < b(\hat{V}_1)$ is bad news, since it rules out values of b more likely to be produced by large values of $\hat{V}_2$. In fact, the lower the value of $b(\hat{V}_1)$, the worse is the news of having won. Hence,

$$\tilde{V}_1 < E(V|\hat{V}_1) = \hat{V}_1, \tag{11.5}$$

and if Bidder 1 had bid $b(\hat{V}_1) = \hat{V}_1$ he would immediately regret having won. If his winning bid were enough below $\hat{V}_1$, however, he would be pleased to win.

Deciding how much to scale down the bid is a hard problem because the amount depends on how much all the other players scale down. In a second-price auction a player calculates the value of $\tilde{V}_1$ using equation (11.4), but that equation hides considerable complexity under the disguise of "$b(\hat{V}_2)$," which is itself calculated as a function of $b(\hat{V}_1)$ using an equation like (11.4).

Oil Tracts and the Winner's Curse

The best known example of the winner's curse is from bidding for offshore oil tracts. Offshore drilling can be unprofitable even if oil is discovered, because something must be paid the government for the mineral rights. Capen, Clapp, & Campbell (1971) suggest that bidders' ignorance of the winner's curse caused overbidding in US government auctions of the 1960s. If the oil companies bid close to what their engineers estimated the tracts were worth, rather than scaling down their bids, the winning companies would lose on their investments. The hundredfold difference in the sizes of the bids in the sealed-bid auctions shown in Table 11.1 lends some plausibility to the view that this happened.

Table 11.1 Bids by serious competitors in oil auctions

Offshore Louisiana 1967 Tract SS 207	Santa Barbara Channel 1968 Tract 375	Offshore Texas 1968 Tract 506	Alaska North Slope 1969 Tract 253
32.5	43.5	43.5	10.5
17.7	32.1	15.5	5.2
11.1	18.1	11.6	2.1
7.1	10.2	8.5	1.4
5.6	6.3	8.1	0.5
4.1		5.6	0.4
3.3		4.7	
		2.8	
		2.6	
		0.7	
		0.7	
		0.4	

Note: All bids are in millions of dollars
Source: Capen, Clapp, & Campbell (1971)

Later studies such as Mead et al. (1984) that actually looked at profitability conclude that the rates of return from offshore drilling were not abnormally low, so perhaps the oil companies did scale down their bids rationally. The spread in bids is surprisingly wide, but that does not mean that the bidders did not properly scale down their estimates. Although expected profits are zero under optimal bidding, realized profits could be either positive or negative. With some probability, one bidder makes a large overestimate which results in too high a bid even after rationally adjusting for the winner's curse. The knowledge of how to bid optimally does not eliminate bad luck; it only mitigates its effects.

Another consideration is irrationality of the other bidders. If bidder Allied has figured out the winner's curse, but bidders Brydox and Central have not, what should Allied do? Its rivals will overbid, which affects Allied's best response. Allied should scale down its bid even further, because the winner's curse is intensified against overoptimistic rivals. If Allied wins against a rival who standardly overbids, Allied has very likely overestimated the value.

Risk aversion affects bidding in a surprisingly similar way. If all the players were equally risk averse, the bids would be lower, because the asset is a gamble, whose value is lower for the risk averse. If Smith is more risk averse than Brown, then Smith should be more cautious for two reasons: the direct reason that the gamble is worth less to Smith, and the indirect reason that when Smith wins against a rival like Brown who regularly bids more, Smith probably overestimated the value. Parallel reasoning holds if the players are risk neutral, but the private value of the object differs among them.

Asymmetric equilibria can even arise when the players are identical. Second-price two-person common-value auctions usually have many asymmetric equilibria besides the symmetric equilibrium we have been discussing (see Milgrom [1981c] and Bikhchandani [1988]). Suppose that Smith and Brown have identical payoff functions, but Smith thinks Brown is going to bid aggressively. The winner's curse is intensified for Smith, who would probably have overestimated if he won against an aggressive bidder like Brown, so Smith bids more cautiously. But if Smith bids cautiously, Brown is safe in bidding aggressively, and there is an asymmetric equilibrium. For this reason, acquiring a reputation for aggressiveness is valuable.

Oddly enough, if there are three or more players the sealed-bid second-price common-value auction has a unique equilibrium, which is symmetric. The open exit auction is different: it has asymmetric equilibria, because after one bidder drops out, the two remaining bidders know that they are alone together in a subgame which is a two-player auction. Regardless of the number of players, first-price sealed-bid auctions do not have this kind of asymmetric equilibrium. Threats in a first-price auction are costly because the high bidder pays his bid even if his rival decides to bid less in response. Thus, a bidder's aggressiveness is not made safer by intimidation of another bidder.

The winner's curse crops up in situations seemingly far removed from auctions. An employer must beware of hiring a worker passed over by other employers. Someone renting an apartment must hope that he is not the first visitor who arrived when the neighboring trumpeter was asleep. A firm considering a new project must worry that the project has been considered and rejected by competitors. The winner's curse can even be applied to political theory, where certain issues keep popping up.

Opinions are like estimates, and one interpretation of different valuations is that everyone gets the same data, but they analyze it differently.

On a more mundane level, as I write this in 1987, four major candidates—Bush, Kemp, Dole, and Other—are running for the Republican nomination for President of the United States. Consider an entrepreneur auctioning off four certificates, each paying one dollar if its particular candidate wins the nomination. If every bidder is rational, the entrepreneur should receive a maximum of one dollar in total revenue from these four auctions, and less if bidders are risk averse. But holding the auction in a bar full of partisans, how much do you think he would actually receive?

11.5 Information in Common-Value Auctions

The Seller's Information

Milgrom & Weber (1982) have found that honesty is the best policy as far as the seller is concerned. If it is common knowledge that he has private information, he should release it before the auction. The reason is not that the bidders are risk averse (though perhaps this strengthens the result), but the "No News is Bad News" result of Section 7.1. If the seller refuses to disclose something, buyers know that the information must be unfavorable, and an unravelling argument tells us that the quality must be the very worst possible.

Quite apart from unravelling, another reason to disclose information is to mitigate the winner's curse, even if the information just reduces uncertainty over the value without changing its expectation. In trying to avoid the winner's curse, bidders lower their bids, so anything which makes it less of a danger raises their bids.

Asymmetric Information Among the Buyers

If bidder Smith knows he has uniformly worse information than bidder Brown (that is, his information partition is coarser than Brown's), then he should stay out of the auction: his expected payoff is negative if Brown expects zero profits.

If Smith's information is not uniformly worse, then he can still benefit by entering the auction. Having independent information, in fact, is more valuable than having good information. Consider a common-value first-price sealed-bid auction with four bidders. Bidders Smith and Black have the same good information, Brown has that same information plus an extra signal, and Jones usually has only a poor estimate, but one different from any other bidder's. Smith and Black should drop out of the auction—they can never beat Brown without overpaying. But Jones will sometimes win, and his expected surplus is positive. If, for example, real estate tracts are being sold, and Jones is quite ignorant of land values, he can still do well if on rare occasions he has inside information concerning the location of a new freeway, even though ordinarily he should refrain from bidding. If Smith and Black both use the same appraisal formula, they will compete each other's profits away, and if Brown uses the formula plus extra private information, he drives their profits negative by taking some of the best deals from them and leaving the worst ones.

In general, a bidder should bid less if there are more bidders or his information is absolutely worse (that is, his information partition is coarser). He should also bid less if parts of his information partition are coarser than those of his rivals, even if his information is not uniformly worse. These considerations are most important in sealed-bid auctions, because in an open-cry auction, information is revealed while other bidders still have time to act on it.

Recommended Reading

Capen, E., R. Clapp, & W. Campbell (1971) "Competitive Bidding in High-Risk Situations" *Journal of Petroleum Technology.* June 1971. 23, 1: 641–53.

McAfee, R. Preston & John McMillan (1987) "Auctions and Bidding" *Journal of Economic Literature.* June 1987. 25, 2: 699–754.

Milgrom, Paul (1987) "Auction Theory" In *Advances in Economic Theory, Fifth World Congress.* Truman Bewley, ed. Cambridge: Cambridge University Press, 1987.

Robinson, Marc (1985) "Collusion and the Choice of Auction" *Rand Journal of Economics.* Spring 1985. 16, 1: 141–5.

Vickrey, William (1961) "Counterspeculation, Auctions, and Competitive Sealed Tenders" *Journal of Finance.* March 1961. 16, 1: 8–37.

Problem 11

An Auction with Stupid Bidders

Smith's value for an object has a private component equal to 1 and also a component common with Jones and Brown. Jones's and Brown's private components equal zero. Each player estimates the common component Z independently, and player i's estimate is either x_i above the true value or x_i below, with equal probability. Jones and Brown are naive and always bid their valuations. The auction is English.

(1) If $x_{Smith} = 0$, what is Smith's dominant strategy if his estimate of Z equals 20?

(2) If $x_i = 8$ for all players and Smith estimates $Z = 20$, what are the probabilities that he puts on different values of Z?

(3) If Smith knows that $Z = 12$ with certainty, what are the probabilities he puts on the different combinations of bids by Jones and Brown?

(4) Why is 8.72 a better upper limit on bids for Smith than 21, if his estimate of Z is 20, and $x_i = 8$ for all three players? (compute the payoffs from the two strategies)

Notes

N11.1 Introduction

- McAfee & McMillan (1987) and Milgrom (1987) are excellent surveys of the literature and theory of auctions. Both articles take some pains to relate the material to models of asymmetric information. Milgrom & Weber (1982) is a classic article that covers many aspects of auctions.
- Auctions look like tournaments in that the winner is the player who chooses the largest

amount for some costly variable, but in auctions the losers generally do not incur costs proportional to their bids. Shubik (1971), however, has suggested an auction in which both the first and the second-highest bidders pay the second price. If both players begin with infinite wealth, the game illustrates why equilibrium might not exist if strategy sets are unbounded. Once one bidder has started bidding against another, both of them do best by continuing to bid so as to win the prize as well as pay the bid. This auction may seem absurd, but it has some similarity to patent races (see Section 13.4) and arms races.

N11.2 Auction Classification and Private-Value Strategies

- Cassady (1967) is an excellent source of institutional detail on auctions. The appendix to his book includes advertisements and sets of auction rules, and he cites numerous newspaper articles. See also *New York Times*, 26 July 1985, p. C23; *New York Times*, 31 July 1985, pp. A1, C20; *Wall Street Journal*, 24 August 1984, pp. 1, 16; and "The Crackdown on Colluding Roadbuilders," *Fortune*, 3 October 1983, p. 79.
- One might think that a second-price open-cry auction would come to the same results as a first-price open-cry auction, because if the price advances by epsilon at each bid, the first and second bids are practically the same. But the second-price auction can be manipulated. If somebody initially bids \$10 for something worth \$80, another bidder could safely bid \$1,000. No one else would bid more, and he would pay only the second price: \$10.
- In one variant of the English auction, the auctioneer announces each new price and a bidder can hold up a card to indicate he is willing to bid that price. This rule is practical to administer in large crowds and it also allows the seller to act strategically during the course of the auction. If, for example, the two highest valuations are 100 and 120, this auction could yield a price of 110, while the usual rules would only allow a price of $100 + \varepsilon$.
- Vickrey (1961) notes that a Dutch auction could be set up as a second-price auction. When the first bidder presses his button, he primes a buzzer that goes off when a second bidder presses a button.
- After the last bid of an open-cry art auction in France, the representative of the Louvre has the right to raise his hand and shout "pre-emption de l'etat," after which he takes the painting at the highest price bid (*The Economist*, 23 May 1987, p. 98). How does that affect the equilibrium strategies? What would happen if the Louvre could resell?
- **Share Auctions.** In a share auction each buyer submits a bid for both a quantity and a price. The bidder with the highest price receives the quantity for which he bid at that price. If any of the product being auctioned remains, the bidder with the second-highest price takes the quantity he bid for, and so forth. The rules of a share auction can allow each buyer to submit several bids, often called a **schedule** of bids. The details of share auctions vary, and they can be either first-price or second-price. Models of share auctions are very complicated; see R. Wilson (1979).
- **Reserve prices** are prices below which the seller refuses to sell. They can increase the seller's revenue, and their effect is to make the auction more like a regular fixed-price market. For discussion, see Milgrom & Weber (1982). They are also useful when buyers collude, a situation of bilateral monopoly (see "At Many Auctions, Illegal Bidding Thrives as a Longtime Practice Among Dealers," *Wall Street Journal*, 19 February 1988, p. 21.)

   In some real-world English auctions, the auctioneer does not announce the reserve price in advance, and he starts the bidding below it. This can be explained as a way to allow bidders to show each other that their valuations are greater than the starting price, even though it may turn out that they are all lower than the reserve price.
- Concerning auctions with risk-averse players, see Maskin & Riley (1984).

N11.4   Common-Value Auctions and the Winner's Curse

- Even if valuations are correlated, the optimal bidding strategies can still be the same as in private-value auctions if the values are independent. If everyone overestimates their values by 10 percent, then a player can still extract no information about his value by seeing other players' valuations.

- "Getting carried away" may be a rational feature of a common-value auction. If a bidder has a high private value, and then learns from the course of the bidding that the common value is larger than he thought, he may well end up paying more than he had planned, although he would not regret it afterwards. Other explanations for why bidders seem to pay too much are the winner's curse and the fact that in every auction all but one or two of the bidders think that the winning bid is greater than the value of the object.

- Milgrom & Weber (1982) use the concept of **affiliated** variables in classifying auctions. Roughly speaking, random variables X and Y are affiliated if a larger value of X means that a larger value of Y is more likely, or at least no less likely. Independent random variables are affiliated.

12 Pricing and Product Differentiation

12.1 Quantities as Strategies: the Cournot Equilibrium Revisited

Introduction

Chapter 12 is about how firms with market power set prices. This section generalizes the Cournot model of firms choosing the quantities they sell, while Section 12.2 sets out the Bertrand model of firms choosing the prices at which they sell. An important point that arises is whether firms produce differentiated products. Section 12.3 develops two location models of product differentiation. Section 12.4 uses the overlapping generations model so well-known from macroeconomics to analyze another source of market power: customer switching costs that make the customers, if not the products, heterogeneous. Section 12.5 shows that even if a firm is a monopolist, if it sells a durable good it suffers competition from its future self.

Cournot Behavior with General Cost and Demand Functions

In the next few sections, sellers compete against each other using simultaneous moves. We will start by generalizing the Cournot duopoly model of Section 3.4 from linear demand and zero costs to a wider class of functions. The two players are firms Allied and Brydox, and their strategies are the quantities q_a and q_b. The payoffs are based on the total cost functions, $c(q_a)$ and $c(q_b)$, and the demand function, $p(q)$, where q is the total quantity. This specification says that only the sum of the outputs matters to the price. The implication is that the firms produce an identical product, because whether it is Allied or Brydox that produces an extra unit, the effect on price is the same.

Let us take the point of view of Allied. In the Cournot-Nash analysis, Allied chooses his output of q_a for a given level of q_b as if his choice did not affect q_b. From his point of view, q_a is a function of q_b, but q_b is exogenous. Allied sees the effect of his output on price as

$$\frac{\partial p}{\partial q_a} = \frac{dp}{dq} \frac{\partial q}{\partial q_a} = \frac{dp}{dq}. \tag{12.1}$$

Allied's payoff function is

$$\pi_a = p(q)q_a - c(q_a). \tag{12.2}$$

To find Allied's reaction function, we differentiate with respect to his strategy to obtain

$$\frac{d\pi_a}{dq_a} = p + \frac{dp}{dq}q_a - \frac{dc}{dq_a} = 0, \tag{12.3}$$

which implies

$$q_a = \frac{\frac{dc}{dq_a} - p}{\frac{dp}{dq}}, \tag{12.4}$$

or, simplifying the notation,

$$q_a = \frac{c' - p}{p'}. \tag{12.5}$$

If particular functional forms for $p(q_a + q_b)$ and $c(q_a)$ are available, equation (12.5) can be solved to find q_a as a function of q_b. More generally, to find the change in Allied's best response for an exogenous change in Brydox's output, differentiate (12.5) with respect to q_b, remembering that q_b exerts not only a direct effect, but possibly an indirect effect.

$$\frac{dq_a}{dq_b} = \frac{(p - c')(p'' + p''\frac{dq_a}{dq_b})}{p'^2} + \frac{c''\frac{dq_a}{dq_b} - p' - p'\frac{dq_a}{dq_b}}{p'}. \tag{12.6}$$

Equation (12.6) can be solved for $\frac{dq_a}{dq_b}$ to obtain the slope of the reaction function,

$$\frac{dq_a}{dq_b} = \frac{(p - c')p'' - p'^2}{2p'^2 - c''p' - (p - c')p''}. \tag{12.7}$$

If both costs and demand are linear, as we assumed in Section 3.4, then $c'' = 0$ and $p'' = 0$, so equation (12.7) becomes

$$\frac{dq_a}{dq_b} = -\frac{p'^2}{2p'^2} = -\frac{1}{2}. \tag{12.8}$$

The general model faces two problems that did not arise in the linear model: nonuniqueness and nonexistence. If demand is concave and costs are convex, which implies that $p'' < 0$ and $c'' > 0$, then all is well as far as existence goes. Since price is greater than marginal cost ($p > c'$), equation (12.7) tells us that the reaction functions are downward sloping, because $2p'^2 - c''p' - (p - c')p''$ is positive and both $(p-c')p''$ and $-p'^2$ are negative. If the reaction curves are downward sloping, they cross and an equilibrium exists, as was shown in Figure 3.1 for the linear case represented by equation (12.8). If the demand curves are not linear, the contorted reaction functions of equation (12.7) might give rise to multiple Cournot equilibria as in Figure 12.1.

Figure 12.1 Multiple Cournot–Nash equilibria

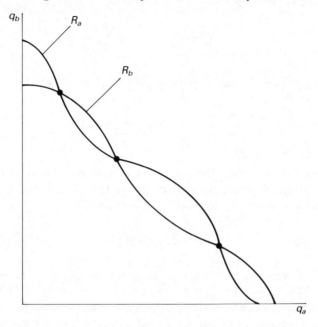

If demand is convex or costs are concave, so that $p'' > 0$ or $c'' < 0$, the reaction functions can be upward sloping, in which case they might never cross and no equilibrium would exist. The problem can also be seen from Allied's payoff function, equation (12.2). If $p(q)$ is convex, the payoff function might not be concave, in which case standard maximization techniques break down. The problems of the general Cournot model teach a lesson to those of us who use no-fat modelling: sometimes simple assumptions, and linearity in particular, generate atypical results.

Many Oligopolists

Let us return to the simpler game in which production costs are zero and demand is linear. For concreteness, we will use the particular inverse demand function

$$p(q) = 120 - q. \tag{12.9}$$

Using (12.9), the payoff function (12.2) becomes

$$\pi_a = 120q_a - q_a^2 - q_b q_a. \tag{12.10}$$

In Section 3.4 we found that the firms picked outputs of 40 apiece given demand function (12.9). This generated a price of 40. With n firms instead of two, the demand function is

$$p\left(\sum_{i=1}^{n} q_i\right) = 120 - \sum_{i=1}^{n} q_i, \tag{12.11}$$

and firm j's payoff function is

$$\pi_j = 120q_j - q_j^2 - q_j \sum_{i \neq j} q_i. \tag{12.12}$$

Differentiating j's payoff function with respect to q_j we obtain

$$\frac{d\pi_j}{dq_j} = 120 - 2q_j - \sum_{i \neq j} q_i = 0. \tag{12.13}$$

To find the equilibrium, make the guess that it is symmetric, so that $q_j = q_i$, $(i = 1, \ldots, n)$, an educated guess, since every player faces a first order condition like (12.13). Under symmetry, equation (12.13) becomes $120 - (n+1)q_j = 0$, so that

$$q_j = 120/(n+1). \tag{12.14}$$

Consider several different values for n. If $n = 1$, we get $q_j = 60$, the monopoly optimum; and if $n = 2$ we get $q_j = 40$, the Cournot output found in Section 3.4. If $n = 5$, we get $q_j = 20$; and as n rises, individual output shrinks to zero. Moreover, the total output of $nq_j = 120n/(n+1)$ gradually approaches 120, the competitive output, and the market price falls to zero, the marginal cost of production. As the number of firms increases, profits fall.

Conjectural Variation

Conjectural variation, an equilibrium concept different in flavor from any that has yet appeared in this book, is a way to quantify the degree of cooperation between oligopolists. Let us continue to specify the strategies as quantities. In a subgame perfect equilibrium, no player wants to deviate by himself, and his beliefs about how the other players would behave would be confirmed whatever nodes were reached. Under conjectural variation, a player believes, for unspecified reasons, that if he deviated the other players would deviate in specified ways. This is most clearly seen in an example. Returning to the two-player model, we can use equation (12.3) to write Allied's self-perceived first order condition as

$$\frac{d\pi_a}{dq_a} = p + \left(\frac{dp}{dq}\right)\left(\frac{dq}{dq_a}\right)q_a - \frac{dc}{dq_a} = 0, \tag{12.15}$$

The difference between first order conditions (12.3) and (12.15) is that (12.15) multiplies q_a by

$$\frac{dq}{dq_a} = 1 + \frac{dq_b}{dq_a}. \tag{12.16}$$

Equation (12.16) says that the expected effect on industry output of an increase in q_a by one unit has two components: the direct increase of one unit, and an indirect increase from Brydox increasing his output in response. First order condition (12.15) must be qualified by "self-perceived" because Allied might be mistaken in his beliefs about Brydox's response. The belief implicit in Nash equilibrium, that Allied's deviation is not followed by a response from Brydox, is the only belief that supports an equilibrium in which one player or the other is not mistaken. But if consistency of beliefs is not required, other beliefs are possible that lead to different behavior. Let q_{-i} denote the total output of all firms except i.

Firm i's **conjectural variation** *is the rate $\frac{dq_{-i}}{dq_i}$ at which he conjectures that the output of other firms would change if i's own output changed.*

CV = 0   In a Cournot–Nash equilibrium, Allied believes that if he deviated by producing more, Brydox would not deviate, so the conjectural variation equals 0.

CV = −1   If Allied believes that an increase in his output is matched by a decrease in Brydox's output, so the total industry output is left unchanged, the conjectural variation is −1. If both firms use this conjectural variation, the industry output is the competitive level; firms ignore the effect of their output in depressing the price. Of course, if both firms use a negative value, their beliefs are inconsistent.

CV = 1   If Allied believes that Brydox would exactly match his output changes, the conjectural variation is 1. With two firms, industry output is at the monopoly level. For an n-player game $CV = n - 1$ achieves that level.

In Stackelberg equilibrium (Section 3.4), the conjectural variation of the Stackelberg follower is between 0 and 1, and takes the value given by a reaction function like equation (12.7).

In the world oil market, fringe producers like Britain face the OPEC cartel. If Britain's conjectural variation equals −1, Britain believes that producing more would make OPEC cut back an equal amount; if 0, that OPEC would ignore Britain; if 0.5, that OPEC would follow with a smaller increase; if 1, that OPEC would match every increase; and if 10, that OPEC would respond by flooding the market. Setting up equations with the appropriate value for the conjectural variations of all the players, we could solve for the equilibrium output. The idea is useful for organizing different models of duopoly and it is simple enough to be empirically estimated. Even without knowing the correct theory, an estimate could be made of how much OPEC actually does respond to Britain.

12.2   Prices as Strategies: the Bertrand Equilibrium

The Bertrand (1883) duopoly model seems to be only slightly different from the

Cournot model, but it reaches radically different conclusions. The Bertrand solution is nothing more than a Nash equilibrium in prices rather than quantities. We will use the same two-player zero-cost linear-demand world as before, but now the strategy spaces are the prices, not the quantities. We will also use the same demand function, equation (12.9), which implies that if p is the lowest price, $q = 120 - p$. In the Cournot model, firms chose quantities but allowed the market price to vary freely; in the Bertrand model, they choose prices and sell as much as they can. The strategies for Allied and Brydox are p_a and p_b. The payoff function for Allied (Brydox's is analogous) is

$$\pi_a = \begin{cases} p_a(120 - p_a) & \text{if } p_a < p_b. \\ \frac{p_a(120 - p_a)}{2} & \text{if } p_a = p_b. \\ 0 & \text{if } p_a > p_b. \end{cases} \qquad (12.17)$$

The Bertrand game has a unique Nash equilibrium: $p_a = p_b = 0$. No other pair of prices could be an equilibrium, because one firm could capture the entire market by slightly undercutting the other's price. The only pair of prices where undercutting is not a temptation is $(0,0)$. Duopoly profits are not just less than monopoly profits, they are zero!

The result might seem less surprising in a homely example. If two wells are located across the street from each other, and a passer-by asks the owners how much it would cost for a drink, what price will he pay? The Bertrand model predicts a price of zero. Nothing else is an equilibrium, because the well-owners would undercut each other, so unless the Nash equilibrium concept is abandoned, the price equals the marginal cost of zero.

Like the surprising outcome of the Prisoner's Dilemma, the Bertrand equilibrium is not so surprising once one thinks about the limitations of the model. What it shows is that duopoly profits do not arise solely from the fact of there being two firms; profits arise from something else like multiple periods or incomplete information.

While it is more natural to think of firms competing in prices rather than quantities, both the Bertrand and the Cournot model are in common use. The Bertrand model can be awkward mathematically because of the discontinuous jump from a market share of zero to 100 percent after a slight price cut. The Cournot model is useful as a simple model that avoids this problem while letting the price fall with entry. But there are also ways to modify the Bertrand model to obtain an intermediate price, and we will look at them next.

Capacity Constraints: the Edgeworth Paradox

Let us start by altering the Bertrand model by constraining each firm to sell no more than $K = 70$ units. The industry capacity of 140 exceeds the competitive output, but do profits continue to be zero?

When capacities are limited we require additional assumptions because of the new possibility that a firm might attract more customers than it can supply. We need to specify a **rationing rule** telling which customers are served at the low price and which must buy from the high-price firm. The rationing rule is unimportant to the payoff of the low-price firm, but crucial to the high-price firm. One possible rule is

Intensity rationing. *The customers that value the product most buy from the firm with the lower price.*

The inverse demand function from equation (12.9) is $p = 120 - q$, and under intensity rationing, the K customers with the strongest demand buy from the low-price firm. Suppose that Brydox is the low-price firm, charging a price of 30 so that 90 consumers wish to buy there. The residual demand facing Allied is then

$$q_a = 120 - p_a - K. \tag{12.18}$$

The demand curve is shown in Figure 12.2a.

Under intensity rationing, the payoff functions are

$$\pi_a = \begin{cases} p_a \cdot Min\{120 - p_a, K\} & \text{if } p_a < p_b. & (12.19a) \\ \frac{p_a(120 - p_a)}{2} & \text{if } p_a = p_b. & (12.19b) \\ 0 & \text{if } p_a > p_b, p_b \geq 50. & (12.19c) \\ p_a(120 - p_a - K) & \text{if } p_a > p_b, p_b < 50. & (12.19d) \end{cases}$$

The appropriate rationing rule depends on what is being modelled. Intensity rationing is appropriate if buyers with more intense demand make greater efforts to obtain low prices. If the intense buyers are wealthy people who are unwilling to wait in rationing lines, then the least intense buyers might end up at the low-price firm—**inverse-intensity rationing**. An intermediate rule is proportional rationing, under which every type of consumer is equally likely to be able to buy at the low price.

Proportional rationing. *Each customer has the same probability of being able to buy from the low-price firm.*

Under proportional rationing, if $K = 70$ and 90 customers wanted to buy from Brydox, 2/9 of each type of customer (for example, the type willing to pay 120) are forced to buy from Allied. The residual demand curve facing Allied, shown in Figure 12.2b and equation (12.20), intercepts the price axis at 120, but slopes down at a rate 4.5 times as fast as market demand because there are only 2/9 $(= \frac{q(p_b) - K}{q(p_b)})$ as many remaining consumers of each type.

$$q_a = (120 - p_a)\left(\frac{120 - p_b - K}{120 - p_b}\right). \tag{12.20}$$

The capacity constraint has a very important effect: (0,0) is no longer a Nash equilibrium in prices. Consider Allied's best response when Brydox charges a price of zero. If Allied raises his price above zero, he retains most of his customers

Figure 12.2   Rationing rules: (a) intensity rationing if $K = 70$; (b) proportional rationing

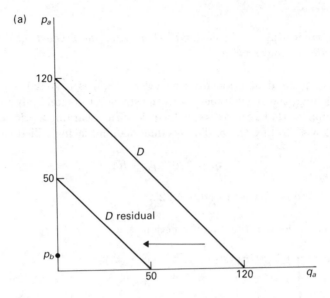

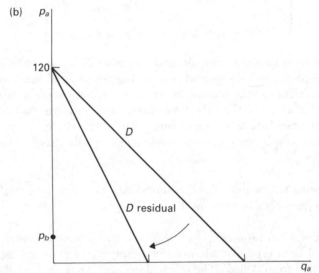

(because Brydox is already producing at capacity), but his profits rise from zero to some positive number, regardless of the rationing rule. In any equilibrium, both players must charge prices within some small amount ϵ of each other, or the one with the lower price would deviate by raising his price. But if the prices are equal, then both players have unused capacity, and each has an incentive to undercut the other. Hence, there is no pure strategy Nash equilibrium. This is known as the **Edgeworth paradox**, after Edgeworth (1897).

A mixed strategy Nash equilibrium does exist, calculated using intensity rationing by Levitan & Shubik (1972) and analyzed in Dasgupta & Maskin (1986b).

Expected profits are positive, because the firms charge positive prices. Under proportional rationing, as under intensity rationing, profits are positive in equilibrium, but the high-price firm does better with proportional rationing. The high-price firm would do best with inverse-intensity rationing, under which the customers with the least intense demand are served at the low-price firm, leaving the ones willing to pay more at the mercy of the high-price firm.

Even if capacity were made endogenous, the outcome would be inefficient, either because firms would charge prices higher than marginal cost (if their capacity is low), or they would invest in excess capacity (even though they price at marginal cost).

Product Differentiation

The Bertrand model without capacity constraints generates zero profits because only slight price discounts are needed to bid away customers. The assumption behind this is that the two firms sell identical goods, so that if Allied's price is slightly higher than Brydox's all the customers go to Brydox. If customers have brand loyalty or bad information on prices, the equilibrium is different and the demand curves facing Allied and Brydox might be

$$q_a = 24 - 2p_a + p_b \qquad (12.21)$$

and

$$q_b = 24 - 2p_b + p_a. \qquad (12.22)$$

The greater the difference in the coefficients on prices in demand curves like these, the less substitutable are the products. As with standard demand curves like (12.9), we have made implicit assumptions about the extreme points of (12.21) and (12.22). These equations only apply if the quantities demanded turn out to be nonnegative, and we might also want to restrict them to prices below some ceiling, since otherwise the demand facing one firm becomes infinite as the other's price rises to infinity. With those restrictions, the payoffs are

$$\pi_a = p_a(24 - 2p_a + p_b) \qquad (12.23)$$

and

$$\pi_b = p_b(24 - 2p_b + p_a). \qquad (12.24)$$

Maximizing Allied's payoff, we obtain the first order condition

$$\frac{d\pi_a}{dp_a} = 24 - 4p_a + p_b = 0, \qquad (12.25)$$

and the reaction function

$$p_a = 6 + p_b/4. \qquad (12.26)$$

Since Brydox has a parallel first order condition, the equilibrium is where $p_a =$

$p_b = 8$. The quantity each firm produces is 16, which is below the 24 each would produce at prices of zero. Figure 12.3 shows that the reaction functions intersect. Allied's demand curve has the elasticity

$$\left(\frac{\partial q_a}{\partial p_a}\right) \cdot \left(\frac{p_a}{q_a}\right) = -2\left(\frac{p_a}{q_a}\right), \tag{12.27}$$

which is less than infinity even when $p_a = p_b$, unlike in the standard Bertrand model.

Figure 12.3   Bertrand reaction functions with differentiated Products

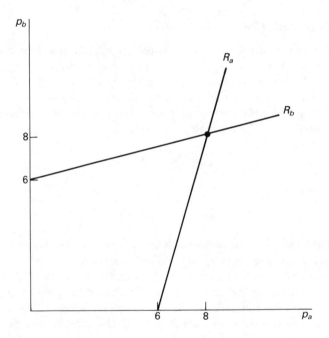

Cournot Equilibrium with Differentiated Products

We can also work out the Cournot equilibrium for demand functions (12.21) and (12.22), but product differentiation does not affect it much. Start by expressing the price in terms of quantities alone, obtaining

$$p_a = 12 - \frac{1}{2}q_a + \frac{1}{2}p_b \tag{12.28}$$

and

$$p_b = 12 - \frac{1}{2}q_b + \frac{1}{2}p_a. \tag{12.29}$$

After substituting from (12.29) into (12.28) and solving for p_a, we obtain

$$p_a = 24 - 2q_a/3 - q_b/3. \tag{12.30}$$

The first order condition for Allied's maximization problem is

$$\frac{d\pi_a}{dq_a} = 24 - 4q_a/3 - q_b/3 = 0, \tag{12.31}$$

which gives rise to the reaction function

$$q_a = 18 - q_b/4. \tag{12.32}$$

We can guess that $q_a = q_b$. It follows from (12.32) that $q_a = 14.4$ and the market price is 9.6. On checking, you would find this to indeed be a Nash equilibrium.

12.3 Location Models

In Section 12.2 we analyzed the Bertrand model with differentiated products using demand functions whose arguments were the prices of both firms. Such a model is suspect because it is not based on primitive assumptions. In particular, the demand functions might not be generated by maximizing any possible utility function. A demand curve with a constant elasticity less than one, for example, is impossible because as the price goes to zero, the amount spent on the commodity goes to infinity. Demand functions (12.21) and (12.22) were also restricted to prices below a certain level, and it would be good to be able to justify that restriction.

Location models construct demand functions like (12.21) and (12.22) from primitive assumptions. In these models, a differentiated product's characteristics are points in a space. If cars differ only in their mileage, the space is a one-dimensional line. If acceleration is also important, the space is a two-dimensional plane. An easy characteristic to think about this way is the location where a product is sold. The product "gasoline sold at the corner of Wilshire and Westwood," is different from "gasoline sold at the corner of Wilshire and Fourth." Depending on where consumers live, they have different preferences over the two, but if prices diverge enough they are willing to switch from one gas station to the other.

Location models form a literature in themselves. We will look at the first two models analyzed in the classic article of Hotelling (1929), a model of price choice and a model of location choice. Figure 12.4 shows what is common to both. Two firms are located at points x_a and x_b along a line running from zero to one, with a constant density of consumers throughout. In the Hotelling Pricing Game, firms choose prices for given locations. In the Hotelling Location Game, prices are fixed and the firms choose the locations.

The Hotelling Pricing Game

(Hotelling [1929])

Players

Sellers Allied and Brydox, located at x_a and x_b, where $x_a < x_b$, and a continuum of buyers indexed by location $x \in [0, 1]$.

Information

Imperfect, symmetric, complete, and certain.

Actions and Events

(1)   The sellers simultaneously choose prices p_a and p_b.
(2)   Each buyer chooses a seller.

Payoffs

Demand is uniformly distributed on the interval $[0,1]$ with a density equal to one (think of each consumer as buying one unit). Production costs are zero. Each consumer always buys, so his problem is to minimize the sum of the price plus the linear transport cost, which is θ per unit distance travelled.

$$\pi_{buyer\ at\ x} = -Min\{\theta|x_a - x| + p_a,\ \theta|x_b - x| + p_b\}. \tag{12.33}$$

$$\pi_a = \begin{cases} 0 & \text{if } p_a - p_b > \theta(x_b - x_a) \tag{12.34a} \\ & \text{(Brydox captures the entire market)} \\[2mm] p_a & \text{if } p_b - p_a > \theta(x_b - x_a) \tag{12.34b} \\ & \text{(Allied captures the entire market)} \\[2mm] p_a(\frac{1}{2\theta}\left[(p_b - p_a) + \theta(x_a + x_b)\right]) & \text{otherwise} \\ & \text{(the market is divided)} \tag{12.34c} \end{cases}$$

Brydox has analogous payoffs.

Figure 12.4   Location models

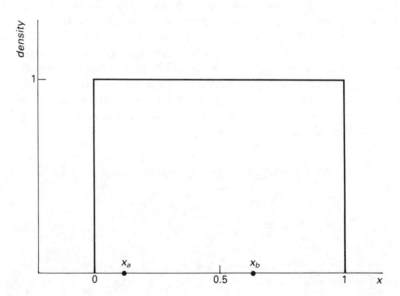

The payoffs need explanation. A buyer's utility depends on the price he pays and the distance he travels. Price aside, Allied is most attractive to the customer at $x = 0$ and least attractive to the customer at $x = 1$. The customer at $x = 0$ will buy from Allied so long as

$$\theta x_a + p_a < \theta x_b + p_b, \tag{12.35}$$

which implies that

$$p_a - p_b < \theta(x_b - x_a), \tag{12.36}$$

which gives us payoff (12.34a). The customer at $x = 1$ will buy from Brydox if

$$\theta(1 - x_a) + p_a > \theta(1 - x_b) + p_b, \tag{12.37}$$

which implies that

$$p_b - p_a < \theta(x_b - x_a), \tag{12.38}$$

which gives us payoff (12.34b).

Very likely, inequalities (12.36) and (12.38) are both satisfied, in which case Customer 0 goes to Allied and Customer 1 goes to Brydox. This is the case represented by payoff (12.34c), and we must find the customer, denoted by x^*, who is at the boundary, indifferent between the two sellers. First, notice that if Allied attracts Customer x_b, he also attracts all $x > x_b$, because going beyond x_b the customers' distances from both sellers increase at the same rate. So we know that if there is an indifferent consumer he is between x_a and x_b. Knowing this, (12.33) tells us that

$$\theta(x^* - x_a) + p_a = \theta(x_b - x^*) + p_b, \tag{12.39}$$

so that

$$p_b - p_a = \theta(2x^* - x_a - x_b), \tag{12.40}$$

and

$$x^* = \frac{1}{2\theta} \left[(p_b - p_a) + \theta(x_a + x_b) \right]. \tag{12.41}$$

Since Allied keeps all the customers between 0 and x^*, equation (12.41) is the demand function facing Allied so long as he does not set his price so far above Brydox's that he loses even Customer 0. The demand facing Brydox equals $(1 - x^*)$. Note that if $p_b = p_a$, then from (12.41), $x^* = \frac{x_a + x_b}{2}$, independent of θ, which is just what we would expect. Demand is linear in the prices of both firms, and looks similar to demand curves (12.21) and (12.22), which we used in Section 12.2 for the Bertrand game with differentiated products.

Now that we have found the demand functions, the Nash equilibrium can be calculated in the same way as in Section 12.2, by setting up the profit functions for each firm, differentiating with respect to the price of each, and solving the two first order conditions for the two prices. Subject to the proviso that the firms are willing to pick prices to satisfy inequalities (12.36) and (12.38), the resulting Nash equilibrium is

$$p_a = \frac{(2 + x_a + x_b)\theta}{3}, \quad p_b = \frac{(4 - x_a - x_b)\theta}{3}. \tag{12.42}$$

From (12.42) one can see that Allied charges a higher price if a large x_a gives it more safe customers or a large x_b makes the number of contestable customers greater. The simplest case is when $x_a = 0$ and $x_b = 1$, when (12.42) tells us that both firms charge a price equal to θ. Profits are positive and increasing in the transportation cost, unless $\theta = 0$, in which case we have returned to the basic Bertrand model.

We cannot rest satisfied with the neat equilibrium given in equation (12.42), because the proviso that firms are willing to charge prices to satisfy (12.36) and (12.38) is often violated. Hotelling did not notice this, and fell into a trap which is very tempting for economists doing game theory. Economists are used to models in which the calculus approach gives an answer that is both the local optimum and the global optimum. But in games like this one, the local optimum is not global, because of the discontinuity in the objective function. D'Aspremont, Gabszewicz, & Thisse (1979) have shown that if x_a and x_b are close together, no pure strategy equilibrium exists, for reasons similar to why none exists in the Bertrand model with capacity constraints. If both firms charge nonrandom prices, neither would deviate to a slightly different price, but one might deviate to a much lower price that would capture every single customer. But if both firms charged that low price, each would deviate by raising his price slightly. It turns out that if Allied and Brydox are located symmetrically around the center of the interval, then if $x_a \geq 0.25$ and $x_b \leq 0.75$ no pure strategy equilibrium exists, although by Theorem 5.8 (Dasgupta & Maskin [1986b]), a mixed strategy equilibrium does exist.

Let us now turn to the choice of location. We will simplify the model by pushing consumers into the background and imposing a single exogenous price for all firms.

The Hotelling Location Game

(Hotelling [1929])

Players

n Sellers.

Information

Imperfect, symmetric, complete, and certain.

Actions and Events

The sellers simultaneously choose locations $x_i \in [0, 1]$.

Payoffs

Consumers are distributed along the interval $[0,1]$ with a uniform density equal to one. The price equals one, and production costs are zero. The sellers are ordered by their location so $x_1 \leq x_2 \leq \ldots \leq x_n$, $x_0 \equiv 0$ and $x_{n+1} \equiv 1$. Seller i attracts half the customers from the gaps on each side of him, so that his payoff is $\pi_1 = x_1 + \frac{x_2 - x_1}{2}$, $\pi_n = \frac{x_n - x_{n-1}}{2} + 1 - x_n$, or, for $i = 2, \ldots, n-1, \pi_i = \frac{x_i - x_{i-1}}{2} + \frac{x_{i+1} - x_i}{2}$.

With **one seller**, the location does not matter in this model, since the customers are captive. If price were a choice variable and demand were elastic, we would expect the monopolist to locate at $x = 0.5$.

With **two sellers**, both firms locate at $x = 0.5$, regardless of whether demand is elastic. This is a stable Nash equilibrium, as can be seen by inspecting Figure 12.4 and imagining best responses to the other firm's location. The best response is always to locate ε closer than one's rival to the center of the line. When both firms do this, they end up splitting the market by both locating exactly at the center.

With **three sellers** the model does not have a Nash equilibrium in pure strategies. Consider any strategy combination in which each player locates at a separate point. Such a strategy combination is not an equilibrium, because the two players nearest the ends would edge in to squeeze the middle player's market share. But if a strategy combination has any two players at the same point, then the third player would be able to acquire a share of at least $(0.5 - \epsilon)$ by moving next to them; and if the third player's share is that large, one of the doubled-up players would deviate by jumping to his other side and capturing his entire market share. The only equilibrium is in mixed strategies. Shaked (1982) has computed the symmetric mixing probability density $m(x)$ to be

$$m(x) = \begin{cases} 2 & \text{if } \frac{1}{4} \leq x \leq \frac{3}{4} \\ 0 & \text{otherwise.} \end{cases} \tag{12.44}$$

Strangely enough, three is a special number. With **more than three sellers**, an equilibrium in pure strategies does exist (Eaton & Lipsey [1975]). Dasgupta & Maskin (1986b), as amended by Simon (1987), have also shown that an equilibrium, possibly in mixed strategies, exists for any number of players n in a space of any dimension m.

Since prices are inflexible, the competitive market does not achieve efficiency. A benevolent social planner or a monopolist who could charge higher prices if he located his outlets closer to more consumers would choose different locations than competing firms. In particular, when two competing firms both locate in the center of the line, consumers are no better off than if there were just one firm. The average distance of a consumer from a seller would be minimized by setting $x_1 = 0.25$ and $x_2 = 0.75$, the locations that would be chosen either by the social planner or the monopolist.

12.4   An Overlapping Generations Model of Consumer Switching Costs

The next model is included in this book for two reasons: to illustrate another source of market power and to demonstrate a general modelling technique, the **overlapping generations model**, in which different cohorts of otherwise identical players enter and leave the game with overlapping "lifetimes." Such models, best-known from the consumption-loans model of Samuelson (1958), are most often used in macroeconomics, but they can also be useful in microeconomics. The article of Klemperer (1987) has stimulated considerable interest in customers who are bound to firms, and we will look at the paper of Farrell & C. Shapiro on this subject, which is an overlapping generations model.

Customer Switching Costs

(Farrell & C. Shapiro [1988])

Players

Firms Allied and Brydox, and a series of customers, each of whom is first called a youngster and then an oldster.

Information

Perfect and certain.

Actions and Events

(1a) Brydox, the initial incumbent, picks the incumbent price p_1^i.
(1b) Allied, the initial entrant, picks the entrant price p_1^e.
(1c) The oldster picks a firm.
(1d) The youngster picks a firm.
(1e) Whichever firm attracted the youngster becomes the incumbent.
(1f) The oldster dies and the youngster becomes an oldster.
(2a) Return to (1a), possibly with new identities for entrant and incumbent.

Payoffs

The discount factor is δ. The customer reservation price is R and the switching cost is c. The per period payoffs in period t (not the total payoffs) are, for $(j = i, e)$,

$$\pi_{firm\ j} = \begin{cases} 0 & \text{if no customers are attracted.} \\ p_t^j & \text{if just oldsters or just youngsters are attracted.} \\ 2p_t^j & \text{if both oldsters and youngsters are attracted.} \end{cases}$$

$$\pi_{oldster} = \begin{cases} R - p_t^i & \text{if he buys from the incumbent.} \\ R - p_t^e - c & \text{if he switches to the entrant.} \end{cases}$$

$$\pi_{youngster} = \begin{cases} R - p_t^i & \text{if he buys from the incumbent.} \\ R - p_t^e & \text{if he buys from the entrant.} \end{cases}$$

Finding all the perfect Nash equilibria of an infinite game like this one is difficult, so we will follow Farrell & Shapiro in limiting ourselves to the much easier task of finding the perfect Markov equilibrium, which is unique. A Markov strategy does not depend directly on the past history of the game, only on the current values of state variables (see note N4.4). Here, a firm's Markov strategy is its price as a function of whether it is the incumbent or the entrant, and not of the entire past history of the game.

Brydox, the initial incumbent, moves first and chooses p^i low enough that Allied is not tempted to choose $p^e < p^i - c$ and steal away the oldsters. Allied's profit is p^i if it chooses $p^e = p^i$ and serves just youngsters, and $2(p^i - c)$ if it chooses $p^e = p^i - c$ and serves both oldsters and youngsters. Brydox chooses p^i to make Allied indifferent between these alternatives, so

$$p^i = 2(p^i - c), \tag{12.45}$$

and

$$p^i = p^e = 2c. \tag{12.46}$$

In equilibrium, Allied and Brydox take turns being the incumbent and they charge the same price.

Because the game lasts forever and the equilibrium strategies are Markov, we can use a trick from dynamic programming to calculate the payoffs from being the entrant and the incumbent. The equilibrium payoff of the current entrant is the immediate payment of p^e plus the discounted value of being the incumbent next period:

$$\pi_e^* = p^e + \delta \pi_i^*. \tag{12.47}$$

The incumbent's payoff can be similarly stated as the immediate payment of p^i plus the discounted value of being the entrant next period:

$$\pi_i^* = p^i + \delta \pi_e^*. \tag{12.48}$$

We could use equation (12.46) to substitute for p^e and p^i, which would leave us with the two equations (12.47) and (12.48) for the two unknowns π_i^* and π_e^*, but an easier way to compute the payoff is to realize that in equilibrium the incumbent and the entrant sell the same amount at the same price, so $\pi_i^* = \pi_e^*$ and equation

(12.48) becomes

$$\pi_i^* = 2c + \delta\pi_i^*. \tag{12.49}$$

It follows that

$$\pi_i^* = \pi_e^* = \frac{2c}{1-\delta}. \tag{12.50}$$

Prices and total payoffs are increasing in the switching cost c, because that is what gives the incumbent market power and prevents ordinary Bertrand competition. The total payoffs are increasing in δ for the usual reason, unrelated to game theory, that future payments increase in value as δ approaches one.

12.5 Durable Monopoly

We commonly think that a firmly established monopoly will earn positive profits, but this is not necessarily the case when the monopolist sells a good that is durable. When the monopolist sells something like a refrigerator to a consumer, that consumer drops out of the market demand curve until the refrigerator wears out. The demand curve is therefore changing over time as a result of the monopolist's choice of price, and the modeller cannot separate out just one period for his model. Demand is not **time separable**, because a rise in price at time t_1 affects the quantity demanded at time t_2.

The reason that a monopoly over a durable good might not be very profitable is that each period the monopolist does have competitors—himself in later periods. If the monopolist sets a high price in the first period, removing the high-demand buyers from the market, he would be tempted to set a lower price in the next period to take advantage of the remaining consumers. But if it were known he would lower the price, the high-demand buyers would not buy at a high price in the first period. The threat of the future low price forces the monopolist to keep his current price low.

Let us formalize this argument. Suppose that a seller has a monopoly on a durable good. His good lasts two periods, and he can produce, price, and sell it in both periods. The buyer must decide what quantity of the good to buy in each period. Because this one buyer is meant to represent the entire market demand, the moves are ordered so that he has no market power, as in the principal–agent models in Section 6.3. Alternatively, the buyer can be viewed as representing a continuum of consumers (see Coase [1972] and Bulow [1982]).

Durable Monopoly

Players

A seller and a buyer.

Information

Perfect and certain.

Actions and Events

(1) The seller picks a sale price, p_1.
(2) The buyer buys quantity, q_1.
(3) The seller picks a sale price, p_2.
(4) The buyer buys an additional quantity, q_2.

Payoffs

Production cost is zero and there is no discounting. The seller's payoff is his revenue, and the buyer's is the sum across periods of his benefits (B) from consumption minus his costs (C). His benefits arise from his being willing to pay as much as

$$R(q_t) = 60 - q_t/2 \tag{12.51}$$

per unit consumed in period t if he consumes a total quantity of q_t in that period.

$$\pi_s = q_1 p_1 + q_2 p_2. \tag{12.52}$$

$$\pi_b = (B_1 - C_1) + (B_2 - C_2),$$

$$= (\frac{60 - R(q_1)}{2} q_1 + R(q_1)q_1 - p_1 q_1)$$

$$+ (\frac{60 - R(q_1 + q_2)}{2} (q_1 + q_2) + R(q_1 + q_2)(q_1 + q_2) - p_2 q_2). \tag{12.53}$$

The benefit for the buyer in period 1 is the amount he would have been willing to pay for renting q_1, which can be found from equation (12.51) using the typical consumer's surplus calculation of the two areas shown in Figure 12.5.

Consider the situation in the second period if the buyer has already bought q_1. The purchase price is the amount per unit that the consumer would pay for a single period of consuming the total amount $q_1 + q_2$. Therefore, the residual demand remaining after the first period's sales is, using (12.51),

$$p_2 = R(q_1 + q_2) = 60 - \frac{(q_1 + q_2)}{2}. \tag{12.54}$$

Solving for the monopoly quantity, q_2^*, the seller maximizes $q_2 p_2$, solving the problem

$$\underset{q_2}{Maximize}\, q_2(60 - q_2/2 - q_1/2), \tag{12.55}$$

which generates the first order condition

$$60 - q_2 - q_1/2 = 0, \tag{12.56}$$

so that

$$q_2^* = 60 - q_1/2, \text{ and } p_2^* = 30 - q_1/4. \tag{12.57}$$

Figure 12.5   Durable Monopoly: buyer benefit per period

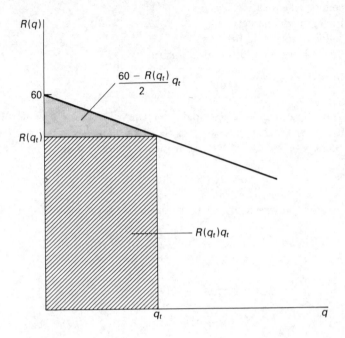

We must now find q_1^*. In period one the buyer looks ahead to the possibility of buying in period two at a lower price. The price he would pay for a unit in period one cannot exceed the benefit in period one plus the expected value of p_2, which from (12.57) is $30 - q_1/4$. The maximum the buyer is willing to pay for quantity q_1 is therefore

$$p_1(q_1) = R(q_1) + p_2 = (60 - q_1/2) + (30 - q_1/4),$$

$$= 90 - \frac{3}{4}q_1. \tag{12.58}$$

Knowing that in the second period he will choose q_2 according to (12.57), the seller combines that with (12.58) to solve

$$\underset{q_1}{Maximize}\, p_1 q_1 + p_2 q_2 = (90 - \frac{3}{4}q_1)q_1 + (30 - q_1/4)(60 - q_1/2),$$

$$= 1800 + 60q_1 - \frac{5}{8}q_1^2, \tag{12.59}$$

which has the first order condition

$$60 - \frac{5}{4}q_1 = 0, \tag{12.60}$$

so that, making use of (12.58) to find p_1,

$$q_1^* = 48, \text{ and } p_1^* = 54. \tag{12.61}$$

It follows from (12.57) that $q_2^* = 36$ and $p_2^* = 18$. The sum of the profits is $\pi_s = 3,240 \ (= 54[48] + 18[36])$.

The purpose of these calculations is to compare the situation with three other market structures: a competitive market, a monopolist who rents instead of selling, and a monopolist who commits to selling only in the first period. The competitive market bids down the price to the marginal cost of zero, so $p_1 = 0$ and $q_1 = 120$ from (12.51), and profits equal zero.

If the monopolist rents instead of selling, then equation (12.51) is like an ordinary demand equation, because the monopolist is effectively selling the good's services separately each period. He could rent a quantity of 60 each period at a rental fee of 30 and his profits would sum to $\pi_s = 3,600$. That is higher than 3,240, so profits are higher from renting than from selling outright. Selling, the problem is that the buyer knows that the seller will be tempted to lower the price once the buyer has bought in the first period. Renting avoids this problem.

If the monopolist can commit to not producing in the second period, he will do just as well as the monopolist who rents, since he can sell a quantity of 60 at a price of 60, the sum of the rents for the two periods. An example is the artist who breaks the plates for his engravings after a production run of announced size. We must also assume that the artist can convince the market that he has broken the plates.

With more than two periods, the difficulties of the durable-goods monopolist become even more striking. In an infinite period model without discounting, if the marginal cost of production is zero, then the equilibrium price for outright sale instead of renting is constant—at zero! Think about this in the context of a model with many buyers. Early consumers foresee that the monopolist has an incentive to cut the price after they buy, in order to sell to the remaining consumers who value the product less. In fact, the monopolist would continue to cut the price and sell more and more units to consumers with weaker and weaker demand until the price fell to marginal cost. Without discounting, even the high-valuation consumers refuse to buy at a high price, because they know they could wait until the price falls to zero. And this is not a trick of infinity: a large number of periods generates a price close to zero.

We can also use the durable monopoly model to think about the durability of the product. If the seller can develop a product so flimsy that it only lasts one period, that is equivalent to renting. A consumer is willing to pay the same price to own a one-hoss shay that he knows will break down in one year as he would pay to rent it for a year. Low durability leads to the same output and profits as renting, which explains why a firm with market power might produce goods that wear out quickly. The explanation is not that the monopolist can use his market power to inflict lower quality on consumers—after all, the price he receives is lower too—but that the lower durability makes it credible to high-valuation buyers that the seller expects their business in the future and will not lower his price.

I hesitate to end the chapter with such a result, since the implication is that this book is less durable than would be socially optimal. But remember that your discount rate for this material should be high, which softens the plight of us durable monopolists. Game theory is becoming more and more important in a variety of disciplines, but its basic structure has by now been fixed. This is a good time to learn the subject and jump aboard while the amount of material is still barely manageable, and that is true even if the books might come down in price in a few years. If you do not have your own copy of the book, buy it now!

Recommended Reading

Bulow, Jeremy (1982) "Durable-Goods Monopolists" *Journal of Political Economy.* April 1982. 90, 2: 314–32.

Bulow, Jeremy, John Geanakoplos, & Paul Klemperer (1985) "Multimarket Oligopoly: Strategic Substitutes and Complements" *Journal of Political Economy.* June 1985. 93, 3: 488–511.

d'Aspremont, Claude, J. Gabszewicz, & Jacques Thisse (1979) "On Hotelling's 'Stability of Competition' " *Econometrica.* September 1979. 47, 5: 1145–50.

Hotelling, Harold (1929) "Stability in Competition" *Economic Journal.* 1929. 39: 41–57.

Problem 12

Duopoly

Firms Allied and Brydox have the total cost curves $c(q_a) = 1,000 + q_a^2$ and $c(q_b) = 2q_b^2$. Demand is $p = 115 - q_a - q_b$.

(1)   What is the equation for Allied's Cournot reaction function?
(2)   What are the outputs and profits in the Cournot equilibrium?
(3)   What are the socially optimal outputs and prices for each firm?

Notes

N12.1   Quantities as Strategies: the Cournot Equilibrium Revisited

- Articles on the existence of a pure strategy equilibrium in the Cournot model include Novshek (1985) and Roberts & Sonnenschein (1976).
- **Merger in a Cournot Model.** A problem with the Cournot model is that a firm's best policy is often to split up into separate firms. Allied gets half the industry profits in a duopoly game. If Allied split into firms *Allied*$_1$ and *Allied*$_2$, it would get two-thirds of the profit in the Cournot triopoly game, even though industry profit falls. A Cournot duopoly would rapidly dissolve into perfect competition!

  This point was made by Salant, Switzer, & Reynolds (1983). It is interesting that nobody noted this earlier, given the intense scrutiny of Cournot models. The insight comes from looking at the problem and asking whether a player could improve his lot if his strategy space were expanded in reasonable ways.

- The idea of conjectural variation is attributed to Bowley (1924) and is discussed in Jacquemin (1985). Varian (1984) defines it differently, as the increase in industry output when Allied increases his output rather than as the increase in non-Allied output. Varian's values are one greater than the conventional values.
- Do not confuse a conjectural variation of -1 with perfect competition, even though both may lead to the efficient output. In perfect competition, individuals do not believe that they affect the rest of the market, but if $CV = -1$ a firm believes that other firms will cut back when it produces more. Perfect competition is more like a game with players so small relative to the market that even though $CV = 0$, as in Nash equilibrium, each player correctly believes that his actions have a trivial effect on the market price.
- More than one reviewer has told me that conjectural variation is an obsolete idea that should be dropped from the book, since the idea behind it is better modelled by a dynamic game. But the reader should know what the term means, and it may be useful in describing reduced form behavior.

N12.2 Prices as Strategies: the Bertrand Equilibrium

- Intensity rationing has also been called **efficient rationing**. Sometimes, however, as in Section 12.2, this rationing rule is inefficient. Some low-intensity customers left facing the high price decide not to buy the product even though their benefit is greater than its marginal cost. The reason intensity rationing has been thought to be efficient is that it is efficient if the rationed-out customers are unable to buy at any price.
- OPEC has tried both price and quantity controls ("OPEC, Seeking Flexibility, May Choose Not to Set Oil Prices, but to Fix Output," *Wall Street Journal*, 8 October 1987, p. 2; "Saudi King Fahd is Urged by Aides To Link Oil Prices to Spot Markets," *Wall*
- *Street Journal*, 7 October 1987, p. 29). Weitzman (1974) is the classic reference on price vs. quantity control by regulators, although he does not use the context of oligopoly. The decision rests partly on enforceability, and OPEC has also hired accounting firms to monitor prices ("Dutch Accountants Take On a Formidable Task: Ferreting Out 'Cheaters' in the Ranks of OPEC," *Wall Street Journal*, 26 February 1985, p. 39).
- Kreps & Scheinkman (1983) show how capacity choice and Bertrand pricing can lead to a Cournot outcome. Two firms face downward-sloping market demand. In the first stage of the game, they simultaneously choose capacities, and in the second stage they simultaneously choose prices (possibly by mixed strategies). If a firm cannot satisfy the demand facing it in the second stage (because of the capacity limit), it uses intensity rationing (the results depend on this). The unique subgame perfect equilibrium is for each firm to choose the Cournot capacity and price.
- Haltiwanger & Waldman (unpub) have suggested a dichotomy applicable to many different games between players who are **responders**, choosing their actions flexibly, and **nonresponders**, who are inflexible. A player might be a nonresponder because he is irrational, because he moves first, or simply because his strategy set is small. The categories are used in a second dichotomy, between games exhibiting **synergism**, in which responders choose to do whatever the majority do (upward sloping reaction curves), and games exhibiting **congestion**, in which responders want to join the minority (downward sloping reaction curves). Under synergism, the equilibrium is more like what it would be if all the players were nonresponders; under congestion, the responders have more influence. Haltiwanger & Waldman apply the dichotomies to network externalities, efficiency wages, and reputation.

    If the reaction functions of two firms are upward sloping, it has been said that the actions are **strategic complements**, and if they are downward sloping, **strategic substitutes** (Bulow, Geanakoplos, & Klemperer [1985]), an unfortunate pair of names since the idea has nothing to do with demand. Gal-Or (1985) notes that if reaction curves slope down (as in Cournot) there is a first-mover advantage, while if they slope up (as in differentiated Bertrand) there is a second-mover advantage.

- Upward sloping reaction functions can result in what has been called the **fat-cat effect** (Fudenberg & Tirole [1986a] p. 23). Consider a two-stage game, in the first stage of which an incumbent firm chooses its advertising level, and in the second stage it plays a Bertrand subgame with an entrant. If the advertising in the first stage gives the incumbent a base of captive customers who have inelastic demand, he will choose a higher price than the entrant. The incumbent has become a "fat cat." The effect is present in many models, including the Hotelling Pricing Game of Section 12.3 and Customer Switching Costs of Section 12.4. In the Hotelling Pricing Game, a firm located so that it has a large "safe" market would choose a higher price. In Customer Switching Costs, a firm that has old customers locked in would choose a higher price than a fresh entrant in the last period of a finitely repeated game.
- Section 12.3 shows how to generate demand curves (12.21) and (12.22) using a location model, but they can also be generated directly by a quadratic utility function. Dixit (1979) states that over the three goods 0, 1, and 2, the utility function

$$U = q_0 + \alpha_1 q_1 + \alpha_2 q_2 - \frac{1}{2}\left(\beta_1 q_1^2 + 2\gamma q_1 q_2 + \beta_2 q_2^2\right) \tag{12.62}$$

(where the constants $\alpha_1, \alpha_2, \beta_1$, and β_2 are positive and $\gamma^2 \leq \beta_1\beta_2$) generates the inverse demand functions

$$p_1 = \alpha_1 - \beta_1 q_1 - \gamma q_2 \tag{12.63}$$

and

$$p_2 = \alpha_2 - \beta_2 q_2 - \gamma q_1. \tag{12.64}$$

N12.3   Location Models

- Greenhut & Ohta (1975) have written a book on location models.
- There is a moral to the story of the mistake in Hotelling (1929): it is worth understanding basic models. Presumably nobody thought very hard about Hotelling's famous article for 50 years after he wrote it, despite its fame and the reward of surefire publication for pointing out an error.
- Location models and switching cost models are attempts to go beyond the notion of a market price. Antitrust cases are good sources for descriptions of the complexities of pricing in particular markets. See, for example, Sultan's 1974 book on electrical equipment in the 1950s, or antitrust opinions such as United States v. Addyston Pipe & Steel Co. et al, 85 Fed 271.
- It is important in location models whether the positions of the players on the line are moveable. See, for example, Lane (1980).
- The location games in this chapter have used a one-dimensional space with end points, i.e., a line segment. Another kind of one-dimensional space is a circle (not to be confused with a disk). The difference is that no point on a circle is distinctive, so no consumer preference can be called extreme. It is, if you like, Peoria vs. Berkeley. The circle might be used for modelling convenience or because it fits a situation: e.g., airline flights spread over the 24 hours of the day. With two players, the Hotelling location game on a circle has a continuum of pure strategy equilibria of two varieties: both players locating in the same spot, and each player separated from the other by 180°. The three-player model also has a continuum of pure strategy equilibria, each player separated from another by 120°, in contrast to the nonexistence of a pure strategy equilibrium when the game is played on a line segment.
- Some characteristics like the color of cars could be modelled as location, but only on a player-by-player basis, because they have no natural ordering. While Smith's ranking of (red=1, yellow=2, blue=10) could be depicted on a line, if Brown's ranking is (red=1, blue=5, yellow=6) we cannot use the same line for him. In the text, the characteristic

was something like physical location, about which people may have different preferences but they agree on what positions are close to what other positions.

N12.4   An Overlapping Generations Model of Consumer Switching Costs

- We assumed that the incumbent chooses its price first, but the alternation of incumbency remains even if we make the opposite assumption. The natural assumption is that prices are chosen simultaneously, but because of the discontinuity in the payoff function, that subgame has no equilibrium in pure strategies.

N12.5   Durable Monopoly

- The proposition that price falls to marginal cost in a durable monopoly with no discounting and infinite time is called the "Coase conjecture," after Coase (1972). Despite being presented non-mathematically, it is really a proposition and not a conjecture, but alliteration was too strong to resist.
- Gaskins (1974) has written a well-known article on the problem of the durable monopolist who foresees that he creates his own future competition because his product is recycled, using the context of the aluminum market.
- Leasing by a durable monopoly was the main issue in the antitrust case *US v. United Shoe Machinery Corporation*, 110 F. Supp. 295 (1953), but not because it increased monopoly profits. The complaint was rather that long-term leasing impeded entry by new sellers of shoe machinery, a curious idea when the proposed alternative was outright sale. More likely, leasing was used as a form of financing for the machinery customers; by leasing, they did not need to borrow as they would to finance a purchase.
- Another way out of the durable monopolist's problem is to give best-price guarantees to customers, promising to refund part of the purchase price if any future customer gets a lower price. Perversely, this hurts consumers, because it stops the seller from being tempted to lower his price. The "most-favored-nation" contract, which is the analogous contract in markets with several sellers, is analyzed by Holt & Scheffman (1987), for example, who demonstrate how it can maintain high prices, and Png & D. Hirshleifer (1987), who show how it can be used to price discriminate between different types of buyers.

13 Industrial Organization

13.1 Predatory Pricing: the Kreps–Wilson Model

This chapter contains applications of dynamic game theory to industrial organization. We will start with an entry deterrence game that applies the Gang of Four model to predatory pricing. If the established firm has an advantage in obtaining credit, predatory pricing can be explained more simply, but explaining the credit advantage, as is attempted in Section 13.2, is not simple. Section 13.3 shows that even with predatory pricing and credit advantages, entry might not be deterred if the entrant hopes to be bought out. Section 13.4 consists of two patent race models and an overview of innovation. Section 13.5 concludes with models of the takeover process.

Let us start by returning to the Chainstore Paradox of Section 4.4. The heart of the paradox is the perfectness problem faced by an incumbent who wishes to threaten a prospective entrant with low post-entry prices. The incumbent can respond to entry in two ways. He can collude with the entrant and share the profits, or he can fight by lowering his price so that both firms make losses—a response commonly known as "predatory pricing." We have already seen that the incumbent would not fight in a perfect equilibrium if the game has complete information. Foreseeing the incumbent's accommodating response, the potential entrant ignores any threats, and enters.

In Kreps & Wilson (1982a), incomplete information allows the threat of predatory pricing to successfully deter entry. They model a monopolist with outlets in N towns who faces an entrant who can enter each town. In our adaption of the model, we will start by assuming that the order in which the towns can be entered is common knowledge, and that if the entrant passes up his chance to enter a town, he cannot enter it later, an assumption relaxed at the end of the analysis. The incomplete information takes the form of a small probability that the monopolist is "tough" and has nothing but *Fight* in his action set: he is an uncontrolled manager who satisfies his passions in squelching entry, rather than maximizing profits.

Predatory Pricing

(Kreps & Wilson [1982a])

Players

The entrant and the monopolist.

Information

Asymmetric, incomplete, and certain.

Actions and Events

(0)   Nature chooses the monopolist to be *Tough* with probability θ or *Standard*, with probability $(1 - \theta)$. Only the monopolist observes Nature's move.
(1)   The entrant chooses *Enter* or *Stay Out* for the first town.
(2)   The monopolist chooses *Collude* or *Fight* if he is standard, *Fight* if he is tough.
(3)   Steps (1) and (2) are repeated for towns $i = 2$ through N.

Payoffs

The discount rate is zero. Table 13.1 gives the payoffs per period, which are the same as in Table 4.1.

Table 13.1 Predatory Pricing

Incumbent

		Collude	Fight
	Enter	**40, 50**	$-10, 0$
Entrant			
	Stay Out	0, 100	**0, 100**

Payoffs to: (Entrant, Incumbent)

In describing the equilibrium, we will denote towns by names such as i_{-30} and i_{-5}, where the numbers are to be taken purely ordinally. The entrant has an opportunity to enter town i_{-30} before i_{-5}, but there are not necessarily 25 towns between them. The actual gap depends on θ but not N.

Part of the equilibrium for Predatory Pricing

Entrant: Enter first at town i_{-10}. If entry has occurred before i_{-10} and been answered with *Collude*, enter every town after the first one entered.

Tough monopolist: Always fight entry.

Standard monopolist: Fight any entry up to i_{-30}. Fight the first entry after i_{-30} with a probability $m(i)$ that diminishes until it reaches zero at i_{-5}. If *Collude* is ever chosen instead, always collude thereafter. If *Fight* was chosen in response to the first attempt at entry, increase the mixing probability $m(i)$ in subsequent towns.

This description, which is illustrated by Figure 13.1, only covers the equilibrium path and small deviations. Note that out-of-equilibrium beliefs do not have to be specified (unlike in the original model of Kreps and Wilson), since whenever a monopolist colludes, in or out of equilibrium, Bayes's Rule says that the entrant must believe him to be *Standard*.

Figure 13.1   Predatory Pricing: strategies over time

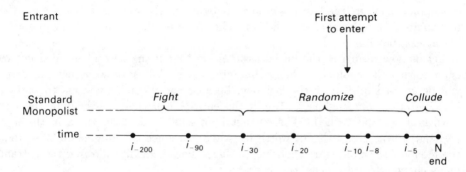

The entrant will certainly stay out until i_{-30}. If no town is entered until i_{-5} and the monopolist is *Standard*, then entry at i_{-5} is undoubtedly profitable. But entry is attempted at i_{-10}, because since $m(i)$ is diminishing in i, the standard monopolist probably would not fight even there.

Out of equilibrium, if an entrant were to enter at i_{-90}, the standard monopolist would be willing to fight, to maintain i_{-10} as the next town to be entered. If he did not, then the entrant, realizing that he could not possibly be facing a tough monopolist, would enter every subsequent town from i_{-89} to i_{-1}. If no town were entered until i_{-5}, the standard monopolist would be unwilling to fight in that town, because too few towns are left to protect. If a town between i_{-30} and i_{-5} has been entered and fought, the monopolist raises the mixing probability that he fights in the next town entered because he has a more valuable reputation to defend. By fighting in the first town he has increased the belief that he is tough and increased the gap until the next town to be entered.

What if the entrant deviated and entered town i_{-20}? The equilibrium calls for a mixed strategy response beginning with i_{-30}, so the standard monopolist must be indifferent between fighting and not fighting. If he fights, he loses current revenue but the entrant's posterior belief that he is tough rises, rising more if the fight occurs late in the game. The entrant knows that in equilibrium the standard monopolist would fight with a probability of, say, 0.9 in town i_{-20}, so fighting there would not much increase the belief that he was tough, but if he fought in town i_{-13}, where the mixing probability has fallen to 0.2, the belief would rise much more. On the other hand, the gain from a given reputation diminishes as fewer towns remain to be protected, so the mixing probability falls over time.

The description of the equilibrium strategies is incomplete because describing what happens after unsuccessful entry becomes rather intricate. Even in the simultaneous-move games of Chapter 3, we saw that games with mixed strategy equilibria have many different possible realizations. In repeated games like Predatory Pricing, the number of possible realizations makes an exact description very complicated indeed. If, for example, the entrant entered town i_{-20} and the monopolist chose *Fight*, the entrant's belief that he was tough would rise, pushing the next town entered to i_{-8} instead of i_{-10}. A complete description of the strategies would say what would happen for every possible history of the game, which is impractical.

Moreover, because of mixing, even the equilibrium path becomes nonunique after i_{-10}, when the first town is entered. When the entrant enters at i_{-10}, the standard monopolist chooses randomly whether to fight, so the entrant's belief that the monopolist is tough increases if he is fought. As a result, the next entry might be not at i_{-9}, but i_{-7}.

Let us now return to the initial assumption that if the entrant decided not to enter town i, he could not change his mind later. We have seen that no towns will be entered until near the last one, because the incumbent wants to protect his reputation for toughness. But if the entrant can change his mind, the last town is never approached. The entrant knows he would take losses in the first $(N - 30)$ towns, and it is not worth his while to reduce the number to 30 so that the monopolist would start choosing *Collude*. Paradoxically, allowing the entrant many chances to enter helps the incumbent.

13.2   Credit and the Age of the Firm: the Diamond Model

Telser (1966) has suggested that predatory pricing would be a credible threat if the incumbent had access to cheaper credit than the entrant, and so could hold out for more periods of losses before going bankrupt. While one might wonder whether this is effective protection against entry—what if the entrant is a large old firm from another industry?—we shall focus on how better-established firms might get cheaper credit.

D. Diamond (unpub) aims to explain why old firms are less likely than young firms to default on debt. His model has both adverse selection, because firms differ in type, and moral hazard, because they take hidden actions. The three types of firms, R, S, and RS, are "born" at time zero and borrow to finance projects at the start of each of T periods. We must imagine that there are overlapping generations of firms, so that at any point in time a variety of ages are coexisting, but the model looks at the lifecycle of only one generation. All the players are risk neutral. Type RS firms can choose independently risky projects with negative expected values or safe projects with low but positive expected values. Although the risky projects are worse in expectation, if they are successful the return is much higher than from safe projects. Type R firms can only choose risky projects, and Type S firms only safe projects. At the end of each period the projects bring in their profits and loans are repaid, after which new loans and projects are chosen for the next period. Lenders cannot tell which project is chosen or what a firm's current profits are, but they can seize the firm's assets if a loan is not repaid, which always happens if the risky project was chosen and turned out unsuccessfully.

The equilibrium path has three parts. The RS firms start by choosing risky projects. Their downside risk is limited by bankruptcy, but if the project is successful the firm keeps large residual profits after repaying the loan. Over time, the number of firms with access to the risky project (the RSs and Rs) diminishes through bankruptcy, while the number of Ss remains unchanged. Lenders can therefore maintain zero profits while lowering their interest rates. When the interest rate falls, the value of a stream of safe investment profits minus interest payments rises relative to the expected value of the few periods of risky returns minus interest payments before bankruptcy. After the interest rate has fallen enough, the second phase of the game begins when the RS firms switch to safe projects at a period we will call t_1. Only the tiny and diminishing group of Type R firms continue to choose risky projects. Since the lenders know that the RS firms switch, the interest rate can fall sharply at t_1. A firm that is older is less likely to be a Type R, so it is charged a lower interest rate. Figure 13.2 shows the path of the interest rate over time.

Figure 13.2   The interest rate over time

Source: Diamond (unpub)

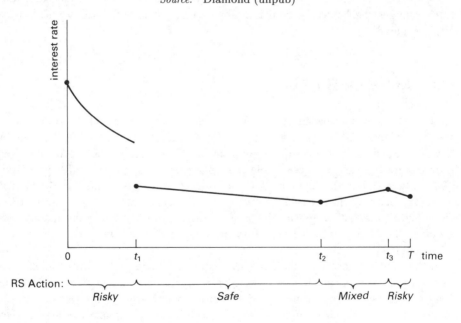

Towards period T, the value of future profits from safe projects declines and even with a low interest rate the RSs are again tempted to choose risky projects. They do not all switch at once, however, unlike in period t_1. In period t_1, if a few RSs had decided to switch to safe projects, the lenders would have been willing to lower the interest rate, which would have made switching even more attractive. If a few firms switch to risky projects at some time t_2, on the other hand, the interest rate rises and switching to risky projects becomes more attractive—a result reminiscent of the Lemons model in Chapter 8. Between t_2 and t_3, the RSs follow a mixed strategy, an increasing number of them choosing risky projects as time passes. The increasing proportion of risky projects causes the interest rate to rise. At t_3, the interest rate is high enough and the end of the game is close enough that the RSs revert to the pure strategy of choosing risky projects. The interest rate declines during this last phase as the number of RSs diminishes because of failed risky projects.

One might ask, in the spirit of no-fat modelling, why the Diamond model has three types of firms rather than two. Types S and RS are clearly needed, but why Type R? A reason is that with three types, bankruptcy is never out-of-equilibrium behaviour, since the failing firm might be a Type R, so Bayes's Rule can always be applied and there is no problem of weird beliefs justifying absurd perfect Bayesian equilibria. The little extra detail in the game description allows simplification of the equilibrium. Also type R is quite realistic.

Both this model and the model of the previous section are Gang of Four models, but they differ in an important respect: it is crucial that the predatory pricing model have a finite number of periods. The Diamond model is different because it is not stationary: as time progresses, some firms of Type R and RS go bankrupt, which changes the lenders' payoff functions. Thus, it is not, strictly speaking, a repeated game.

13.3   Entry for Buyout

Even if entry costs exceeded operating revenues, entry might still be profitable if it is bought out. To see this, let us start by thinking about how entry might be deterred in a model with complete information. The incumbent needs some way to precommit himself to unprofitable post-entry pricing. Spence (1977) and Dixit (1980) suggest that the incumbent could enlarge his initial capacity so that the post-entry price would naturally drop to below average cost. The post-entry price would still be above average variable cost, so having already sunk the capacity cost the incumbent fights entry without further expense. The entrant's capacity cost is not yet sunk, so he refrains from entry.

In the model below with the extensive form of Figure 13.3, the incumbent has the additional option of buying out the entrant. An incumbent who fights entry bears two costs: the loss from selling at a price below average total cost, and the opportunity cost of not earning monopoly profits. He can make the first a sunk cost, but not the second. The entrant, foreseeing that the incumbent will buy him out, enters despite knowing that the duopoly price will be less than average total cost. The incumbent faces a second perfectness problem, for while he may try to deter entry by threatening not to buy out the entrant, the threat is not credible.

Figure 13.3 Entry for Buyout

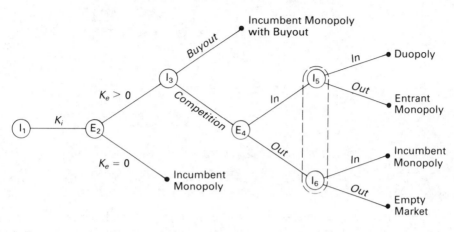

Entry for Buyout

(Rasmusen [1988a])

Players

The incumbent and the entrant.

Information

Imperfect, symmetric, complete, and certain.

Actions and Events

(1) The incumbent selects capacity K_i.
(2) The entrant decides whether to enter or stay out, choosing a capacity $K_e \geq 0$.
(3) If the entrant picks a positive capacity, the incumbent decides whether to buy him out at price B.
(4) If the entrant has been bought out, the incumbent selects output $q_i \leq K_i + K_e$.
(5) If the entrant has not been bought out, each player decides whether to stay in the market or exit.
(6) If a player has remained in the market, he selects the output $q_i \leq K_i$ or $q_e \leq K_e$.

Payoffs

Each unit of capacity costs a, the constant marginal cost is c, a firm that stays in the market incurs fixed cost F, and there is no discounting. There is only one period of production.

If no entry occurs, $\pi_{inc} = [p(q_i) - c]q_i - aK_i - F$ and $\pi_{ent} = 0$.

If entry occurs and is bought out, $\pi_{inc} = [p(q_i) - c]q_i - aK_i - B - F$ and $\pi_{ent} = B - aK_e$.

Otherwise,

$$\pi_{incumbent} = \begin{cases} [p(q_i, q_e) - c]q_i - aK_i - F & \text{if the incumbent stays.} \\ -aK_i & \text{if the incumbent exits.} \end{cases}$$

$$\pi_{entrant} = \begin{cases} [p(q_i, q_e) - c]q_e - aK_e - F & \text{if the entrant stays.} \\ -aK_e & \text{if the entrant exits.} \end{cases}$$

Two things have yet to be specified: the buyout price B and the price function $p(q_i, q_e)$. To specify them requires particular solution concepts for bargaining and duopoly, which Chapters 10 and 12 have shown are not uncontroversial. Here, they are subsidiary to the main point and can be chosen according to the taste of the modeller. We have "blackboxed" the pricing and bargaining subgames in order not to deflect attention to subsidiary parts of the model. The numerical example below will name specific functions for those subgames, but other numerical examples could use different functions to illustrate the same points.

Numerical Example

Assume that the market demand curve is

$$p = 100 - q_i - q_e. \tag{13.1}$$

Let the cost per unit of capacity be $a = 10$, the marginal cost of output be $c = 10$, and the fixed cost be $F = 601$. Assume that output follows Cournot behavior and the bargaining solution splits the surplus equally, in accordance with the Nash bargaining solution and Rubinstein (1982) (see Sections 10.2 and 10.4).

If the incumbent faced no threat of entry, he would behave as a simple monopolist, choosing a capacity equal to the output which solves

$$\underset{q_i}{Maximize} \, (100 - q_i)q_i - 10q_i - 10q_i. \tag{13.2}$$

Problem (13.2) has the first order condition

$$80 - 2q_i = 0, \tag{13.3}$$

so the monopoly capacity and output would both equal 40, yielding a net operating revenue of 1,399 $(= [p - c]q_i - F)$, well above the capacity cost of 400.

We will not go into details, but under the parameters assumed the incumbent chooses the same output and capacity of 40 even if entry is possible but buyout is not. If the potential entrant were to enter, he could do no better than to choose

$K_e = 30$, which costs 300. With capacities $K_i = 40$ and $K_e = 30$, Cournot behavior leads the two firms to solve

$$\underset{q_i}{Maximize}\, (100 - q_i - q_e)q_i - 10q_i \ \ s.t. \ \ q_i \leq 40 \tag{13.4}$$

and

$$\underset{q_e}{Maximize}\, (100 - q_i - q_e)q_e - 10q_e \ \ s.t. \ \ q_e \leq 30, \tag{13.5}$$

which have first order conditions

$$90 - 2q_i - q_e = 0 \tag{13.6}$$

and

$$90 - q_i - 2q_e = 0. \tag{13.7}$$

The Cournot outputs both equal 30, yielding a price of 40 and net revenues of $R_i^d = R_e^d = 299$ $(= [p - c]q_i - F)$. The entrant's profit net of capacity cost would be -1 $(= R_e^d - 30a)$, less than the zero from not entering.

What if both entry and buyout are possible, but the incumbent still chooses $K_i = 40$? If the entrant chooses $K_e = 30$ again, then $R_e^d = R_i^d = 299$, just as above. If he buys out the entrant, the incumbent, having increased his capacity to 70, produces a monopoly output of 45. Half of the surplus from buyout is

$$B = (1/2) \left[\underset{q_i}{Maximum}\, \{[p(q_i) - c]q_i | q_i \leq 70\} - F - (R_e^d + R_i^d) \right] \tag{13.8}$$
$$= (1/2)[(55 - 10)45 - 601 - (299 + 299)] = 413.$$

The entrant is bought out for his Cournot revenue of 299 plus the 413 which is his share of the buyout surplus, a total buyout price of 712. Since 712 exceeds the entrant's capacity cost of 300, buyout induces entry which would otherwise have been deterred. Nor can the incumbent deter entry by picking a different capacity. Choosing any K_i greater than 30 leads to the same Cournot output of 60 and the same buyout price of 712. Choosing K_i less than 30 allows the entrant to make a profit even without being bought out.

Realizing that entry cannot be deterred, the incumbent would choose a smaller initial capacity. A Cournot player whose capacity is less than 30 would produce a quantity equal to his capacity. Since buyout will occur, if a firm starts with a capacity less than 30 and adds one unit, the marginal cost of capacity is 10 and the marginal benefit is the increase (for the entrant) or decrease (for the incumbent) in the buyout price. If it is the entrant who adds a unit of capacity, R_e^d rises by at least $(40 - 10)$, the lowest possible Cournot price minus the marginal cost of output. Moreover, R_i^d falls because the entrant's extra output lowers the market price, so under our bargaining solution the buyout price rises by more than 15 $(= [40 - 10]/2)$ and the entrant should add extra capacity up to $K_e = 30$. A parallel argument shows why the incumbent should build a capacity of at least 30. Increasing the capacities any further leaves the buyout price unchanged, because the duopoly net revenues are unaffected, so both firms choose exactly 30.

The industry capacity equals 60 when buyout is allowed, but after the buyout only 45 is used. Industry profits in the absence of possible entry would have been 999 ($= 1,399 - 400$), but with buyout they are 824 ($= 1,424 - 600$), so buyout has decreased industry profits by 175. Consumer surplus has risen from 800 ($= 0.5[100-p(q|K = 40)][q|K = 40]$) to 1,012.5 ($= 0.5[100-p(q|K = 60)][q|K = 60]$), a gain of 212.5, so buyout raises total welfare in this example. The increase in output outweighs the inefficiency of the entrant's investment in capacity, an outcome that depends on the particular parameters we chose.

This model is a bundle of paradoxes. In the numerical example, allowing the incumbent to buy out the entrant raised total welfare, even though it solidified monopoly power and resulted in wasteful excess capacity. Under other parameters, the effect of excess capacity dominates, and allowing buyout would lower welfare— but only because it encourages entry, of which we usually approve. Adding more potential entrants would also have perverse effects. If the incumbent's excess capacity can deter one entrant, it can deter any number. We have seen that a single entrant might enter anyway, for the sake of the buyout price. But if there are many potential entrants, it is easier to deter entry. Buying out a single entrant would not do the incumbent much good, so he would only be willing to pay a small buyout price, and the small price would discourage any entrant from being the first. The game becomes complicated, but clearly the multiplicity of potential entrants makes entry more difficult for any of them.

13.4   Innovation and Patent Races

Market Power as a Precursor of Innovation

In a second-best sense, at least, market power is not always inimical to social welfare. Although restrictive monopoly output is inefficient, the profits thereby generated encourage innovation, an important source of both market power and economic growth. The importance of innovation, however, is diminished because of imitation, which can so severely restrict the rewards to innovation as to entirely prevent it. An innovator generally incurs some research cost, but a discovery instantly imitated would yield zero net revenues in many markets—those, for example, with constant returns to scale and Bertrand pricing. Table 13.2 shows how the payoffs might look if the firm that innovates incurs a cost equal to one, but imitation is costless and results in Bertrand competition. Innovation is a dominated strategy.

Table 13.2 Imitation with Bertrand pricing

		Brydox	
		Innovate	*Imitate*
	Innovate	$-1, -1$	$-1, 0$
Allied			
	Imitate	$0, -1$	**0, 0**

Payoffs to: (Allied, Brydox)

Under different assumptions, innovation can occur even with costless imitation. The key is what happens in the product market. Suppose there are two firms in the industry and they behave according to some rule such as Cournot or complete collusion that can be represented ordinally by the payoffs in Table 13.3 (a version of Section 3.3's Chicken). Although the firm that innovates pays the entire cost and keeps only half the benefit, imitation is no longer dominant. Allied imitates if Brydox innovates, but not if Brydox imitates. If Allied could move first, he would bind himself not to innovate, perhaps by disbanding his research laboratory.

Table 13.3 Imitation with profits in the product market

Brydox

		Innovate	Imitate
	Innovate	1, 1	**1, 2**
Allied			
	Imitate	**2, 1**	0, 0

Payoffs to: (Allied, Brydox)

Without a first-mover advantage, the game has two pure strategy Nash equilibria, (*Innovate*, *Imitate*) and (*Imitate*, *Innovate*), and a symmetric Nash equilibrium in mixed strategies in which each firm innovates with probability 0.5. The mixed strategy equilibrium is inefficient, since sometimes both firms innovate and sometimes neither.

History might provide a focal point or explain why one player moves first. Japan was for many years incapable of doing basic scientific research, and does relatively little even today. The United States therefore had to innovate rather than imitate in the past, and today continues to do much more basic research than Japan.

Much of the literature on innovation compares the relative merits of monopoly and competition. One reason a monopoly might innovate more is because it can capture more of the benefits, capturing the entire benefit if perfect price discrimination is possible (otherwise, some of the benefit goes to consumers). In addition, the monopoly avoids a second inefficiency: entrants innovating solely to steal the old innovator's rents without much increasing consumer surplus. The welfare aspects of innovation theory—and, indeed, all aspects—are intricate, and the interested reader is referred to the surveys by Kamien & Schwartz (1982) and Reinganum (forth).

Patent Races

One way that governments respond to imitation is by issuing patents: exclusive rights to make, use, or sell an innovation. If a firm patents its discovery, other firms cannot imitate, or even make the discovery independently. Research effort therefore has a discontinuous payoff: if the researcher is the first to make a discovery, he receives the patent; if he is second, nothing. This makes patents examples of the tournaments discussed in Section 7.4, although if no player exerts any effort, none of them will get the reward, unlike in standard tournaments. Patents are also special because they lose their value if consumers find a substitute and stop buying the patented product. Moreover, the effort in tournaments is usually exerted over a fixed time period, while research usually has an endogenous time period, ending when the discovery is made. Because of this endogeneity, we call the competition a **patent race.**

We will consider two models. On the technical side, the first model shows how to derive an entire mixed strategy probability distribution, instead of just the single number derived in Section 3.3. On the substantive side, it shows how patent races lead to inefficiency.

Patent Race for a New Market

Players

Three identical firms, Allied, Brydox, and Central.

Information

Imperfect, symmetric, complete, and certain.

Actions and Events

Each firm simultaneously chooses research spending $x_i \geq 0$, $(i = a, b, c)$.

Payoffs

Firms are risk neutral and the discount rate is zero. Innovation occurs at time $T(x_i)$ where $T' < 0$. The value of the patent is V, and if several players innovate simultaneously they share its value.

$$
\pi_i = \begin{cases} V - x_i & \text{if } T(x_i) < T(x_j), \ (\forall j \neq i) & \text{(Firm } i \text{ gets the patent)} \\ \frac{V}{1+m} - x_i & \text{if } T(x_i) = T(x_k), & \text{(Firm } i \text{ shares the patent with} \\ & & m = 1 \text{ or } 2 \text{ other firms)} \\ -x_i & \text{if } T(x_i) > T(x_j) \text{ for some } j & \text{(Firm } i \text{ does not get the patent)} \end{cases}
$$

The game does not have any pure strategy Nash equilibria, because the payoff functions are discontinuous. A slight difference in research by one player can make a big difference in the payoffs, as shown in Figure 13.4 for fixed values of x_b and x_c. The research levels shown in Figure 13.4 are not equilibrium values. If Allied chose any research level x_a less than V, Brydox would respond with $x_a + \varepsilon$ and win the patent. If Allied chose $x_a = V$, then Brydox and Central would respond with $x_b = 0$ and $x_c = 0$, which would make Allied want to switch to $x_a = \varepsilon$.

There does exist a symmetric mixed strategy equilibrium, which we can derive. Let the probability with which firm i chooses a research level less than or equal to x be denoted by $M_i(x)$. In a mixed strategy equilibrium a player is indifferent

Figure 13.4   The Payoffs in "Patent Race for a New Market."

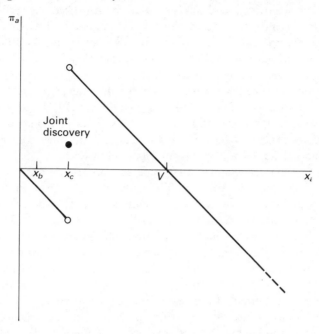

between any of the pure strategies among which he is mixing. Since we know that the pure strategies $x_a = 0$ and $x_a = V$ yield zero payoffs, so must the expected payoff for every strategy mixed between, if Allied mixes over the support $[0, V]$. The expected payoff from the pure strategy x_a is the expected value of winning minus the cost of research. Letting x stand for nonrandom and X for random variables, this is

$$V \cdot Prob(x_a \geq X_b, x_a \geq X_c) - x_a = 0, \tag{13.9}$$

which can be rewritten as

$$V \cdot Prob(X_b \leq x_a) Prob(X_c \leq x_a) - x_a = 0, \tag{13.10}$$

or

$$V \cdot M_b(x_a) M_c(x_a) - x_a = 0. \tag{13.11}$$

We can rearrange equation (13.11) to obtain

$$M_b(x_a) M_c(x_a) = \frac{x_a}{V}. \tag{13.12}$$

If all three firms choose the same mixing distribution M, we obtain

$$M(x) = \left(\frac{x}{V}\right)^{1/2} \quad \text{for } 0 \leq x \leq V. \tag{13.13}$$

What is noteworthy about a patent race is not the nonexistence of a pure strategy equilibrium, but the overexpenditure on research. All three players have expected payoffs of zero, because the patent value V is completely dissipated in the race. As in Brecht's *Threepenny Opera*, "When all race after happiness/Happiness comes in last." To be sure, the innovation is made earlier than it would have been by a monopolist, but hurrying the innovation is not worth the cost, from society's point of view, a result that would persist even if the discount rate were positive. The patent race is an example of **rent-seeking** (see Posner [1975] and Tullock [1967]), in which players dissipate the value of monopoly rents in the struggle to acquire them. Rogerson (1982) uses a game very similar to Patent Race for a New Market to analyze competition for a government monopoly franchise.

The second patent race we will analyze is asymmetric because one player is an incumbent and the other an entrant. The aims are to discover which firm spends more and to explain why firms acquire valuable patents they do not use. A typical story of a sleeping innovation, though not in this case patented, is that of synthetic caviar. In 1976, Romanoff Caviar Co. said that they had developed synthetic caviar as a "defensive marketing weapon" that they would not introduce in the US unless the Soviet Union introduced a synthetic caviar they claimed to have developed. The new product would sell for one quarter of the old price and *Business Week* said that the reason Romanoff did not introduce it was to avoid cannibalizing its old market (*Business Week*, 28 June 1976, p. 51). The game theoretic aspects of this situation put the claims of all its players in doubt, but this dubious validity makes it all the more typical of sleeping patent stories.

However difficult to validate, there do exist good theoretical models of sleeping patents, the best known of which is Gilbert & Newbery (1982), in which the incumbent firm does research and acquires a sleeping patent, while the entrant does no research. We will look at a slightly more complicated model, which does not reach such an extreme result.

Patent Race for an Old Market

Players

An incumbent and an entrant.

Information

Imperfect, symmetric, complete, and uncertain.

Actions and Events

(1)   The firms simultaneously choose research spending x_i and x_e, which result in research achievements $f(x_i)$ and $f(x_e)$, where $f' > 0$ and $f'' < 0$.

(2)   Nature chooses which player wins the patent using a function g that maps research achievement to a probability between zero and one.

$$Prob(\text{incumbent wins patent}) = g[f(x_i) - f(x_e)], \qquad (13.14)$$

where $g' > 0$, $g(0) = 0.5$, and $0 \le g \le 1$.

(3)   The winner of the patent decides whether to spend Z to implement it.

Payoffs

The old patent yields revenue y and the new patent yields v. The payoffs are shown in Table 13.4.

Table 13.4 Patent Race for an Old Market: payoffs

Outcome	$\pi_{incumbent}$	$\pi_{entrant}$
The entrant wins and implements	$-x_i$	$v - x_e - Z$
The incumbent wins and implements	$v - x_i - Z$	$-x_e$
Neither player implements	$y - x_i$	$-x_e$

The entrant will do no research unless he plans to implement, so we will disregard his strongly dominated strategy, $(x_e > 0, \textit{no implementation})$. The incumbent wins with probability g and the entrant with probability $1 - g$, so from Table 13.4 the expected payoff functions are

$$\pi_{incumbent} = (1 - g[f(x_i) - f(x_e)])(-x_i) + g[f(x_i) - f(x_e)]Max\{v - x_i - Z, y - x_i\} \qquad (13.15)$$

and

$$\pi_{entrant} = (1 - g[f(x_i) - f(x_e)])(v - x_e - Z) + g[f(x_i) - f(x_e)](-x_e). \qquad (13.16)$$

On differentiating and simplifying the notation, we obtain the first order conditions

$$\frac{d\pi_i}{dx_i} = -(1 - g[f_i - f_e]) - g' f_i'(-x_i) + g' f_i' Max\{v - x_i - Z, y - x_i\} - g[f_i - f_e] = 0 \qquad (13.17)$$

and

$$\frac{d\pi_e}{dx_e} = -(1 - g[f_i - f_e]) + g' f_e'(v - x_e - Z) - g[f_i - f_e] + g' f_e' x_e = 0. \qquad (13.18)$$

Equating (13.17) and (13.18), which both equal zero, we obtain

$$-(1-g) + g'f_i'x_i + g'f_i'Max\{v - x_i - Z, y - x_i\} - g =$$
$$-(1-g) + g'f_e'(v - x_e - Z) - g + g'f_e'x_e, \quad (13.19)$$

which simplifies to

$$f_i'[x_i + Max\{v - x_i - Z, y - x_i\}] = f_e'[v - x_e - Z + x_e], \quad (13.20)$$

or

$$\frac{f_i'}{f_e'} = \frac{v - Z}{Max\{v - Z, y\}}. \quad (13.21)$$

We can use equation (13.21) to show that different parameters generate two qualitatively different outcomes.

Outcome 1   *The entrant and incumbent spend equal amounts, and each implements if successful.* This happens if there is a big gain from patent implementation, that is, if

$$v - Z \geq y, \quad (13.22)$$

so that equation (13.21) becomes

$$\frac{f_i'}{f_e'} = \frac{v - Z}{v - Z} = 1, \quad (13.23)$$

which implies that $x_i = x_e$.

Outcome 2   *The incumbent spends more and does not implement if he is successful (he acquires a sleeping patent).* This happens if the gain from implementation is small, that is, if

$$v - Z < y, \quad (13.24)$$

so that equation (13.21) becomes

$$\frac{f_i'}{f_e'} = \frac{v - Z}{y} < 1, \quad (13.25)$$

which implies that $f_i' < f_e'$. Since we assumed that $f'' < 0$, f' is decreasing in x, and it follows that $x_i > x_e$.

This model shows that the presence of another player can stimulate the incumbent to do research he otherwise would not, and that he may or may not implement the discovery. The incumbent has at least as much incentive for research as the entrant, because a large part of a successful entrant's payoff comes at the incumbent's expense. The benefit to the incumbent is the maximum of the benefit from implementing and the benefit from stopping the entrant, but the entrant's benefit can only come from implementing. Contrary to the popular belief that sleeping patents are bad, here they can help society by eliminating wasteful implementation.

13.5   Takeovers and Greenmail

The Free Rider Problem

Game theory is well suited to modelling takeovers because the takeover process depends crucially on information and includes a number of sharply delineated actions and events. Suppose that under its current mismanagement, a firm has a value per share of v, but no shareholder has enough shares to justify the expense of a proxy fight to throw out the current managers, although doing so would raise the value to $(v + x)$. An outside bidder makes a tender offer, for any shares offered, conditional upon obtaining a majority. Any bid p between v and $(v + x)$ can make both the bidder and the shareholders better off. But do the shareholders accept such an offer?

We will see that they do not. Quite simply, the only reason the bidder makes a tender offer is that the value would rise higher than his bid, so no shareholder should accept any bid he makes.

The Free Rider Problem in Takeovers

(Grossman & Hart [1980])

Players

A bidder and a continuum of shareholders, with amount m of shares.

Information

Imperfect, symmetric, complete, and certain.

Actions and Events

(1)   The bidder offers p per share for the m shares.
(2)   Each shareholder decides whether to accept the bid (denote by θ the fraction that accept).
(3)   If $\theta \geq 0.5$, the bid price is paid out, and the value of the firm rises from v to $(v + x)$ per share.

Payoffs

If $\theta < 0.5$, the takeover fails, the bidder's payoff is zero, and the shareholder's payoff is v per share. Otherwise,

$$\pi_{bidder} = \{ \theta m(v + x - p) \quad \text{if } \theta \geq 0.5.$$

$$\pi_{shareholder} = \begin{cases} p & \text{if the shareholder accepts.} \\ v + x & \text{if the shareholder rejects.} \end{cases}$$

In any iterated dominant strategy equilibrium, the bidder's payoff equals zero. Bids above $(v + x)$ are dominated strategies, since the bidder could not possibly profit from them. But if the bid is any lower, an individual shareholder should hold out for the new value of $(v + x)$ rather than accepting p. To be sure, when they all do that, the offer fails and they end up with v, but no individual wants to accept if he thinks the offer will succeed. The only equilibria are the many strategy combinations that lead to a failed takeover, or a bid of $p = (v + x)$ accepted by a majority, which succeeds but yields a payoff of zero to the bidder. If organizing an offer has even the slightest cost, the bidder would not do it.

The free rider problem is clearest where there is a continuum of shareholders, so that the decision of any individual does not affect the success of the tender offer. If there were, instead, nine players with one share each, then in one asymmetric equilibrium five of them tender at a price just slightly above the old market price and four hold out. Each of the five tenderers knows that if he held out, the offer would fail and his payoff would be zero. This is an example of the discontinuity problem of Section 7.7.

In practice, the free rider problem is not quite so severe even with a continuum of shareholders. If the bidder can quietly buy a sizeable number of shares without driving up the price (something severely restricted in the United States by the Williams Act), then his capital gains on those shares can make a takeover profitable even if he makes nothing from shares bought in the public offer. Dilution tactics such as freeze-out mergers (see Macey & McChesney [1985]) also help the bidder. In a freeze-out, the bidder buys 51 percent of the shares and merges the new acquisition with another firm he owns, at a price below its full value. If dilution is strong enough, the shareholders are willing to sell at a price less than $v + x$.

Still another takeover tactic is the two-tier tender offer, a nice application of the Prisoner's Dilemma. Suppose that the underlying value of the firm is 30, which is the initial stock price. A monopolistic bidder offers a price of 10 for 51 percent of the stock and 5 for the other 49 percent, conditional upon 51 percent tendering. It is then a weakly dominant strategy to tender, even though all the shareholders would be better off refusing to sell.

Greenmail

Greenmail occurs when managers buy out some shareholders at an inflated stock price to stop them from taking over. Opponents of greenmail explain this using the Corrupt Managers model. Suppose that a little dilution is possible, or the bidder owns some shares to start with, so that he can take over the firm but would lose most of the gains to the other shareholders. The managers are willing to pay the bidder a large amount of greenmail to keep their jobs, and both manager and bidder prefer greenmail to an actual takeover, despite the fact that the other shareholders are considerably worse off. The most common objection to this model is that it fails to explain why the corporate charter does not prohibit greenmail.

Managers often use what we might call the Noble Managers model to justify greenmail. In this model, current management knows the true value of the firm, which is greater than both the current stock price and the takeover bid. They pay greenmail to protect the shareholders from selling their mistakenly undervalued shares. This implies either that shareholders are irrational or that the stock price rises after greenmail because shareholders know that the greenmail signal (giving up the benefits of a takeover) is more costly for a firm which really is not worth more than the takeover bid.

Shleifer & Vishny (1986) have constructed a more sophisticated model in which greenmail is in the interest of the shareholders. The idea is that greenmail encourages potential bidders to investigate the firm, eventually leading to a takeover at a higher price than the initial offer. Greenmail is costly, but for that very reason it is an effective signal that the managers think a better offer could come along later. (Like Noble Managers, this assumes that the manager acts in the interests of the shareholders.) I will present a numerical example in the spirit of Shleifer & Vishny rather than following them exactly, since their exposition is not directed towards the behavior of the stock price.

The story behind the model is that a manager has been approached by a bidder, and he must decide whether to pay him greenmail in the hopes that other bidders—"white knights"—will appear. The manager has better information than the market as a whole about the probability of other bidders appearing, and some other bidders can only appear after they undertake costly investigation, which they will not do if they think the takeover price will be bid up by competition with the first bidder. The manager pays greenmail to encourage new bidders by getting rid of their competition.

Greenmail to Attract White Knights

(Shleifer & Vishny [1986])

Players

The manager, the market, and bidder Brydox. (Bidders Raider and Allied do not make decisions.)

Information

Asymmetric, incomplete, and certain.

Actions and Events

The game tree is shown in Figure 13.5. After each time t, the market picks a share price p_t.

(0)   Unobserved by any player, Nature picks the state to be (A), (B), (C), or (D), with probabilities 0.1, 0.3, 0.1, and 0.5.

(1)   Unless the state is (D), the Raider appears and offers a price of 15. The manager's information partition becomes $\{(A), (B, C), (D)\}$; everyone else's becomes $\{(A, B, C), (D)\}$.

(2)   The manager decides whether to pay greenmail and extinguish the Raider's offer at a cost of 5 per share.

(3)   If the state is (A), Allied appears and offers a price of 25 if greenmail was paid, and 30 otherwise.

(4)   If the state is (B), Brydox decides whether to buy information at a cost of 8 per share. If he does, then he can make an offer of 20 if the Raider has been paid greenmail, or 27 if he must compete with the Raider.

(5)   Shareholders accept the best offer outstanding, which is the final value of a share. If no offer is outstanding, the final value is 5 if greenmail was paid, 10 otherwise.

Payoffs

The manager maximizes the final value.

The market minimizes the absolute difference between p_t and the final value.

If he buys information, Brydox receives 23 ($= 31 - 8$) minus the value of his offer; otherwise he receives zero.

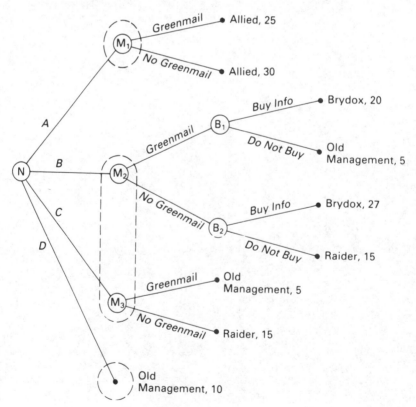

Figure 13.5   Greenmail to Attract White Knights

The payoffs specify that the manager should maximize the final value of the firm, rather than a weighted average of the prices p_0 through p_5. This assumption is reasonable because the only shareholders to benefit from a high value of p_t are those that sell their stock at t. The manager cannot say: "The stock is overvalued: Sell!", because the market would learn the overvaluation too, and refuse to buy.

The prices 15, 20, 27, and 30 are assumed to be the results of blackboxed bargaining games between the manager and the bidders. Assuming that the value of the firm to Brydox is 31 ensures that he will not buy information if he foresees that he would have to compete with the Raider. Since Brydox has a dominant strategy— buy information if the Raider has been paid greenmail and not otherwise—our focus will be on the market price and the decision of whether to pay greenmail. This model is also not designed to answer the question of why the Raider appears. His behavior is exogenous. As the model stands, his expected profit is positive since he is sometimes paid greenmail, but if he actually had to buy the firm he would regret it in states B and C, since the final value of the firm would be 10.

We will see that in equilibrium the manager pays greenmail in states (B) and (C), but not in (A) or (D). Table 13.5 shows the equilibrium path of the market price.

Table 13.5: Greenmail to Attract White Knights: the equilibrium price

State	Prob.	p_0	p_1	p_2	p_3	p_4	p_5	Final management
(A)	0.1	14.5	19	30	30	30	30	Allied
(B)	0.3	14.5	19	16.25	16.25	20	20	Brydox
(C)	0.1	14.5	19	16.25	16.25	5	5	Old Management
(D)	0.5	14.5	10	10	10	10	10	Old Management

The market's optimal strategy amounts to estimating the final value. Before the market receives any information, its prior beliefs estimate the final value to be 14.5 ($= 0.1[30] + 0.3[20] + 0.1[5] + 0.5[10]$). If state (D) is ruled out by the arrival of the Raider, the price rises to 19 ($= 0.2[30] + 0.6[20] + 0.2[5]$). If the Raider does not appear, it becomes common knowledge that the state is (D), and the price falls to 10.

If the state is (A), the manager knows it and refuses to pay greenmail in expectation of Allied's offer of 30. Observing the lack of greenmail, the market deduces that the state is (A), and the price immediately rises to 30.

If the state is (B) or (C) the manager does pay greenmail and the market, ruling out (A), uses Bayes's Rule to assign probabilities of 0.75 to (B) and 0.25 to (C). The price falls from 19 to 16.25 ($= 0.75[20] + 0.25[5]$).

It is clear that the manager should not pay greenmail in states (A) or (D), when the manager knows that Brydox is not around to investigate. What if the manager deviates in the information set (B,C) and refuses to pay greenmail? The market would initially believe that the state was (A), so the price would rise to $p_2 = 30$. But the price would fall again after Allied failed to make an offer and the market realized that the manager had deviated. Brydox would refuse to enter at Time 3, and the Raider's offer of 15 would be accepted. The payoff of 15 would be less than the expected payoff of 16.25 from paying greenmail.

The model does not say that greenmail is always good for the shareholders, only that it can be good *ex ante*. If the true state turns out to be (C), then greenmail was a mistake, *ex post*, but since state (B) is more likely, the manager is correct to pay greenmail in information set (B,C). What is noteworthy is that greenmail is optimal even though it drives down the stock price from 19 to 16.25. Greenmail communicates the bad news that Allied is not around, but makes the best of that misfortune by attracting Brydox.

Recommended Reading

Gilbert, Richard & David Newbery (1982) "Preemptive Patenting and the Persistence of Monopoly" *American Economic Review.* June 1982. 72, 3: 514–26.

Kreps, David & Robert Wilson (1982a) "Reputation and Imperfect Information" *Journal of Economic Theory.* August 1982. 27, 2: 253–79.

Rasmusen, Eric (1988a) "Entry for Buyout" *Journal of Industrial Economics.* March 1988. 36, 3: 281–300.

Problem 13

A Patent Race with Two Firms

Adopt the model of Patent Race for a New Market except with two firms, Allied and Brydox, instead of three. Let $T(x) = 2000 - x$ and $V = 100$.

(1)   If Allied moves first, what strategy does it choose?
(2)   What is the value of $M(x)$ in a symmetric equilibrium when both firms move simultaneously?
(3)   What is the expected innovation time for each firm?
(4)   What is the expected innovation time for the industry?
       (Hint: find the probability that x_a equals a particular value y and $x_b \leq x_a$, do the same for x_b, and integrate.)

Notes

N13.1   Predatory Pricing: the Kreps–Wilson Model

- Books on the new theoretical organization include Fudenberg & Tirole (1986a), Jacquemin (1985), Krouse (unpub), the collection edited by Stiglitz & Mathewson (1986), Schmalensee & Willig (forth), and Tirole (1988). Currently the most advanced textbook on the theory of regulation is Spulber (forth).
- Kreps & Wilson (1982a) call the types "strong" and "weak" rather than "tough" and "standard". I am uncomfortable calling the fighting monopolist "strong," because he has a smaller set of possible actions and he is often modelled as irrational. The terms "strong" and "weak" also fail to emphasize that one type is rare.
- Kreps & Wilson (1982a) do not simply assume that one type of monopolist always chooses *Fight*. They make the more elaborate but primitive assumption that his payoff function makes fighting a dominant strategy. Table 13.6 shows a set of payoffs for the tough monopolist which generate this result.

Table 13.6 Tough monopolist

Incumbent

		Collude	*Fight*
	Enter	20, 10	−10, 40
Entrant			
	Stay Out	0, 100	**0, 100**

Payoffs to: (Entrant, Incumbent)

Under the Kreps-Wilson assumption, the tough monopolist would actually choose to collude in the early periods of the game in some perfect Bayesian equilibria. Such an equilibrium could be supported by out-of-equilibrium beliefs that the authors point out are absurd: if the monopolist fights in the early periods, the entrant believes he must be a standard monopolist.

N13.3 Entry for Buyout

- Dynamic models like Entry for Buyout frequently contain subgames where the modeller must choose between having two players move simultaneously or in sequence. Usually having them move in sequence is simpler, as in the Switching Costs model of Section 12.4, but sometimes, as here in the decision of whether to stay or exit, the first player to move would have an unwarranted first-mover advantage.

N13.4 Innovation and Patent Races

- The idea of the patent race is described by Barzel (1968), although his model showed the same effect of overhasty innovation even without patents.
- The Brecht quotation is from Act III, scene 7 of the *Threepenny Opera*, translated by John Willett (Berthold Brecht, *Collected Works*, London: Eyre Methuen, 1987).
- Reinganum (1985) has shown that an important element of patent races is whether increased research hastens the arrival of the patent or just affects whether it is acquired. If more research hastens the innovation, then the incumbent might spend less than the entrant because the incumbent is enjoying a stream of profits from his present position that the new innovation destroys.
- **Uncertainty in Innovation.** Patent Race for an Old Market is only one way to model innovation under uncertainty. A more common way to model uncertainty, used by Loury (1979) and Dasgupta & Stiglitz (1980), uses continuous time with discrete discoveries and specifies that discoveries arrive as a Poisson process with parameter $\lambda(X)$, where X is research expenditure, $\lambda' > 0$, and $\lambda'' < 0$. Then

$$Prob(\text{invention at } t) = \lambda e^{-\lambda(X)t};$$
$$Prob(\text{invention before } t) = 1 - e^{-\lambda(X)t}. \tag{13.26}$$

A little algebra gives us the current value of the firm, V_0, as a function of the innovation rate, the interest rate, the post-innovation value V_1, and the current revenue flow R_0. The return on the firm equals the current cash flow plus the probability of a capital gain.

$$rV_0 = R_0 - X + \lambda(V_1 - V_0), \tag{13.27}$$

which implies

$$V_0 = \frac{\lambda V_1 + R_0 - X}{\lambda + r}. \tag{13.28}$$

Expression (13.28) is frequently useful in modelling.

Mathematical Appendix

This appendix has three purposes: to remind some readers of the definitions of terms they have seen before, to give other readers an idea of what the terms mean, and to list a few theorems for reference. In accordance with these limited purposes, some terms such as "boundary point" are left undefined. For fuller exposition, see Rudin (1964) on real analysis, Debreu's *Theory of Value* (1959), and Chiang (1984) and Takayama (1985) on mathematics for economists. Intriligator (1971) and Varian (1984) both have good mathematical appendices and are strong in discussing optimization, and Kamien & Schwartz (1981) covers maximizing by choice of functions. Border (1985) is a recent reference on fixed point theorems.

Notation

$\sum$ Summation. $\sum_{i=1}^{3} x_i = x_1 + x_2 + x_3$.

Π Product. $\Pi_{i=1}^{3} x_i = x_1 x_2 x_3$.

$|\ |$ Absolute value. If $x \geq 0$ then $|x| = x$ and if $x < 0$ then $|x| = -x$.

$|$ "Such that," "given that," or "conditional upon." $\{x | x < 3\}$ denotes the set of real numbers less than three. $Prob(x | y < 5)$ denotes the probability of x given that y is less than 5.

$:$ "Such that." $\{x : x < 3\}$ denotes the set of real numbers less than three. The colon is a synonym for $|$.

$\mathbf{R}^n$ The set of n-dimensional vectors of real numbers (integers, fractions, and the least upper bounds of any subsets thereof).

$\{\ \}$ A set of elements. The set $\{3, 5\}$ consists of two elements, 3 and 5.

$\in$ "Is an element of." $a \in \{2, 5\}$ means that a takes either the value 2 or 5.

$\subset$ Set inclusion. If $X = \{2, 3, 4\}$ and $Y = \{2, 4\}$, then $Y \subset X$ because Y is a subset of X.

[a, b] A closed interval. The interval $[0, 1000]$ is the set $\{x | 0 \leq x \leq 1000\}$. Square brackets are also used as delimiters.

(a, b) An open interval. The interval $(0, 1000)$ is the set $\{x | 0 < x < 1000\}$. $(0, 1000]$ would be a half-open interval, the set $\{x | 0 < x \leq 1000\}$. Parentheses are also used as delimiters.

× The Cartesian product. $X \times Y$ is the set of points $\{x, y\}$, where $x \in X$ and $y \in Y$.

ϵ An arbitrarily small positive number. If my payoff from both *Left* and *Right* equals 10, I am indifferent between them; if my payoff from *Left* is changed to $10 + \epsilon$, I prefer *Left*.

~ We say that $X \sim F$ if the random variable X is distributed according to distribution F.

∃ "There exists..."

∀ "For all..."

≡ "Equals by definition."

→ If f maps space X into space Y then $f : X \rightarrow Y$.

$\frac{df}{dx}, \frac{d^2f}{dx^2}$ The first and second derivatives of a function. If $f(x) = x^2$ then $\frac{df}{dx} = 2x$ and $\frac{d^2f}{dx^2} = 2$.

f', f'' The first and second derivatives of a function. If $f(x) = x^2$ then $f' = 2x$ and $f'' = 2$. Primes are also used on variables (not functions) for other purposes: x' and x'' might denote two particular values of x.

$\frac{\partial f}{\partial x}, \frac{\partial^2 f}{\partial x \partial y}$ Partial derivatives of a function. If $f(x, y) = x^2 y$ then $\frac{\partial f}{\partial x} = 2xy$ and $\frac{\partial^2 f}{\partial x \partial y} = 2x$.

y_{-i} The set y minus element i. If $y = \{y_1, y_2, y_3\}$, then $y_{-2} = \{y_1, y_3\}$.

$Max(\ ,\ )$ The maximum of two numbers. $Max(x, y)$ denotes the greater of x and y.

$Min(\ ,\ )$ The minimum of two numbers. $Min(5, 3) = 3$.

$Sup\ X$ The supremum (least upper bound) of set X. If $X = \{x | 0 \leq x < 1000\}$, then $sup\ X = 1000$. The supremum is useful because sometimes, as here, no maximum exists.

$Inf\ X$ The infimum (greatest lower bound) of set X. If $X = \{x | 0 \leq x < 1000\}$, then $inf\ X = 0$.

Argmax The argument that maximizes a function. If $e^* = argmax\ EU(e)$, then e^* is the value of e that maximizes the function $EU(e)$. The argmax of $f(x) = x - x^2$ is $1/2$.

Maximum The greatest value that a function can take. $Maximum(x - x^2) = 1/4$.

The Greek Alphabet

A	α	alpha	N	ν	nu
B	β	beta	Ξ	ξ	xi
Γ	γ	gamma	O	o	omicron
Δ	δ	delta	Π	π	pi
E	ϵ or ε	epsilon	P	ρ	rho
Z	ζ	zeta	Σ	σ	sigma
H	η	eta	T	τ	tau
Θ	θ	theta	Υ	υ	upsilon
I	ι	iota	Φ	ϕ	phi
K	κ	kappa	X	χ	chi
Λ	λ	lambda	Ψ	ψ	psi
M	μ	mu	Ω	ω	omega

Glossary

almost always   See "generically."

closed   A closed set in $\mathbf{R}^n$ includes its boundary points. The set $\{x : 0 \le x \le 1000\}$ is closed.

compact   If set X in $\mathbf{R}^n$ is closed and bounded, then X is compact.

complete lattice   A lattice is complete if the infimum and supremum of each of its subsets are in the lattice.

complete metric space   All compact metric spaces and all Euclidean spaces are complete.

concave function   The continuous function $f(x)$ is concave if for all elements w and z of X, $f(0.5w + 0.5z) \ge 0.5f(w) + 0.5f(z)$. If f maps $\mathbf{R}$ into $\mathbf{R}$ and f is concave, then if $f' > 0$, $f'' \le 0$, while if $f' < 0$, $f'' \ge 0$. See Figure A.1.

continuous function   Let $d(x, y)$ represent the distance between points x and y. The function f is continuous if for every $\epsilon > 0$ there exists a $\delta(\epsilon) > 0$ such that $d(x, y) < \delta(\epsilon)$ implies $d(f(x), f(y)) < \epsilon$.

continuum   A continuum is a closed interval of the real line, or a set that can be mapped one-to-one onto such an interval.

contraction   The mapping $f(x)$ is said to be a contraction if there exists a number $c < 1$ such that for the metric d of the space X,

$$d(f(x), f(y)) \le cd(x, y), \text{ for all } x, y \in X. \qquad (A.1)$$

convex function   The continuous function $f(x)$ is convex if for all elements w and z of X, $f(0.5w + 0.5z) \le 0.5f(w) + 0.5f(z)$. See Figure A.1. Convex functions are only loosely related to convex sets.

convex set   If set X is convex, then if you take any two of its elements w and z and a real number $t : 0 \le t \le 1$, then $tw + (1 - t)z$ is also in X.

Figure A.1   Concave and convex functions

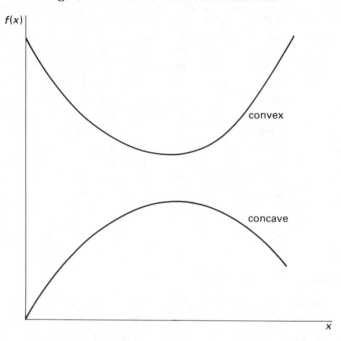

correspondence   A correspondence is a mapping that maps each point to one or more other points, as opposed to a function, which only maps to one.

Euclidean space.   Euclidean n-space is the set of vectors $\mathbf{R}^n$ with the standard inner product and norm.

function   If f maps each point in X to exactly one point in Y, f is called a function. The two mappings in Figure A.1 are functions, but the mapping in Figure A.2 is not.

generically   If a fact is true on set X generically, "except on a set of measure zero," or "almost always," then it is false only on a subset of points Z that have the property that if a point is randomly chosen using a density function with support X, a point in Z is chosen with probability zero. This implies that if the fact is false on $z \in \mathbf{R}^n$ and z is perturbed by adding a random amount ϵ, the fact is true on $z + \epsilon$ with probability one.

isotone   A mapping f is isotone if $a \geq b$ implies $f(a) \geq f(b)$.

Lagrange multiplier   The Lagrange multiplier λ is the marginal value of relaxing a constraint in an optimization problem. If the problem is

$$\{ \underset{x}{Maximize}\ x^2 \text{ subject to } x \leq 5 \}, \text{ then } \lambda = 2x^* = 10.$$

lattice   A lattice is a partially ordered set (the $\geq$ mapping is defined) where for any two elements a and b, the values $inf(a, b)$ and $sup(a, b)$ are also in the set.

lower semicontinuous correspondence   The correspondence ϕ is lower semicontinuous at the point x_0 if

$$x_n \to x_0, \ y_0 \in \phi(x_0), \text{ implies } \exists y_n \in \phi(x_n) \text{ such that } y_n \to y_0, \qquad \text{(A.2)}$$

which means that associated with every x sequence leading to x_0 is a y sequence leading to its image. See Figure A.2. This idea is not as important as upper semicontinuity.

measure zero   See "generically."

metric   The function $d(w, z)$ defined over elements of set X is a metric if (1) $d(w, z) > 0$ if $w \neq z$ and $d(w, z) = 0$ if $w = z$; (2) $d(w, z) = d(z, w)$; and (3) $d(w, z) \leq d(w, y) + d(y, z)$ for points $w, y, z \in X$.

metric space   Set X is a metric space if it is associated with a metric that defines the distance between any two of its elements.

one-to-one   The mapping $f : X \to Y$ is one-to-one if every point in set X is mapped into one point in set Y.

onto   The mapping $f : X \to Y$ is onto Y if every point in Y is mapped onto by some point in X.

open   In the space $\mathbf{R}^n$, an open set is one that does not include all its boundary points. The set $\{x : 0 < x < 1000\}$ is open. In more general spaces, an open set is a member of a topology.

quasi-concave   The continuous function f is quasi-concave if for $w \neq z$, $f(0.5w + 0.5z) > min[f(w), f(z)]$, or, equivalently, if the set $\{x \in X | f(x) > b\}$ is convex for any number b. Every concave function is quasi-concave, but not every quasi-concave function is concave.

strict   The word "strict" is used in a variety of contexts to mean that a relationship does not hold with equality or is not arbitrarily close to being violated. If function f is concave and $f' > 0$, then $f'' \leq 0$, but if f is strictly concave, then $f'' < 0$. The opposite of "strictly" is "weakly." The word "strong" is sometimes used as a synonym for "strict."

support   The support of a probability distribution $F(x)$ is the closure of the set of values of x such that the density is positive. If each output between 0 and 20 has a positive probability density, and no other output does, then the support of the output distribution is [0,20].

topology   Besides denoting a field of mathematics, a topology is a collection of subsets of a space called "open sets" that includes (1) the entire space and the empty set, (2) the intersection of any finite number of open sets, and (3) the union of any number of open sets. In a metric space, the metric "induces" a topology by defining an open set. Imposing a topology on a space is something like defining which elements are close to each other, which is easy to do for $\mathbf{R}^n$ but not for every space (e.g., spaces consisting of functions or of game trees).

uniform distribution.   A variable is uniformly distributed over support X if each point in X has equal probability. The density function for support $X =$

$[\alpha, \beta]$ is

$$f(x) = \begin{cases} 0 & x < \alpha \\ \frac{1}{\beta - \alpha} & \alpha \leq x \leq \beta \\ 0 & x > \beta, \end{cases} \tag{A.3}$$

and the cumulative density function is

$$F(x) = \begin{cases} 0 & x < \alpha \\ \frac{x - \alpha}{\beta - \alpha} & \alpha \leq x \leq \beta \\ 1 & x > \beta. \end{cases} \tag{A.4}$$

upper semicontinuous correspondence   The correspondence $\phi : X \to Y$ is upper semicontinuous at point x_0 if

$$x_n \to x_0, \ y_n \in \phi(x_n), \ y_n \to y_0, \ \text{implies} \ y_0 \in \phi(x_0), \tag{A.5}$$

which means that every sequence of points in $\phi(x)$ leads to a point also in $\phi(x)$. See Figure A.2. An alternative definition, appropriate only if Y is compact, is that ϕ is upper semicontinuous if the set of points $\{x, \phi(x)\}$ is closed.

Figure A.2   Upper semicontinuity

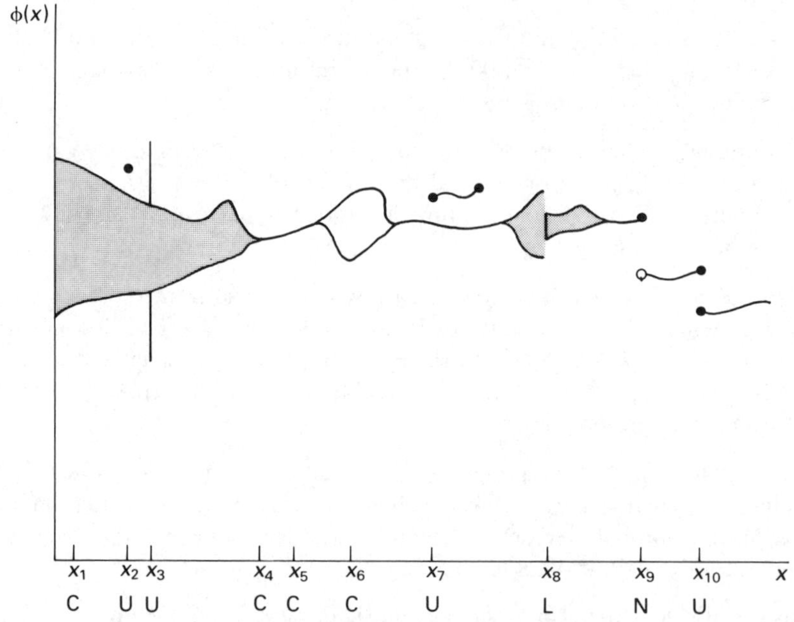

Note: Points at which the correspondence is just upper semicontinuous are labelled U; just lower semicontinuous, L; both, C; and neither, N.

vector   A vector is an ordered set of real numbers, a point in $\mathbf{R}^n$. The point $(2.5, 3, -4)$ is a vector in $\mathbf{R}^3$.

weak   The word "weak" is used in a variety of contexts to mean that a relationship might hold with equality or be on a borderline. If f is concave and $f' > 0$, then $f'' \leq 0$, but to say that f is weakly concave, while technically adding nothing to the meaning, emphasizes that $f'' = 0$ under some or all parameters. The opposite of "weak" is "strict" or "strong."

Fixed Point Theorems

Theorem A.1   Brouwer's Fixed Point Theorem

A continuous function mapping a compact, convex set into itself has a fixed point. If $f : B \rightarrow B$ is a continuous function defined for $x \in B \subset \mathbf{R^n}$, where B is compact and convex, then there exists an x^* such that $f(x^*) = x^*$.

Brouwer's theorem (Rudin [1964] p. 203) is the most basic of the fixed point theorems.

Theorem A.2   Kakutani Fixed Point Theorem (Kakutani [1941])

Let $\phi(x) : B \rightarrow B$ be an upper semicontinuous correspondence defined for $x \in B \subset \mathbf{R^n}$. If B is compact and convex, and for all x it is true that $\phi(x)$ is nonempty and convex, then ϕ has a fixed point: there exists $x^* : x^* \in \phi(x^*)$.

Theorems 5.5, 5.6, 5.7, and 5.8 are proved using the Kakutani fixed point theorem.

Theorem A.3   Contraction Mapping Theorem

A contraction $f(x)$ of a complete metric space X into itself has a unique fixed point x^* such that $f(x^*) = x^*$.

The contraction mapping theorem (Rudin [1964] p. 220) is special because it guarantees a unique fixed point, but it rarely applies.

Theorem A.4   Tarski Fixed Point Theorem

Any isotone mapping from a complete lattice into itself has a fixed point.

The Tarski fixed point theorem is little-known. Its usefulness is that existence of a fixed point does not depend on the strategy set being compact or the best responses being continuous. For details, see Lippman, Mamer, & McCardle (1987) or Tarski (1955).

Theorem A.5   Fan's Lemma (Fan [1952])

Let $S \subset \mathbf{R}^n$ be convex and compact. Let $F : S \times S \rightarrow \mathbf{R}^1$ be continuous, and *concave in its first argument, and suppose that $F(x, y) = 0$ for any $x = y$. Then there exists an x^* such that*

$$\underset{y \in S}{Maximum} \;\; F(y, x^*) = 0. \tag{A.6}$$

Fan's lemma comes up occasionally, and Dasgupta & Maskin (1986a) use it to prove Theorem 5.10. The set S can represent a strategy set, and F can represent a payoff function. Fan has done other work in this area, so one must be careful of confusing his propositions: Border (1985) refers to a different result as "Fan's Lemma."

Answers to Problems

Problem 1 A Discoordination Game

(1)

Woman

Prize Fight   *Ballet*

	Prize Fight	*Ballet*
Prize Fight	2, 1	−1, 10
Ballet	−3, 8	1, 2

Man (row player label)

Payoffs to: (Man, Woman)

(2) (*Ballet, Ballet*).

(3) No—it has a first-mover disadvantage.

(4) (*Prize Fight, Ballet*) and (*Ballet, Prize Fight*) are not Nash because the man would deviate; (*Prize Fight, Prize Fight*) and (*Ballet, Ballet*) are not, because the woman would.

Problem 2 Joint Ventures

(1) Diagram.

(2) (*Low* if defective parts, *Low* if not defective parts, *Low*).

(3) One.

(4) $0.72 = (1)(0.7)/(0.7 + 0.27)$.

Problem 3   Mixed Strategies in the Battle of the Sexes

(1) $\pi_m = p_m(2p_w + -1[1 - p_w]) + (1 - p_m)(-5p_w + 1[1 - p_w])$.
(2) $\frac{d\pi_m}{dp_m} = (2p_w - [1 - p_w]) - (-5p_w + [1 - p_w]) = 0$.
(3) $p_w^* = 2/9, p_m^* = 7/9, \pi_m = -1/3, \pi_w = -1/3$.
(4)   *(Prize Fight, Ballet)* for (M, W). 0.604.
(5)   *Prob(Large)* $= 2/5$, $\pi = 1/5$.
(6)   Pareto optimality produces a pure strategy focal point in Pure Coordination.

Problem 4   Repeated Entry Deterrence

(1)   Diagram.
(2)   Let E_1, E_2, F_1, etc. denote the realized actions:
       $\{(E, E|(E_1, C_1), S|(E_1, F_1); E|S_1), (F|E_1, C|(E_1, F_1), C|(E_1, C_1), F|(S_1, E_2))\}$.
(3) $\{(\text{Always enter}),(\text{Always collude})\}$.
(4) $\{(\text{Stay out, then enter}), (\text{Fight, then collude})\}$.

Problem 5   Limit Pricing

(1)   Nature moves, then Allied, then Brydox.
(2)   Either Brydox would never enter, so Allied would wish to deviate from $Low|20$
       or Brydox would enter after $High$, so Allied would wish to deviate from
       $High|30$.
(3) $\{(Low|20, Low|30), (Out|Low, In|High), Prob(20|High) = 0\}$.
(4) $Prob(20|High) = 1$.

Problem 6   Worker Effort

(1)   Yes. He wants to be insured.
(2)   Yes. He needs the incentive.
(3)   The wage would vary more across states.
(4)   Pay equal amounts whether or not a mistake is made, but zero if a slight
       mistake is made.

Problem 7   Defense Contractors

(1)   The plane is ordered if $m_a + m_b \leq 50$. If the plane is ordered, Allied is paid
       $50 - m_b$, and Brydox $50 - m_a$.
(2)   Allied gets \$40 million and Brydox gets \$30 million.
(3)   Zero.
(4)   Yes. Unpopular transfers, and collusion.

Problem 8   Worker Ability

(1)   *Bads* are fully insured; *Goods* are partially insured and must separate.

(2) Under optimal pooling, workers are fully insured, but *Goods* prefer partial insurance.

(3) Make the proportion of *Goods* very large.

(4) Wilson is pooling; Reactive is separating.

Problem 9   Warranties

(1) Signalling.

(2) Allied: (P=11, *Warranty*), Other: (P=2, *No Warranty*), Consumer: (*Buy from Either*). $11 = 0.9(10) + 0.1(20)$. $11 - 1 - 0.1(20) > 2 - 1, 11 - 1 - 0.8(20) < 2 - 1$.

(3) If $Prob(Low\ quality|Warranty) = 1$, then the high-quality firm still would offer a warranty, with P=18 (= $0.2[10] + 0.8[20]$). If $Prob(Low\ quality|No\ Warranty) = 1$, then the low-quality firm still would offer no warranty, as in (2).

(4) Allied, Brydox: (P=12.5, *Warranty*), Other: (P=2, *No Warranty*), Consumer: (*Buy from Either*). $12.5 = 0.5[0.9(10) + 0.1(20)] + 0.5[0.6(10) + 0.4(20)]$. Beliefs: $\Pr(\text{Brydox}|P \neq 12.5, Warranty) = 1$, $\Pr(\text{Other}|P \neq 2, No\ Warranty) = 1$.

Problem 10   Bilateral Monopoly

(1) $\pi_s = (p - 10)q$, $\pi_b = 210q - q^2 - pq$.

(2) The lower part of the MB curve, between $q = 50$ and $q = 100$.

(3) Not in a straight line.

(4) $p = 110, q = 50, \pi_s = 5000, \pi_b = 2500$ and $p = 10, q = 100, \pi_s = 0, \pi_b = 10,000$.

(5) $\pi_s = 200\sqrt{\pi_b} - 2\pi_b$. $\pi_b = 5625, \pi_s = 3750, p = 60, q = 75$.

Problem 11   An Auction with Stupid Bidders

(1) Bid up to 21.

(2) $Prob(Z = 12) = Prob(Z = 28) = 0.5$.

(3) $Prob(4, 4) = Prob(4, 20) = Prob(20, 4) = Prob(20, 20) = 0.25$.

(4) $\pi(8.72) = 0.5 \cdot 0.25(13 - 4) = 1.125$. $\pi(21) = 0.5 \cdot 0.25(13 - 4) + 0.5 \cdot 0.75(13 - 20) + 0.5 \cdot 0.25(29 - 20) = -0.375$.

Problem 12   Duopoly

(1) $q_a = 28.75 - 0.25q_b$

(2) $q_a = 25, q_b = 15, \pi_a = 250, \pi_b = 675$.

(3) Set $MC_a = MC_b = p$. Then $q_a = 32.86, q_b = 16.43, p = 65.72$.

Problem 13   A Patent Race with Two Firms

(1) $x_a = 0$ or $x_a = 100$.

(2) $F(x) = \frac{x}{V}$.

(3) $ET(x^*) = 2000 - \int_0^{100} x F'(x) dx = 1950$.

(4) $ET(Max(x_a^*, x_b^*)) = 2000 - 2 \int_0^{100} x F'(x) F(x) dx = 1933$.

References and Citation Index

Forthcoming and unpublished articles and books have been assigned, instead of years, (forth) and (unpub). The page numbers where a reference is mentioned in the text are listed after the reference. The date of first publication, which may differ from the date of the printing cited, follows the author's name.

Akerlof, George (1970) "The Market for Lemons: Quality Uncertainty and the Market Mechanism" *Quarterly Journal of Economics.* August 1970. 84, 3: 488–500. **182, 199, 200n, 201n**

Akerlof, George (1976) "The Economics of Caste and of the Rat Race and Other Woeful Tales" *Quarterly Journal of Economics.* November 1976. 90, 4: 599–617. **199n**

Akerlof, George (1980) "A Theory of Social Custom, of which Unemployment may be One Consequence" *Quarterly Journal of Economics.* June 1980. 94, 4: 749–75. **198**

Akerlof, George (1983) "Loyalty Filters" *American Economic Review.* March 1983. 73, 1: 54–63. **170, 198**

Akerlof, George & Janet Yellen, eds. (1986) *Efficiency Wage Models of the Labor Market.* Cambridge: Cambridge University Press, 1986. **177n**

Alchian, Armen. See Klein et al. (1978). **15, 153, 154n**

Alchian, Armen & Harold Demsetz (1972) "Production, Information Costs and Economic Organization" *American Economic Review.* December 1972. 62, 5: 777–95. **154n**

Antle, Rick & Abby Smith (1986) "An Empirical Investigation of the Relative Performance Evaluation of Corporate Executives" *Journal of Accounting Research.* Spring 1986. 24, 1: 1–39. **177n**

Arrow, Kenneth (1985) "The Economics of Agency" In *Principals and Agents: The Structure of Business*, John Pratt & Richard Zeckhauser, eds. Boston: Harvard Business School Press, 1985. 37–51. **154n**

Arrow, Kenneth & Debreu, Gerard (1954) "Existence of an Equilibrium for a Competitive Economy" *Econometrica.* July 1954. 22, 3: 265–90. **129n**

Aumann, Robert (1964a) "Markets with a Continuum of Traders" *Econometrica.* January/April 1964. 32, 1/2: 39–50. **81n**

Aumann, Robert (1964b) "Mixed and Behavior Strategies in Infinite Extensive Games" *Annals of Mathematics Studies*, No. 52. Princeton: Princeton University Press, 1964. 627–50. **81n**

Aumann, Robert (1974) "Subjectivity and Correlation in Randomized Strategies" *Journal of Mathematical Economics.* March 1974. 1, 1: 67–96. **75, 80**

Aumann, Robert (1976) "Agreeing to Disagree" *Annals of Statistics.* November 1976. 4, 6: 1236–9. **66n**

Aumann, Robert (1981) "Survey of Repeated Games" In *Essays in Game Theory and Mathematical Economics in Honor of Oscar Morgenstern*, Aumann, Robert, ed., Mannheim: Bibliographisches Institut, 1981. **101n, 103n**

Aumann, Robert (1987) "Correlated Equilibrium as an Expression of Bayesian Rationality" *Econometrica.* January 1987. 55, 1: 1–18.

Aumann, Robert & Sergiu Hart (forth) *Handbook of Game Theory with Economic Applications.* New York: North-Holland. **5**

Axelrod, Robert (1984) *The Evolution of Cooperation.* New York: Basic Books, 1984. **119, 128n**

Axelrod, Robert & William Hamilton (1981) "The Evolution of Cooperation" *Science.* March 1981. 211, 4489: 1390–96. **129n**

Bagchi, Arunabha (1984) *Stackelberg Differential Games in Economic Models.* Berlin: Springer-Verlag, 1984. **81n**

Bagehot, Walter (1971) "The Only Game in Town" *Financial Analysts Journal.* March/April 1971. 27, 2: 12–22. **202n**

Baiman, Stanley (1982) "Agency Research in Managerial Accounting: A Survey" *Journal of Accounting Literature.* Spring 1982. 1: 154–213. **154n**

Baker, George, Michael Jensen, & Kevin J. Murphy (1988) "Compensation and Incentives: Practice vs. Theory" *Journal of Finance, Papers and Proceedings.* July 1988. 43, 3:593–616. **169**

Baldwin, B. & G. Meese (1979) "Social Behavior in Pigs Studied by Means of Operant Conditioning" *Animal Behavior.* 1979. 27: 947–57. **32, 40n**

Bamberg, Gunter & Klaus Spremann, eds. (1987) *Agency Theory, Information, and Incentives.* Berlin: Springer-Verlag, 1987. **5**

Banks, Jeffrey & Joel Sobel (1987) "Equilibrium Selection in Signaling Games" *Econometrica.* May 1987. 55, 3: 647–61. **128n**

Baron, David (forth) "Design of Regulatory Mechanisms and Institutions" In Schmalensee & Willig (forth). **176n, 177n**

Baron, David & David Besanko (1984) "Regulation, Asymmetric Information, and Auditing" *Rand Journal of Economics.* Winter 1984. 15, 4: 447–70. **178n**

Barzel, Yoram (1968) "Optimal Timing of Innovations" *Review of Economics and Statistics.* August 1968. 50, 3: 348–55. **307n**

Basil Blackwell (1985) *Guide for Authors.* Oxford: Basil Blackwell, 1985. **17n**

Becker, Gary (1968) "Crime and Punishment: An Economic Approach" *Journal of Political Economy.* March/April 1968. 76, 2: 169–217.

Becker, Gary & George Stigler (1974) "Law Enforcement, Malfeasance and Compensation of Enforcers" *Journal of Legal Studies.* January 1974. 3, 1: 1–18. **166**

Benoit, Jean-Pierre & Vijay Krishna (1985) "Finitely Repeated Games" *Econometrica.* July 1985. 17, 4: 317–20. **101n**

Bernanke, Benjamin (1983) "Nonmonetary Effects of the Financial Crisis in the Propagation of the Great Depression" *American Economic Review*. June 1983. 73, 3: 257–76. **197**

Bernheim, B. Douglas (1984a) "Rationalizable Strategic Behavior" *Econometrica*. July 1984. 52, 4: 1007–28. **37, 40n**

Bernheim, B. Douglas (1984b) "Strategic Deterrence of Sequential Entry into an Industry" *Rand Journal of Economics*. Spring 1984. 15, 1: 1–11. **173**

Bernheim, B. Douglas, Bezalel Peleg, & Michael Whinston (1987) "Coalition-Proof Nash Equilibria I: Concepts" *Journal of Economic Theory*. June 1987. 42, 1: 1–12. **101n**

Bernheim, B. Douglas & Michael Whinston (1986) "Common Agency" *Econometrica*. July 1986. 54, 4: 923–42. **157n**

Bernheim, B. Douglas & Michael Whinston (1987) "Coalition-Proof Nash Equilibria II: Applications" *Journal of Economic Theory*. June 1987. 42, 1: 13–29. **101n**

Bertrand, Joseph (1883) "Recherches sur la théorie mathématique de la richesse" *Journal des savants*. September 1883. 48: 499–508. **263**

Besanko, David. See Baron & Besanko (1984). **178n**

Bewley, Truman, ed. (1987) *Advances in Economic Theory, Fifth World Congress*. Cambridge: Cambridge University Press, 1987.

Bikhchandani, Sushil (1988) "Reputations in Repeated Second Price Auctions" *Journal of Economic Theory*. October 1988. 46, 1: 97–119. **254**

Binmore, Ken (1987) "Modelling Rational Players, Part I" *Economics and Philosophy*. October 1987. 3, 2: 179–214. **128n**

Binmore, Ken (unpub) "Nash Bargaining Theory I" LSE ICERD Discussion Paper No. 80/90. 1980.

Binmore, Ken & Partha Dasgupta, eds. (1986) *Economic Organizations as Games*. Oxford: Basil Blackwell, 1986. **5**

Binmore, Ken, Ariel Rubinstein, & Asher Wolinsky (1986) "The Nash Bargaining Solution in Economic Modelling" *Rand Journal of Economics*. Summer 1986. 17, 2: 176–88. **242n**

Blanchard, Olivier (1979) "Speculative Bubbles, Crashes, and Rational Expectations" *Economics Letters*. 1979. 3, 4: 387–9. **103n**

Bond, Eric (1982) "A Direct Test of the 'Lemons' Model: The Market for Used Pickup Trucks" *American Economic Review*. September 1982. 72, 4: 836–40. **200n**

Border, Kim (1985) *Fixed Point Theorems with Applications to Economics and Game Theory*. Cambridge: Cambridge University Press, 1985. **309, 316**

Border, Kim & Joel Sobel (1987) "Samurai Accountant: A Theory of Auditing and Plunder" *Review of Economic Studies*. October 1987. 54 (4), 180: 525–40. **178n**

Bowersock, G. (1985) "The Art of the Footnote" *American Scholar*. Winter 1983/84. 52: 54–62. **18n**

Bowley, Arthur (1924) *The Mathematical Groundwork of Economics*. Oxford: Clarendon Press, 1924. **281n**

Boyd, Robert & Jeffrey Lorberbaum (1987) "No Pure Strategy is Evolutionarily Stable in the Repeated Prisoner's Dilemma Game" *Nature*. May 1987. 327, 6117: 58–9. **123**

Boyd, Robert & Peter Richerson (1985) *Culture and the Evolutionary Process.* Chicago: University of Chicago Press, 1985. **129n**

Brams, Steven (1980) *Biblical Games: A Strategic Analysis of Stories in the Old Testament.* Cambridge, Mass.: MIT Press, 1980.

Brams, Steven & D. Marc Kilgour (1988) *Game Theory and National Security.* Oxford: Basil Blackwell, 1988. **41n**

Bulow, Jeremy (1982) "Durable-Goods Monopolists" *Journal of Political Economy.* April 1982. 90, 2: 314–32. **276, 280**

Bulow, Jeremy, John Geanakoplos, & Paul Klemperer (1985) "Multimarket Oligopoly: Strategic Substitutes and Complements" *Journal of Political Economy.* June 1985. 93, 3: 488–511. **280, 281n**

Caillaud, B., Roger Guesnerie, P. Rey, & Jean Tirole (1988) "Government Intervention in Production and Incentives Theory: a Review of Recent Contributions" *Rand Journal of Economics.* Spring 1988. 19, 1: 1–26. **175, 176n, 177n**

Campbell, Richmond & Lanning Sowden (1985) *Paradoxes of Rationality and Cooperation: Prisoner's Dilemma and Newcomb's Problem.* Vancouver: University of British Columbia Press, 1985. **38n**

Campbell, W. See Capen et al. (1971). **253, 256**

Capen, E., R. Clapp, & W. Campbell (1971) "Competitive Bidding in High-Risk Situations" *Journal of Petroleum Technology.* June 1971. 23, 1: 641–53. **253, 256**

Cass, David & Karl Shell (1983) "Do Sunspots Matter?" *Journal of Political Economy.* April 1983. 91, 2: 193–227. **75**

Cassady, Ralph (1967) *Auctions and Auctioneering.* Berkeley: California University Press, 1967. **250, 256, 257n**

Chammah, Albert. See Rapoport & Chammah (1965). **120**

Chiang, Alpha (1984) *Fundamental Methods of Mathematical Economics.* New York: McGraw Hill, third edition. **309**

Cho, In-Koo (1987) "A Refinement of Sequential Equilibrium" *Econometrica.* November 1987. 55, 6: 1367–90. **128n**

Cho, In-Koo & David Kreps (1987) "Signalling Games and Stable Equilibria" *Quarterly Journal of Economics.* May 1987. 102, 2: 179–221. **114, 128n, 208**

Clapp, R. See Capen et al. (1971). **253, 256**

Coase, Ronald (1972) "Durability and Monopoly" *Journal of Law and Economics.* April 1972. 15, 1: 143–9. **276, 283n**

Cooter, Robert & Peter Rappoport (1984) "Were the Ordinalists Wrong about Welfare Economics?" *Journal of Economic Literature.* June 1984. 22, 2: 507–30. **38n**

Copeland, Thomas & Dan Galai (1983) "Information Effects on the Bid-Ask Spread" *Journal of Finance.* December 1983. 38, 5: 1457–69. **202n**

Copeland, Thomas & J. Fred Weston (1983) *Financial Theory and Corporate Policy.* Reading, Mass.: Addison-Wesley, 1983. **91, 156n, 218**

Cornell, Bradford & Richard Roll (1981) "Strategies for Pairwise Competitions in Markets and Organizations" *Bell Journal of Economics.* Spring 1981. 12, 1: 210–16. **129n**

Cournot, Augustin (1838) *Recherches sur les principes mathématiques de la théorie des richesses.* Paris: M. Rivière & Cie., 1938. Translated in *Researches into the Mathematical Principles of Wealth.* New York: A.M. Kelly, 1960. **77**

Cox, D. & David Hinkley (1974) *Theoretical Statistics.* London: Chapman and Hall, 1974. **155n**

Cramton, Peter (1984) "Bargaining with Incomplete Information: An Infinite Horizon Model with Two-Sided Uncertainty" *Review of Economic Studies.* October 1984. 51(4), 167: 579–93. **242n**

Crawford, Robert. See Klein et al. (1978). **15, 153, 154n**

Crawford, Vincent & Joel Sobel (1982) "Strategic Information Transmission" *Econometrica.* November 1982. 50, 6: 1431–52. **76**

Crocker, Keith. See Masten & Crocker (1985). **154n**

Dalkey, Norman (1953) "Equivalence of Information Patterns and Essentially Determinate Games" In Kuhn & Tucker (1953), 217–43.

Dasgupta, Partha. See Binmore & Dasgupta (1986). **5**

Dasgupta, Partha, Peter Hammond, & Eric Maskin (1979) "The Implementation of Social Choice Rules; Some General Rules on Incentive Compatibility" *Review of Economic Studies.* April 1979. 46(2), 143: 185–216. **176n, 273**

Dasgupta, Partha & Eric Maskin (1986a) "The Existence of Equilibrium in Discontinuous Economic Games, I: Theory" *Review of Economic Studies.* January 1986. 53(1), 172: 1–26. **125, 126, 127, 129n, 316**

Dasgupta, Partha & Eric Maskin (1986b) "The Existence of Equilibrium in Discontinuous Economic Games, II: Applications" *Review of Economic Studies.* January 1986. 53(1), 172: 27–41. **126, 127, 193, 266, 272**

Dasgupta, Partha & Joseph Stiglitz (1980) "Uncertainty, Industrial Structure, and the Speed of R&D" *Bell Journal of Economics.* Spring 1980. 11, 1: 1–28. **307n**

d'Aspremont, Claude, J. Gabszewicz, & Jacques Thisse (1979) "On Hotelling's 'Stability of Competition' " *Econometrica.* September 1979. 47, 5: 1145–50. **272, 280**

David, Paul (1985) "CLIO and the Economics of QWERTY" *AEA Papers and Proceedings.* May 1985. 75, 2: 332–7. **40n**

Davis, Morton (1970) *Game Theory: A Nontechnical Introduction.* New York: Basic Books, 1970. **5**

Davis, Philip & Reuben Hersh (1981) *The Mathematical Experience.* Boston: Birkhauser, 1981. **17n**

Debreu, Gerard (1952) "A Social Equilibrium Existence Theorem" *Proceedings of the National Academy of Sciences.* 1952. 38, 10: 886–93. **129n**

Debreu, Gerard (1959) *Theory of Value: An Axiomatic Analysis of Economic Equilibrium.* New Haven: Yale University Press, 1959. **309**

Debreu, Gerard. See Arrow & Debreu (1954). **129n**

Debreu, Gerard & Herbert Scarf (1963) "A Limit Theorem on the Core of an Economy" *International Economic Review.* September 1963. 4, 3: 235–46. **14**

Demsetz, Harold. See Alchian & Demsetz (1972). **154n**

Diamond, Douglas (1984) "Financial Intermediation and Delegated Monitoring" *Review of Economic Studies.* July 1984. 51(3), 166: 393–414. **178n**

Diamond, Douglas (unpub) "Reputation Acquisition in Debt Markets" University of Chicago mimeo. February 1985. **197, 288, 289**

Diamond, Peter & Michael Rothschild, eds. (1978) *Uncertainty in Economics: Readings and Exercises.* New York: Academic Press, 1978. **5**

DiMona, Joseph. See Haldeman & DiMona (1978). **101n**

Dixit, Avinash (1979) "A Model of Duopoly Suggesting a Theory of Entry Barriers" *Bell Journal of Economics.* Spring 1979. 10, 1: 20–32. **282n**

Dixit, Avinash (1980) "The Role of Investment in Entry Deterrence" *Economic Journal.* March 1980. 90: 95–106. **290**

Dresher, Melvin, Albert Tucker, & Philip Wolfe, eds. (1957) *Contributions to the Theory of Games.* Vol III. Annals of Mathematics Studies, No. 39. Princeton: Princeton University Press, 1957.

Eaton, B. Curtis & Richard Lipsey (1975) "The Principle of Minimum Differentiation Reconsidered: Some New Developments in the Theory of Spatial Competition" *Review of Economic Studies.* January 1975. 42 (1), 129: 27–49. **273**

Edgeworth, Francis (1881) *Mathematical Psychics.* London: Kegan Paul, 1881.

Edgeworth, Francis (1897) "La teoria pura del monopolio" *Giornale degli economisti.* 1925. 40: 13–31. Translated in Edgeworth, Francis, *Papers Relating to Political Economy*, Vol. I. London: Macmillan, 1925, 111–42. **266**

Engers, Maxim (1987) "Signalling with Many Signals" *Econometrica.* May 1987. 55, 3: 663–74. **222n**

Engers, Maxim & Luis Fernandez (1987) "Market Equilibrium with Hidden Knowledge and Self-Selection" *Econometrica.* March 1987. 55, 2: 425–39. **194, 202n, 219**

Fama, Eugene (1980) "Banking in the Theory of Finance" *Journal of Monetary Economics.* 1980. 6: 39–57. **154n, 220**

Fama, Eugene & Michael Jensen (1983a) "Agency Problems and Residual Claims" *Journal of Law and Economics.* June 1983. 26, 2: 327–49. **154n**

Fama, Eugene & Michael Jensen (1983b) "Separation of Ownership and Control" *Journal of Law and Economics.* June 1983. 26, 2: 301–25. **154n**

Fan, Ky (1952) "Fixed Point and Minimax Theorems in Locally Convex Topological Linear Spaces" *Proceedings of the National Academy of Sciences.* 1952. 38: 121–6. **315**

Farrell, Joseph (unpub) "Monopoly Slack and Competitive Rigor: A Simple Model" MIT mimeo. February 1983. **167**

Farrell, Joseph (unpub) "Credible Neologisms in Games of Communication" GTE mimeo. June 1985. **128n**

Farrell, Joseph (1987) "Cheap Talk, Coordination, and Entry" *Rand Journal of Economics.* Spring 1987. 18, 1: 34–9. **76, 80**

Farrell, Joseph & Garth Saloner (1985) "Standardization, Compatibility, and Innovation" *Rand Journal of Economics.* Spring 1985. 16, 1: 70–83. **40n**

Farrell, Joseph & Carl Shapiro (1988) "Dynamic Competition with Switching Costs" *Rand Journal of Economics.* Spring 1988. 19, 1: 123–37. **274**

Fernandez, Luis. See Engers & Fernandez (1987). **194, 202n, 220**

Fernandez, Luis & Eric Rasmusen (unpub) "Equilibrium in Screening Markets when the Uninformed Traders can React to New Offers" Oberlin College mimeo. March 1988. **194**

Forgo, F. See Szep & Forgo (1978). **5**

Fowler, H. (1965) *A Dictionary of Modern English Usage,* second edition. New York: Oxford University Press, 1965. **18n**

Fowler, H. & F. Fowler (1931) *The King's English,* third edition. Oxford: Clarendon Press, 1949. **18n**

Franks, Julian, Robert Harris, & Colin Mayer (1988) "Means of Payment in Takeovers: Results for the UK and US" In *Corporate Takeovers: Causes and Consequences.* Alan Auerbach, ed. Chicago: University of Chicago Press, 1988.

Freixas, Xavier, Roger Guesnerie, & Jean Tirole (1985) "Planning under Incomplete Information and the Ratchet Effect" *Review of Economic Studies.* April 1985. 52 (2), 169: 173–91. **177n**

Friedman, James (1986) *Game Theory with Applications to Economics.* New York: Oxford University Press, 1986. **5, 129n**

Friedman, Milton (1953) *Essays in Positive Economics.* Chicago: University of Chicago Press, 1953. **17n**

Fudenberg, Drew & David Levine (1986) "Limit Games and Limit Equilibria" *Journal of Economic Theory.* April 1986. 38, 2: 261–79. **82n, 102n**

Fudenberg, Drew & Eric Maskin (1986) "The Folk Theorem in Repeated Games with Discounting or with Incomplete Information" *Econometrica.* May 1986. 54, 3: 533–54. **99, 103n, 119**

Fudenberg, Drew & Jean Tirole (1983) "Sequential Bargaining with Incomplete Information" *Review of Economic Studies.* April 1983. 50(2), 161: 221–47. **237, 242n**

Fudenberg, Drew & Jean Tirole (1986a) *Dynamic Models of Oligopoly.* Chur, Switzerland: Harwood Academic Publishers, 1986. **5, 282n, 306n**

Fudenberg, Drew & Jean Tirole (1986b) "A Theory of Exit in Duopoly" *Econometrica.* July 1986. 54, 4: 943–60. **80, 81n**

Fudenberg, Drew & Jean Tirole (unpub) "Moral Hazard and Renegotiation in Agency Contracts" MIT mimeo. February 1988. **141, 157**

Gabszewicz, J. See d'Aspremont et al. (1979). **272, 280**

Galai, Dan. See Copeland & Galai (1983). **202n**

Galbraith, John Kenneth (1954) *The Great Crash, 1929.* Boston: Houghton Mifflin, 1954. **179n**

Gal-Or, Esther (1985) "First Mover and Second Mover Advantages" *International Economic Review.* October 1985. 26, 3: 649–53. **281n**

Gaskins, Darius (1974) "Alcoa Revisited: The Welfare Implications of a Second-Hand Market" *Journal of Economic Theory.* March 1974. 7, 3: 254–71. **283n**

Gaver, Kenneth & Jerold Zimmerman (1977) "An Analysis of Competitive Bidding on BART Contracts" *Journal of Business.* July 1977. 50, 3: 279–95. **178n**

Geanakoplos, John. See Bulow et al. (1985). **280, 281n**

Geanakoplos, John & Heraklis Polemarchakis (1982) "We Can't Disagree Forever" *Journal of Economic Theory.* October 1982. 28, 1: 192–200. **66n**

Ghemawat, Pankaj & Barry Nalebuff (1985) "Exit" *Rand Journal of Economics.* Summer 1985. 16, 2: 184–94. **81n**

Gibbard, Allan (1973) "Manipulation of Voting Schemes: A General Result" *Econometrica.* July 1973. 41, 4: 587–601. **176n**

Gilbert, Richard & David Newbery (1982) "Preemptive Patenting and the Persistence of Monopoly" *American Economic Review.* June 1982. 72, 3: 514–26. **298, 306**

Gillies, Donald (1953) "Locations of Solutions" In *Report of an Informal Conference on the Theory of n-Person Games.* Princeton Mathematics mimeo. 1953. 11–12. **13**

Gjesdal, Froystein (1982) "Information and Incentives: The Agency Information Problem" *Review of Economic Studies.* July 1982. 49 (3), 157: 373–90. **155n**

Glicksberg, Irving (1952) "A Further Generalization of the Kakutani Fixed Point Theorem with Application to Nash Equilibrium Points" *Proceedings of the American Mathematical Society.* February 1952. 3, 1: 170–74. **125**

Gonik, Jacob (1978) "Tie Salesmen's Bonuses to their Forecasts" *Harvard Business Review.* May/June 1978. 56: 116–23. **176n**

Green, Jerry & Jean-Jacques Laffont (1979) *Incentives in Public Decision-Making.* Amsterdam: North Holland, 1979. **5, 176n**

Greenhut, Melvin & Hiroshi Ohta (1975) *Theory of Spatial Pricing and Market Areas.* Durham, N.C.: Duke University Press, 1975. **282n**

Grossman, Gene & Michael Katz (1983) "Plea Bargaining and Social Welfare" *American Economic Review.* September 1983. 73, 4: 749–57. **221n**

Grossman, Sanford & Oliver Hart (1980) "Takeover Bids, the Free-Rider Problem, and the Theory of the Corporation" *Bell Journal of Economics.* Spring 1980. 11, 1: 42–64. **173, 301**

Grossman, Sanford & Oliver Hart (1983) "An Analysis of the Principal Agent Problem" *Econometrica.* January 1983. 51, 1: 7–45. **141, 155n, 156n, 301**

Grossman, Sanford & Motty Perry (1986) "Perfect Sequential Equilibria" *Journal of Economic Theory.* June 1986. 39, 1: 97–119. **128n**

Groves, Theodore (1973) "Incentives in Teams" *Econometrica.* July 1973. 41, 4: 617–31. **179n**

Guasch, J. Luis & Andrew Weiss (1980) "Wages as Sorting Mechanisms in Competitive Markets with Asymmetric Information: A Theory of Testing" *Review of Economic Studies.* July 1980. 47(4), 149: 653–64. **202n**

Guesnerie, Roger. See Caillaud et al. (1988), **177n**; and Freixas et al. (1985), **175, 176n, 177n**

Gul, Faruk (forth) "Bargaining Foundations of Shapley Value" *Econometrica.* **242n**

Guth, Wener, Rold Schmittberger, & Bernd Schwarze (1982) "An Experimental Analysis of Ultimatum Bargaining" *Journal of Economic Behavior and Organization.* December 1982. 3, 4: 367–88. **229**

Guyer, Melvin. See Rapoport & Guyer (1966). **39n**

Guyer, Melvin & Henry Hamburger (1968) "A Note on 'A Taxonomy of 2×2 Games" *General Systems.* 1968. 13: 205–8. **39n**

Haldeman, H. R. & Joseph DiMona (1978) *The Ends of Power.* New York: Times Books, 1978. **101n**

Haller, Hans (1986) "Noncooperative Bargaining of $N \geq 3$ Players" *Economic Letters.* 1986. 22: 11–13. **242n**

Halmos, Paul (1970) "How to Write Mathematics" *L'enseignement mathématique.* May/June 1970. 16, 2: 123–52. **18n**

Haltiwanger, John & Michael Waldman (unpub) "Responders versus Nonresponders: A New Perspective of Heterogeneity" UCLA Economics Working Paper No.436. February 1987. **281n**

Hamburger, Henry. See Guyer & Hamburger (1968). **39n**

Hamilton, William. See Axelrod & Hamilton (1981). **128n**

Hammond, Peter. See Dasgupta et al. (1979). **176n, 273**

Harrington, Joseph (1987) "Collusion in Multiproduct Oligopoly Games under a Finite Horizon" *International Economic Review*. February 1987. 28, 1: 1–14. **101n**

Harris, Milton & Bengt Holmstrom (1982) "A Theory of Wage Dynamics" *Review of Economic Studies*. July 1982. 49 (3), 157: 315–34. **54**

Harris, Robert. See Franks et al. (1988).

Harsanyi, John (1967) "Games with Incomplete Information Played by 'Bayesian' Players, I: The Basic Model" *Management Science*. November 1967. 14, 3: 159–82. **14, 55, 109**

Harsanyi, John (1968a) "Games with Incomplete Information Played by 'Bayesian' Players, II: Bayesian Equilibrium Points" *Management Science*. January 1968. 14, 5: 320–34.

Harsanyi, John (1968b) "Games with Incomplete Information Played by 'Bayesian' Players, III: The Basic Probability Distribution of the Game" *Management Science*. March 1968 14, 7: 486–502.

Harsanyi, John (1973) "Games with Randomly Disturbed Payoffs: A New Rationale for Mixed Strategy Equilibrium Points" *International Journal of Game Theory*. 1973. 2, 1: 1–23. **81n**

Harsanyi, John (1977) *Rational Behavior and Bargaining Equilibrium in Games and Social Situations*. New York: Cambridge University Press, 1977. **5, 242n**

Harsanyi, John and Reinhard Selten (1988) *A General Theory of Equilibrium Selection in Games*. Cambridge, Mass.: MIT Press, 1988. **38n**

Hart, Oliver. See Grossman & Hart (1980), **301**; (1983), **141, 155n, 156n**

Hart, Oliver & Bengt Holmstrom (1987) "The Theory of Contracts" In Bewley (1987). **154n, 155n, 156n**

Hart, Sergiu. See Aumann & Hart (forth). **5**

Haywood, O. (1954) "Military Decisions and Game Theory" *Journal of the Operations Research Society of America*. November 1954. 2, 4: 365–85. **39n**

Henry, O. (1945) *Best Stories of O. Henry*. Garden City, NY: The Sun Dial Press, 1945. **40n**

Herodotus. *The Persian Wars*. George Rawlinson, trans. New York: Modern Library, 1947. **38n**

Hersh, Reuben. See Davis & Hersh (1981). **17n**

Hess, James (1983) *The Economics of Organization*. Amsterdam: North Holland, 1983. **5, 154n**

Hines, W. (1987) "Evolutionary Stable Strategies: A Review of Basic Theory" *Theoretical Population Biology*. April 1987. 31, 2: 195–272. **129n**

Hinkley, David. See Cox & Hinkley (1974). **155n**

Hirshleifer, David. See Png & Hirshleifer (1987). **283n**

Hirshleifer, David and Eric Rasmusen (forth) "Cooperation in a Repeated Prisoner's Dilemma with Ostracism" *Journal of Economic Behavior and Organization.* **102n**

Hirshleifer, David & Sheridan Titman (unpub) "Share Tendering Strategies and the Success of Hostile Takeover Bids" UCLA AGSM Finance Working Paper No. 15–88. July 1988. **128n**

Hirshleifer, Jack (1982) "Evolutionary Models in Economics and Law: Cooperation versus Conflict Strategies" *Research in Law and Economics.* 1982. 4: 1–60. **129n**

Hirshleifer, Jack (1987) "On the Emotions as Guarantors of Threats and Promises" In *The Latest on the Best: Essays on Evolution and Optimality,* John Dupre, ed., Cambridge, Mass.: MIT Press, 1987. **104n**

Hirshleifer, Jack & Juan Martinez-Coll (1988) "What Strategies can Support the Evolutionary Emergence of Cooperation?" *Journal of Conflict Resolution.* June 1988. 32, 2: 367–98. **123**

Hirshleifer, Jack & John Riley (1979) "The Analytics of Uncertainty and Information: An Expository Survey" *Journal of Economic Literature.* December 1979. 17, 4: 1375–421.

Hirshleifer, Jack & John Riley (unpub) *The Economics of Uncertainty and Information.* UCLA mimeo. August 1987. **5**

Hofstadter, Douglas (1983) "Computer Tournaments of the Prisoner's Dilemma Suggest how Cooperation Evolves" *Scientific American.* May 1983. 248, 5: 16–26. **127, 128n**

Holmes, Oliver (1881) *The Common Law.* Boston: Little, Brown and Co., 1923. **152**

Holmstrom, Bengt (1979) "Moral Hazard and Observability" *Bell Journal of Economics.* Spring 1979. 10, 1: 74–91. **149, 155n**

Holmstrom, Bengt (1982) "Moral Hazard in Teams" *Bell Journal of Economics.* Autumn 1982. 13, 2: 324–40. **173, 175, 179n**

Holmstrom, Bengt. See Harris & Holmstrom (1982), **54**; and Hart & Holmstrom (1987), **154n, 155n, 156n**

Holmstrom, Bengt & Roger Myerson (1983) "Efficient and Durable Decision Rules with Incomplete Information" *Econometrica.* November 1983. 51, 6: 1799–819. **153, 157n**

Holt, Charles & David Scheffman (1987) "Facilitating Practices: The Effects of Advance Notice and Best-Price Policies" *Rand Journal of Economics.* Summer 1987. 18, 2: 187–97. **283n**

Hotelling, Harold (1929) "Stability in Competition" *Economic Journal.* 1929. 39: 41–57. **269, 272, 280, 282n**

Hughes, Patricia (1986) "Signalling by Direct Disclosure Under Asymmetric Information" *Journal of Accounting and Economics.* June 1986. 8, 2: 119–42. **222n**

Hurwicz, Leonid, David Schmeidler, & Hugo Sonnenschein, eds. (1985) *Social Goals and Social Organization: Essays in Memory of Elisha Pazner.* Cambridge: Cambridge University Press. 1985. **176n**

Hwang, Chuan-Yang (unpub) "Underpricing of New Issues" UCLA AGSM mimeo. January 1988. **218**

Intriligator, Michael (1971) *Mathematical Optimization and Economic Theory.* Englewood Cliffs, N.J.: Prentice-Hall, 1971. **309**

Isoda, Kazuo. See Nikaido & Isoda (1955). **125**

Jacquemin, Alex (1985) *The New Industrial Organization.* Cambridge, Mass.: MIT Press, 1987. Translated from *Sélection et pouvoir dans la nouvelle économie industrielle.* Louvain-la-Neuve: Cabay Libraire-Editeur, 1985. **129n, 281n, 306n**

Jarrell, Gregg & Sam Peltzman (1985) "The Impact of Product Recalls on the Wealth of Sellers" *Journal of Political Economy.* June 1985. 93, 3: 512–36. **104n**

Jensen, Michael. See Baker et al. (1988), **169**; Fama & Jensen (1983a), **154n**; and Fama & Jensen (1983b), **154n**

Jensen, Michael & William Meckling (1976) "Theory of the Firm: Managerial Behavior, Agency Costs and Ownership Structure" *Journal of Financial Economics.* October 1976. 3, 4: 305–60. **154n**

Joskow, Paul (1985) "Vertical Integration and Longterm Contracts: The Case of Coal-Burning Electric Generating Plants" *Journal of Law, Economics and Organization.* Spring 1985. 1, 1: 33–80. **154n**

Joskow, Paul (1987) "Contract Duration and Relationship-Specific Investments: Empirical Evidence from Coal Markets" *American Economic Review.* March 1987. 77, 1: 168–85. **154n**

Kakutani, Shizuo (1941) "A Generalization of Brouwer's Fixed Point Theorem" *Duke Mathematical Journal.* September 1941. 8, 3: 457–9. **315**

Kamien, Morton & Nancy Schwartz (1981) *Dynamic Optimization: The Calculus of Variations and Optimal Control in Economics and Management.* New York: North Holland, 1981. **309**

Kamien, Morton & Nancy Schwartz (1982) *Market Structure and Innovation.* Cambridge: Cambridge University Press, 1982. **295**

Karlin, Samuel (1959) *Mathematical Methods and Theory in Games, Programming and Economics.* Reading, Mass.: Addison-Wesley, 1959. **82n**

Katz, Lawrence (1986) "Efficiency Wage Theory: A Partial Evaluation" In *NBER Macroeconomics Annual 1986,* Stanley Fischer, ed. Cambridge, Mass.: MIT Press, 1986. **177n**

Katz, Michael (unpub) "Game-Playing Agents: Contracts as Precommitments" Princeton Working Paper. January 1987. **157n**

Katz, Michael. See Grossman & Katz (1983). **222n**

Katz, Michael. See Moskowitz et al. (1980).

Katz, Michael & Carl Shapiro (1985) "Network Externalities, Competition, and Compatibility" *American Economic Review.* June 1985. 75, 3: 424–40. **40n**

Keynes, John Maynard (1933) *Essays in Biography.* New York: Harcourt, Brace and Co., 1933. **17**

Keynes, John Maynard (1936) *The General Theory of Employment, Interest and Money.* London: Macmillan, 1947. **41n**

Kihlstrom, Richard & Michael Riordan (1984) "Advertising as a Signal" *Journal of Political Economy.* June 1984. 92, 3: 427–50. **221n**

Kilgour, D. Marc. See Brams & Kilgour (1988). **41n**

Klein, Benjamin, Robert Crawford, & Armen Alchian (1978) "Vertical Integration,

Appropriable Rents, and the Competitive Contracting Process" *Journal of Law and Economics.* October 1978. 21, 2: 297–326. **15, 153, 154n**

Klein, Benjamin & Keith Leffler (1981) "The Role of Market Forces in Assuring Contractual Performance" *Journal of Political Economy.* August 1981. 89, 4: 615–41. **97, 99, 105n, 177n**

Klein, Benjamin & Lester Saft (1985) "The Law and Economics of Franchise Tying Contracts" *Journal of Law and Economics.* May 1985. 28, 2: 345–61. **157n**

Klemperer, Paul (1987) "The Competitiveness of Markets with Switching Costs" *Rand Journal of Economics.* Spring 1987. 18, 1: 138–50. **274**

Klemperer, Paul. See Bulow et al. (1985). **280, 281n**

Kohlberg, Elon & Jean-Francois Mertens (1986) "On the Strategic Stability of Equilibria" *Econometrica.* September 1986. 54, 5: 1003–7. **82n, 128n**

Kreps, David. See Cho & Kreps (1987). **114n, 128n, 208**

Kreps, David, Paul Milgrom, John Roberts, & Robert Wilson (1982) "Rational Cooperation in the Finitely Repeated Prisoners' Dilemma" *Journal of Economic Theory.* August 1982. 27, 2: 245–52. **14, 118, 127**

Kreps, David & Jose Scheinkman (1983) "Quantity Precommitment and Bertrand Competition Yield Cournot Outcomes" *Bell Journal of Economics.* Autumn 1983. 14, 2: 326–37. **281n**

Kreps, David & A. Michael Spence (1984) "Modelling the Role of History in Industrial Organization and Competition" In *Issues in Contemporary Microeconomics and Welfare.* George Feiwel, ed. London: Macmillan, 1984. **15, 17n**

Kreps, David & Robert Wilson (1982a) "Reputation and Imperfect Information" *Journal of Economic Theory.* August 1982. 27, 2: 253–79. **285, 286, 306, 306n**

Kreps, David & Robert Wilson (1982b) "Sequential Equilibria" *Econometrica.* July 1982. 50, 4: 863–94. **14, 109, 110, 127**

Krishna, Vijay. See Benoit & Krishna (1985). **101n**

Krouse, Clement (unpub) *The Theory of Industrial Organization.* Book draft, University of California, Santa Barbara. January 1988. **306n**

Kuhn, Harold (1953) "Extensive Games and the Problem of Information" In Kuhn & Tucker (1953). **80n**

Kuhn, Harold & Albert Tucker, eds. (1950) *Contributions to the Theory of Games.* Vol I. Annals of Mathematics Studies, No. 24. Princeton: Princeton University Press, 1950.

Kuhn, Harold & Albert Tucker, eds. (1953) *Contributions to the Theory of Games.* Vol II. Annals of Mathematics Studies, No. 28. Princeton: Princeton University Press, 1953.

Kydland, Finn & Edward Prescott (1977) "Rules Rather than Discretion: The Inconsistency of Optimal Plans" *Journal of Political Economy.* June 1977. 85, 3: 473–91. **100n**

Lachman, Judith (1984) "Knowing and Showing Economics and Law" *Yale Law Journal.* July 1984. 93, 8: 1587–624.

Laffont, Jean-Jacques. See Green & Laffont (1979). **5, 176n**

Laffont, Jean-Jacques & Jean Tirole (1986) "Using Cost Observation to Regulate Firms" *Journal of Political Economy.* June 1986. 94, 3: 614–41. **155n, 176n**

Lakatos, Imre (1976) *Proofs and Refutations: The Logic of Mathematical Discovery.* Cambridge: Cambridge University Press, 1976. **15, 17n**

Lane, W. (1980) "Product Differentiation in a Market with Endogenous Sequential Entry" *Bell Journal of Economics.* Spring 1980. 11, 1: 237–60. **282n**

Layard, Richard & George Psacharopoulos (1974) "The Screening Hypothesis and the Returns to Education" *Journal of Political Economy.* September/October 1974. 82, 5: 985–98. **211, 221n**

Lazear, Edward & Sherwin Rosen (1981) "Rank-Order Tournaments as Optimum Labor Contracts" *Journal of Political Economy.* October 1981. 89, 5: 841–64. **175, 177n**

Leffler, Keith. See Klein & Leffler (1981). **97, 100, 105n, 177n**

Leibenstein, Harvey (1950) "Bandwagon, Snob and Veblen Effects in the Theory of Consumers' Demand" *Quarterly Journal of Economics.* May 1950. 64, 2: 183–207. **201n**

Leland, Hayne & David Pyle (1977) "Informational Asymmetries, Financial Structure, and Financial Intermediation" *Journal of Finance.* May 1977. 32, 2: 371–87. **218, 219**

Levering, Robert. See Moskowitz et al. (1980).

Levine, David. See Fudenberg & Levine (1986). **82n, 102n**

Levitan, Richard & Martin Shubik (1972) "Price Duopoly and Capacity Constraints" *International Economic Review.* February 1972. 13, 1: 111–22. **266**

Levmore, Saul (1982) "Self-Assessed Valuation for Tort and Other Law" *Virginia Law Review.* April 1982. 68, 4: 771–861. **176n**

Lippman, Steven, John Mamer, & Kevin McCardle (1987) "Comparative Statics in Non-Cooperative Games via Transfinitely Iterated Play" *Journal of Economic Theory.* April 1987. 41, 2: 288–303. **124, 315**

Lippman, Steven & Richard Rumelt (1982) "Uncertain Imitability: An Analysis of Interfirm Differences in Efficiency under Competition" *Bell Journal of Economics.* Autumn 1982. 13, 2: 418–38.

Lipsey, Richard. See Eaton & Lipsey (1975). **273**

Locke, E. (1949) "The Finan-Seer" *Astounding Science Fiction.* October 1949. 44, 2: 132–40.

Lorberbaum, Jeffrey. See Boyd & Lorberbaum (1987). **123**

Loury, Glenn (1979) "Market Structure and Innovation" *Quarterly Journal of Economics.* August 1979. 93, 3: 395–410. **307n**

Luce, R. Duncan & Howard Raiffa (1957) *Games and Decisions: Introduction and Critical Survey.* New York: Wiley, 1957. **5, 7, 40n, 101n, 103n, 242n**

Luce, Duncan & Albert Tucker, eds. (1959) *Contributions to the Theory of Games.* Vol IV. Annals of Mathematics Studies, No. 40. Princeton: Princeton University Press, 1959.

McAfee, R. Preston & John McMillan (1986) "Bidding for Contracts: A Principal-Agent Analysis" *Rand Journal of Economics.* Autumn 1986. 17, 3: 326–38. **157n**

McAfee, R. Preston & John McMillan (1987) "Auctions and Bidding" *Journal of Economic Literature.* June 1987. 25, 2: 699–754. **256, 256n**

McCardle, Kevin. See Lippman et al. (1987). **124, 315**

Macaulay, Stewart (1963) "Non-Contractual Relations in Business" *American Sociological Review.* February 1963. 28, 1: 55–70. **104n**

McChesney, Fred. See Macey & McChesney (1985). **101n, 302**

McCloskey, Donald (1985) "Economical Writing" *Economic Inquiry.* April 1985. 24, 2: 187–222. **17n**

McCloskey, Donald (1987) *The Writing of Economics.* New York: Macmillan. 1987. **17n**

Macey, Jonathan & Fred McChesney (1985) "A Theoretical Analysis of Corporate Greenmail" *Yale Law Journal.* November 1985. 95, 1: 13–61. **101n, 302**

McGee, John (1958) "Predatory Price Cutting: The Standard Oil (N.J.) Case" *Journal of Law and Economics.* October 1958. 1: 137–69. **85n**

Machina, Mark. (1982) " 'Expected Utility' Analysis without the Independence Axiom" *Econometrica.* March 1982. 50, 2: 277–323. **66n**

McLennan, Andrew (1985) "Justifiable Beliefs in Sequential Equilibrium" *Econometrica.* July 1985. 53, 4: 889–904. **114, 128n**

McMillan, John. (1986) *Game Theory in International Economics.* Chur, Switzerland: Harwood Academic Publishers, 1986. **5**

McMillan, John. See McAfee & McMillan (1986), **157n**; and McAfee & McMillan (1987), **256, 256n**

Mamer, John. See Lippman et al. (1987). **124, 315**

Marschak, Jacob & Roy Radner (1972) *Economic Theory of Teams.* New Haven: Yale University Press, 1972. **179n**

Martinez-Coll. See Hirshleifer & Martinez-Coll (1988). **123**

Maskin, Eric. See Dasgupta & Maskin (1986a), **125, 126, 127, 129n, 316**; Dasgupta & Maskin (1986b), **126, 127, 193, 266, 272**; Dasgupta et al. (1979), **176n, 273**; and Fudenberg & Maskin (1986), **99, 103n, 119**

Maskin, Eric & John Riley (1984) "Optimal Auctions with Risk Averse Buyers" *Econometrica.* November 1984. 52, 6: 1473–518. **257n**

Maskin, Eric & John Riley (1985) "Input vs. Output Incentive Schemes" *Journal of Public Economics.* 1985. 28: 1–23.

Maskin, Eric & Jean Tirole (1987) "Correlated Equilibria and Sunspots" *Journal of Economic Theory.* December 1987. 43, 2: 364–73. **75**

Masten, Scott & Keith Crocker (1985) "Efficient Adaptation in Long-Term Contracts: Take-or-Pay Provisions for Natural Gas" *American Economic Review.* December 1985. 75, 5: 1083–93. **154n**

Mathewson, G. Frank. See Stiglitz & Mathewson (1986). **306n**

Mathewson, G. Frank & Ralph Winter (1985) "The Economics of Franchise Contracts" *Journal of Law and Economics.* October 1985. 28, 3: 503–26. **157n**

Matthews, Steven & John Moore (1987) "Monopoly Provision of Quality and Warranties: An Exploration in the Theory of Multidimensional Screening" *Econometrica.* March 1987. 55, 2: 441–67. **222n**

Mayer, Colin. See Franks et al. (1988).

Maynard Smith, John (1974) "The Theory of Games and the Evolution of Animal Conflicts" *Journal of Theoretical Biology.* September 1974. 47, 1: 209–21. **81n**

Maynard Smith, John (1982) *Evolution and the Theory of Games.* Cambridge: Cambridge University Press, 1982. **129n**

Mead, Walter, Asbjorn Moseidjord, & Philip Sorenson (1984) "Competitive Bidding under Asymmetrical Information: Behavior and Performance in Gulf of Mexico Drainage Lease Sales 1959–1969" *Review of Economics and Statistics.* August 1984. 66, 3: 505–8. **254**

Meckling, William. See Jensen & Meckling (1976). **154n**

Meese, G. See Baldwin & Meese (1979). **32, 40n**

Mertens, Jean-Francois. See Kohlberg & Mertens (1986). **82n, 128n**

Mertens, Jean-Francois & S. Zamir (1985) "Formulation of Bayesian Analysis for Games with Incomplete Information" *International Journal of Game Theory.* 1985. 14, 1: 1–29. **66n**

Milgrom, Paul (1981a) "An Axiomatic Characterization of Common Knowledge" *Econometrica.* January 1981. 49, 1: 219–22. **66n**

Milgrom, Paul (1981b) "Good News and Bad News: Representation Theorems and Applications" *Bell Journal of Economics.* Autumn 1981. 12, 2: 380–91. **153, 156n, 176n, 252**

Milgrom, Paul (1981c) "Rational Expectations, Information Acquisition, and Competitive Bidding" *Econometrica.* July 1981. 49, 4: 921–43. **254**

Milgrom, Paul (1987) "Auction Theory" In Bewley (1987). **256, 256n**

Milgrom, Paul. See Kreps et al. (1982). **14, 118, 127**

Milgrom, Paul & John Roberts (1982a) "Limit Pricing and Entry under Incomplete Information: An Equilibrium Analysis" *Econometrica.* March 1982. 50, 2: 443–59.

Milgrom, Paul & John Roberts (1982b) "Predation, Reputation, and Entry Deterrence" *Journal of Economic Theory.* August 1982. 27, 2: 280–312.

Milgrom, Paul & John Roberts (1986) "Price and Advertising Signals of Product Quality" *Journal of Political Economy.* August 1986. 94, 4: 796–821. **221n**

Milgrom, Paul & Robert Weber (1982) "A Theory of Auctions and Competitive Bidding" *Econometrica.* September 1982. 50, 5: 1089–122. **250, 255, 256n, 257n, 258n**

Mirrlees, James (1974) "Notes on Welfare Economics, Information and Uncertainty" In *Essays on Economic Behavior under Uncertainty,* Balch, McFadden, and Wu, eds. Amsterdam: North Holland, 1974. **202n**

Monteverde, K. & David Teece (1982) "Supplier Switching Costs and Vertical Integration in the Automobile Industry" *Bell Journal of Economics.* Spring 1982. 13, 1: 206–13. **154n**

Mookherjee, Dilip (1984) "Optimal Incentive Schemes with Many Agents" *Review of Economic Studies.* July 1984. 51(3), 166: 433–46.

Mookherjee, Dilip & Ivan Png (unpub) "Optimal Auditing, Insurance, and Redistribution" UCLA AGSM Business Economics Working Paper No.86–5. June 1986. **178n**

Moore, John. See Matthews & Moore (1987). **223n**

Moreaux, Michel (1985) "Perfect Nash Equilibria in Finite Repeated Game and Uniqueness of Nash Equilibrium in the Constituent Game" *Economics Letters.* 1985. 17, 4: 317–20. **101n**

Morgenstern, Oskar. See von Neumann & Morgenstern (1944). **13, 17n, 53, 82n**

Moseidjord, Asbjorn. See Mead et al. (1984). **253**

Moskowitz, Milton, Michael Katz, & Robert Levering, eds. (1980) *Everybody's Business: An Almanac.* San Francisco: Harper and Row, 1980.

Mulherin, J. Harold (1986) "Complexity in Long-term Contracts: An Analysis of Natural Gas Contractual Provisions" *Journal of Law, Economics, and Organization.* Spring 1986. 2, 1: 105–18.

Murphy, Kevin J. (1986) "Incentives, Learning, and Compensation: A Theoretical and Empirical Investigation of Managerial Labor Contracts" *Rand Journal of Economics.* Spring 1986. 17, 1: 59–76. **154n**

Murphy, Kevin J. See Baker et al. (1988). **169**

Myerson, Roger (1978) "Refinements of the Nash Equilibrium Concept" *International Journal of Game Theory.* 1978. 7, 2: 73–80. **128n**

Myerson, Roger (1979) "Incentive Compatibility and the Bargaining Problem" *Econometrica.* January 1979. 47, 1: 61–73. **176n**

Myerson, Roger. See Holmstrom & Myerson (1983). **153, 157n**

Nalebuff, Barry. See Ghemawat & Nalebuff (1985). **81n**

Nalebuff, Barry & John Riley (1985) "Asymmetric Equilibria in the War of Attrition" *Journal of Theoretical Biology.* April 1985. 113, 3: 517–27. **81n**

Nalebuff, Barry & David Scharfstein (1987) "Testing in Models of Asymmetric Information" *Review of Economic Studies.* April 1987. 54(2), 178: 265–78. **199, 202n**

Nalebuff, Barry & Joseph Stiglitz (1983) "Prizes and Incentives: Towards a General Theory of Compensation and Competition" *Bell Journal of Economics.* Spring 1983. 14, 1: 21–43. **175, 177, 178n**

Nash, John (1950a) "The Bargaining Problem" *Econometrica.* January 1950. 18, 2: 155–62. **13, 229, 241**

Nash, John (1950b) "Equilibrium Points in n-Person Games" *Proceedings of the National Academy of Sciences, USA.* January 1950. 36, 1: 48–9. **13**

Nash, John (1951) "Non-Cooperative Games" *Annals of Mathematics.* September 1951. 54, 2: 286–95. **13, 124, 127**

Nelson, Philip (1974) "Advertising as Information" *Journal of Political Economy.* July/August 1974. 84, 4: 729–54. **221n**

Newbery, David. See Gilbert & Newbery (1982). **298, 306**

Nikaido, Hukukane & Kazuo Isoda (1955) "Note on Noncooperative Convex Games" *Pacific Journal of Mathematics.* 1955. 5, 2: 807–15. **125**

Novshek, William (1985) "On the Existence of Cournot Equilibrium" *Review of Economic Studies.* January 1985. 52(1), 168: 85–98. **280n**

Ohta, Hiroshi. See Greenhut & Ohta (1975). **282n**

Ordeshook, Peter (1986) *Game Theory and Political Theory: An Introduction.* Cambridge: Cambridge University Press, 1986. **5, 41n**

Owen, Guillermo (1982) *Game Theory,* second edition. New York: Academic Press, 1982. **5**

Pearce, David (1984) "Rationalizable Strategic Behavior and the Problem of Perfection" *Econometrica.* July 1984. 52, 4: 1029–50. **40n**

Peleg, Bezalel. See Bernheim et al. (1987). **101n**

Peltzman, Sam. See Jarrell & Peltzman (1985). **105n**

Perry, Motty (1986) "An Example of Price Formation in Bilateral Situations: A Bargaining Model with Incomplete Information" *Econometrica.* March 1986. 54, 2: 313–21. **242n, 243n**

Perry, Motty. See Grossman & Perry (1986). **128n**

Phlips, Louis (1983) *The Economics of Price Discrimination.* Cambridge: Cambridge University Press, 1983. **179n**

Png, Ivan (1983) "Strategic Behaviour in Suit, Settlement, and Trial" *Bell Journal of Economics.* Autumn 1983. 14, 2: 539–50. **65**

Png, Ivan. See Mookherjee & Png (unpub). **178n**

Png, Ivan & David Hirshleifer (1987) "Price Discrimination through Offers to Match Price" *Journal of Business.* July 1987. 60, 3: 365–83. **283n**

Polemarchakis, Heraklis. See Geanakoplos & Polemarchakis (1982). **66n**

Popkin, Samuel (1979) *The Rational Peasant: The Political Economy of Rural Society in Vietnam.* Berkeley: University of California Press, 1979. **152**

Porter, Robert (1983a) "Optimal Cartel Trigger Price Strategies" *Journal of Economic Theory.* April 1983. 29, 2: 313–38. **103n, 173**

Porter, Robert (1983b) "A Study of Cartel Stability: The Joint Executive Committee, 1880–1886" *Bell Journal of Economics.* Autumn 1983. 14, 2: 301–14. **103n**

Posner, Richard (1975) "The Social Costs of Monopoly and Regulation" *Journal of Political Economy.* August 1975. 83, 4: 807–27. **298**

Prescott, Edward. See Kydland & Prescott (1977). **100n**

Psacharopoulos, George. See Layard & Psacharopoulos (1974). **211, 221n**

Pyle, David. See Leland & Pyle (1977). **219, 220**

Radner, Roy (1980) "Collusive Behavior in Oligopolies with Long but Finite Lives" *Journal of Economic Theory.* April 1980. 22, 2: 136–56. **101n**

Radner, Roy (1985) "Repeated Principal-Agent Games with Discounting" *Econometrica.* September 1985. 53, 5: 1173–98. **170**

Radner, Roy. See Marschak & Radner (1972). **179n**

Raff, Daniel & Lawrence Summers (1987) "Did Henry Ford Pay Efficiency Wages?" *Journal of Labor Economics.* October 1987. 5, 4: 57–86. **196**

Raiffa, Howard. See Luce & Raiffa (1957). **5, 7, 40n, 101n, 103n, 242n**

Rapoport, Anatol (1960) *Fights, Games and Debates.* Ann Arbor: University of Michigan Press, 1960.

Rapoport, Anatol (1970) *N-Person Game Theory: Concepts and Applications.* Ann Arbor: University of Michigan Press, 1970. **5**

Rapoport, Anatol & Albert Chammah (1965) *Prisoner's Dilemma: A Study in Conflict and Cooperation.* Ann Arbor: University of Michigan Press, 1965. **120**

Rapoport, Anatol & Melvin Guyer (1966) "A Taxonomy of 2x2 Games" *General Systems.* 1966. 11: 203–14. **39n**

Rappoport, Peter. See Cooter & Rappoport (1984). **38n**

Rasmusen, Eric (1987) "Moral Hazard in Risk-Averse Teams" *Rand Journal of Economics.* Autumn 1987. 18, 3: 428–35. **173, 178n, 179n**

Rasmusen, Eric (1988a) "Entry for Buyout" *Journal of Industrial Economics.* March 1988. 36, 3: 281–300. **291, 306**

Rasmusen, Eric (1988b) "Mutual Banks and Stock Banks" *Journal of Law and Economics.* October 1988. 31, 2: 399–422. **154n**

Rasmusen, Eric (unpub) "A New Version of the Folk Theorem" UCLA AGSM Business Economics Working Paper No. 87-6. February 1987. **103n, 104n**

Rasmusen Eric. See Fernandez & Rasmusen (unpub), **194**; and D. Hirshleifer & Rasmusen (forth), **103n**

Rasmusen, Eric & Todd Zenger (unpub) "Firm Size and Agent Sorting" UCLA AGSM Business Economics Working Paper No. 87–13. August 1987. **179n**

Reinganum, Jennifer (1985) "Innovation and Industry Evolution" *Quarterly Journal of Economics.* February 1985. 100, 1: 81–99. **307n**

Reinganum, Jennifer (1988) "Plea Bargaining and Prosecutorial Discretion" *American Economic Review.* September 1988. 78, 4: 713–28. **221n**

Reinganum, Jennifer (forth) "The Timing of Innovation: Research, Development and Diffusion" In Schmalensee & Willig (forth). **295**

Reinganum, Jennifer & Nancy Stokey (1985) "Oligopoly Extraction of a Common Property Natural Resource: The Importance of the Period of Commitment in Dynamic Games" *International Economic Review.* February 1985. 26, 1: 161–74. **101n**

Rey, P. See Caillaud et al. (1988). **175, 176n, 177n**

Reynolds, Robert. See Salant et al. (1983). **280n**

Richerson, Peter. See Boyd & Richerson (1985). **129n**

Riley, John (1979a) "Evolutionary Equilibrium Strategies" *Journal of Theoretical Biology.* January 1979. 76, 2: 109–23. **129n**

Riley, John (1979b) "Informational Equilibrium" *Econometrica.* March 1979. 47, 2: 331–59. **194**

Riley, John (1980) "Strong Evolutionary Equilibrium and the War of Attrition" *Journal of Theoretical Biology.* February 1980. 82, 3: 383–400. **81n**

Riley, John. See Hirshleifer & Riley (1979); Hirshleifer and Riley (unpub), **5**; Maskin & Riley (1984), **257n**; Maskin & Riley (1985); and Nalebuff & Riley (1985), **81n**

Riordan, Michael. See Kihlstrom & Riordan (1984). **222n**

Roberts, John. See Kreps et al. (1982), **14, 118, 127**; Milgrom & Roberts (1982a, 1982b); and Milgrom & Roberts (1986), **222n**

Roberts, John & Hugo Sonnenschein (1976) "On the Existence of Cournot Equilibrium without Concave Profit Functions" *Journal of Economic Theory.* August 1976. 13, 1: 112–17. **280n**

Robinson, Marc (1985) "Collusion and the Choice of Auction" *Rand Journal of Economics.* Spring 1985. 16, 1: 141–5. **251, 256**

Rogerson, William (1982) "The Social Costs of Monopoly and Regulation: A Game-Theoretic Analysis" *Bell Journal of Economics.* Autumn 1982. 13, 2: 391–401. **298**

Roll, Richard. See Cornell & Roll (1981). **129n**

Rosen, J. (1965) "Existence and Uniqueness of Equilibrium Points for Concave n-Person Games" *Econometrica.* July 1965. 33, 3: 520–34. **129n**

Rosen, Sherwin (1986) "Prizes and Incentives in Elimination Tournaments" *American Economic Review.* September 1986. 76, 4: 701–15. **177n**

Rosen, Sherwin. See Lazear & Rosen (1981). **175, 177n**

Ross, Steven (1977) "The Determination of Financial Structure: The Incentive-Signalling Approach" *Bell Journal of Economics.* Spring 1977. 8, 1: 23–40. **221n**

Roth, Alvin (1984) "The Evolution of the Labor Market for Medical Interns and Residents: A Case Study in Game Theory" *Journal of Political Economy.* December 1984. 92, 6: 991–1016. **179n**

Roth, Alvin, ed. (1985) *Game Theoretic Models of Bargaining.* Cambridge: Cambridge University Press, 1985.

Rothkopf, Michael (1980) "TREES: A Decision-Maker's Lament" *Operations Research.* January/February 1980. 28, 1: 3.

Rothschild, Michael. See Diamond & Rothschild (1978). **5**

Rothschild, Michael & Joseph Stiglitz (1970) "Increasing Risk: I. A Definition" *Journal of Economic Theory.* June 1970. 2, 2: 225–43. **156n**

Rothschild, Michael & Joseph Stiglitz (1976) "Equilibrium in Competitive Insurance Markets: An Essay on the Economics of Imperfect Information" *Quarterly Journal of Economics.* November 1976. 90, 4: 629–49. **189, 199, 216**

Rubin, Paul (1978) "The Theory of the Firm and the Structure of the Franchise Contract" *Journal of Law and Economics.* April 1978. 21, 1: 223–33. **157n**

Rubinstein, Ariel (1982) "Perfect Equilibrium in a Bargaining Model" *Econometrica.* January 1982. 50, 1: 97–109. **234, 235, 241, 242n, 292**

Rubinstein, Ariel (1985a) "A Bargaining Model with Incomplete Information about Time Preferences" *Econometrica.* September 1985. 53, 5: 1151–72. **242n**

Rubinstein, Ariel (1985b) "Choice of Conjectures in a Bargaining Game with Incomplete Information" In Roth (1985). **128n, 242n**

Rubinstein, Ariel. See Binmore et al. (1986). **242n**

Rudin, Walter (1964) *Principles of Mathematical Analysis.* New York: McGraw-Hill, 1964. **309, 315**

Rumelt, Richard. See Lippman & Rumelt (1982).

Saft, Lester. See Klein & Saft (1985). **157n**

Salant, Stephen, Sheldon Switzer, & Robert Reynolds (1983) "Losses from Horizontal Merger: The Effects of an Exogenous Change in Industry Structure on Cournot-Nash Equilibrium" *Quarterly Journal of Economics.* May 1983. 98, 2: 185–99. **280n**

Saloner, Garth. See Farrell & Saloner (1985). **40n**

Salop, Steven & Joseph Stiglitz (1977) "Bargains and Ripoffs; A Model of Monopolistically Competitive Price Dispersion" *Review of Economic Studies.* October 1977. 44(3), 138: 493–510. **195**

Samuelson, Paul (1958) "An Exact Consumption-Loan Model of Interest with or without the Social Contrivance of Money" *Journal of Political Economy.* December 1958. 66, 6: 467–82. **274**

Samuelson, William (1984) "Bargaining under Asymmetric Information" *Econometrica.* July 1984. 52, 4: 995–1005. **242n**

Savage, Leonard (1954) *The Foundations of Statistics.* New York: Wiley, 1954. **66n**

Scarf, Herbert. See Debreu & Scarf (1963). **14**

Scharfstein, David. See Nalebuff & Scharfstein (1987). **199, 202n**

Scheffman, David. See Holt & Scheffman (1987). **283n**

Scheinkman, Jose. See Kreps & Scheinkman (1983). **281n**

Schelling, Thomas (1960) *The Strategy of Conflict.* Cambridge, Mass.: Harvard University Press, 1960. **14, 36, 37**

Schelling, Thomas (1966) *Arms and Influence.* New Haven: Yale University Press, 1966. **41n**

Schelling, Thomas (1978) *Micromotives and Macrobehavior*. New York: W. W. Norton, 1978. **41n**

Scherer, Frederick (1980) *Industrial Market Structure and Economic Performance*, second edition. Chicago: Rand McNally, 1980.

Schmalensee, Richard (1982) "Product Differentiation Advantages of Pioneering Brands" *American Economic Review*. June 1982. 72, 3: 349–65. **99**

Schmalensee, Richard & Robert Willig, eds. (forth) *The Handbook of Industrial Organization*. New York: North-Holland. **5, 176n, 306n**

Schmeidler, David. See Hurwicz et al. (1985). **176n**

Schmittberger, Rold. See Guth et al. (1982). **229**

Schwartz, Nancy. See Kamien & Schwartz (1981), **309**; and Kamien & Schwartz (1982), **295**

Schwarze, Bernd. See Guth et al. (1982). **229**

Selten, Reinhard (1965) "Spieltheoretische Behandlung eines Oligopolmodells mit Nachfragetragheit" *Zeitschrift für die gesamte Staatswissenschaft*. 1965. 121: 301–24, 667–89. **14**

Selten, Reinhard (1975) "Reexamination of the Perfectness Concept for Equilibrium Points in Extensive Games" *International Journal of Game Theory*. 1975. 4, 1: 25–55. **109, 127**

Selten, Reinhard (1978) "The Chain-Store Paradox" *Theory and Decision*. April 1978. 9, 2: 127–59. **88, 100, 101n**

Selten, Reinhard. See Harsanyi & Selten (1988). **38n**

Shaked, Avner (1982) "Existence and Computation of Mixed Strategy Nash Equilibrium for 3-Firms Location Problem" *Journal of Industrial Economics*. September/December 1982. 31, 1/2: 93–6. **273**

Shaked, Avner & John Sutton (1984) "Involuntary Unemployment as a Perfect Equilibrium in a Bargaining Model" *Econometrica*. November 1984. 52, 6: 1351–64. **242n**

Shapiro, Carl (1982) "Consumer Information, Product Quality and Seller Reputation" *Bell Journal of Economics*. Spring 1982. 13, 1: 20–35.

Shapiro, Carl (1983) "Premiums for High Quality Products as Returns to Reputation" *Quarterly Journal of Economics*. November 1983. 98, 4: 659–79. **102n**

Shapiro, Carl. See Farrell & Shapiro (1988), **214**; and Katz & Shapiro (1985), **40n**

Shapiro, Carl & Joseph Stiglitz (1984) "Equilibrium Unemployment as a Worker Discipline Device" *American Economic Review*. June 1984. 74, 3: 433–44. **105n, 166, 175**

Shapley, Lloyd (1953a) "Open Questions" In *Report of an Informal Conference on the Theory of n-Person Games*. Princeton Mathematics mimeo. 1953. 15. **13**

Shapley, Lloyd (1953b) "A Value for n-Person Games" In Kuhn & Tucker (1953), 307–17. **13**

Shavell. Steven (1979) "Risk Sharing and Incentives in the Principal and Agent Relationship" *Bell Journal of Economics*. Spring 1979. 10, 1: 55–73. **155n**

Shell, Karl. See Cass & Shell (1983). **75**

Shleifer, Andrei & Robert Vishny (1986) "Greenmail, White Knights, and Shareholders' Interest" *Rand Journal of Economics*. Autumn 1986. 17, 3: 293–309. **128n, 303**

Shubik, Martin (1971) "The Dollar Auction Game: A Paradox in Noncooperative

Behavior and Escalation" *Journal of Conflict Resolution.* March 1971. 15, 1: 109–11. **257n**

Shubik, Martin (1982) *Game Theory in the Social Sciences: Concepts and Solutions.* Cambridge, Mass.: MIT Press, 1982. **5, 65n, 242n**

Shubik, Martin. See Levitan & Shubik (1972). **266**

Simon, Leo (1987) "Games with Discontinuous Payoffs" *Review of Economic Studies.* October 1987. 54(4), 180: 569–98. **129n, 273**

Slade, Margaret (1987) "Interfirm Rivalry in a Repeated Game: An Empirical Test of Tacit Collusion" *Journal of Industrial Economics.* June 1987. 35, 4: 499–516. **103n**

Slatkin, Montgomery (1980) "Altruism in Theory" [review of Scott Boorman & Paul Levitt, *The Genetics of Altruism*] *Science.* 7 November 1980. 210: 633–4. **15**

Smith, Abby. See Antle & Smith (1986). **177n**

Smith, Adam (1776) *An Inquiry into the Nature and Causes of the Wealth of Nations.* Chicago: University of Chicago Press, 1976. **166**

Sobel, Joel. See Banks & Sobel (1987), **128n**; Border & Sobel (1987), **178n**; and Crawford & Sobel (1982), **76**

Sobel, Joel & Ichiro Takahashi (1983) "A Multi-Stage Model of Bargaining" *Review of Economic Studies.* July 1983. 50(3), 162: 411–26. **242n**

Sonnenschein, Hugo (1983) "Economics of Incentives: An Introductory Account" In *Technology, Organization, and Economic Structure: Essays in Honor of Prof. Isamu Yamada.* Ryuzo Sato & Martin Beckmann, eds. Berlin: Springer-Verlag, 1983. **154n**

Sonnenschein, Hugo. See Hurwicz et al. (1985), **176n**; and Roberts & Sonnenschein (1976), **280n**

Sorenson, Philip. See Mead et al. (1984). **253**

Sowden, Lanning. See Campbell & Sowden (1985). **38n**

Spence, A. Michael (1973) "Job Market Signalling" *Quarterly Journal of Economics.* August 1973. 87, 3: 355–74. **205, 221n**

Spence, A. Michael (1977) "Entry, Capacity, Investment, and Oligopolistic Pricing" *Bell Journal of Economics.* Autumn 1977. 8, 2: 534–44. **290**

Spence, A. Michael. See Kreps & Spence (1984). **15, 17n**

Spremann, Klaus. See Bamberg & Spremann (1987). **5**

Spulber, Daniel (1981) "Capacity, Output, and Sequential Entry" *American Economic Review.* June 1981. 71, 3: 503–14

Spulber, Daniel (forth) *Regulation and Markets.* Cambridge, Mass.: MIT Press. **306n**

Stackelberg, Heinrich von (1934) *Marktform und Gleichgewicht.* Berlin: J. Springer, 1934. Translated by Alan Peacock as *The Theory of the Market Economy.* London: William Hodge, 1952. **82n**

Staten, Michael & John Umbeck (1982) "Information Costs and Incentives to Shirk: Disability Compensation of Air Traffic Controllers" *American Economic Review.* December 1982. 72, 5: 1023–37. **154n**

Staten, Michael & John Umbeck (1986) "A Study of Signaling Behavior in Occupational Disease Claims" *Journal of Law and Economics.* October 1986. 29, 2: 263–86. **221n**

Stigler, George (1964) "A Theory of Oligopoly" *Journal of Political Economy.* February 1964. 72, 1: 44–61. **103n**

Stigler, George. See Becker & Stigler (1974). **166**

Stiglitz, Joseph (1977) "Monopoly, Non-linear Pricing and Imperfect Information: The Insurance Market" *Review of Economic Studies.* October 1977. 44, 138: 407–30. **179n**

Stiglitz, Joseph (1987) "The Causes and Consequences of the Dependence of Quality on Price" *Journal of Economic Literature.* March 1987. 25, 1: 1–48. **99, 105n, 201n, 221n**

Stiglitz, Joseph. See Dasgupta & Stiglitz (1980), **307**; Nalebuff & Stiglitz (1983), **175, 177, 178n**; Rothschild & Stiglitz (1970), **156n**; Rothschild & Stiglitz (1976), **189, 199, 216**; Salop & Stiglitz (1977), **195**; and Shapiro & Stiglitz (1984), **105n, 166, 175**

Stiglitz, Joseph & G. Frank Mathewson, eds. (1986) *New Developments in the Analysis of Market Structure.* Cambridge, Mass.: MIT Press, 1986. **306n**

Stiglitz, Joseph & Andrew Weiss (1981) "Credit Rationing in Markets with Imperfect Information" *American Economic Review.* June 1981. 71, 3: 393–410. **197, 199**

Stiglitz, Joseph & Andrew Weiss (unpub) "Sorting Out the Differences between Screening and Signaling Models" Princeton University mimeo. August 1983. **221n**

Stokey, Nancy. See Reinganum & Stokey (1985). **101n**

Straffin, Philip (1980) "The Prisoner's Dilemma" *UMAP Journal.* 1: 101–3. **38n**

Strunk, William & E. B. White (1959) *The Elements of Style.* New York: Macmillan, 1959. **18n**

Sugden, Robert (1986) *The Economics of Rights, Co-operation and Welfare.* Oxford: Basil Blackwell, 1986. **88, 129n**

Sultan, Ralph (1974) *Pricing in the Electrical Oligopoly*, Vol. I: *Competition or Collusion.* Cambridge, Mass.: Harvard University Press, 1974.

Summers, Lawrence. See Raff & Summers (1987). **196**

Sutton, John (1986) "Non-Cooperative Bargaining Theory: An Introduction" *Review of Economic Studies.* October 1986. 53(5), 176: 709–24. **241**

Sutton, John. See Shaked & Sutton (1984). **242n**

Switzer, Sheldon. See Salant et al. (1983). **280n**

Szep, J. & F. Forgo (1985) *Introduction to the Theory of Games.* Dordrecht: D. Reidel, 1985. **5, 129n**

Takahashi, Ichiro. See Sobel & Takahashi (1983). **242n**

Takayama, Akira (1985) *Mathematical Economics*, second edition. Cambridge: Cambridge University Press, 1985. **309**

Tan, Tommy & Sergio Werlang (unpub) "On Aumann's Notion of Common Knowledge: An Alternative Approach" Chicago Graduate School of Business Working Paper No. 85–26. January 1986. **66n**

Tarski, Alfred (1955) "A Lattice-Theoretical Fixpoint Theorem and its Applications" *Pacific Journal of Mathematics.* June 1955. 5, 2: 285–309. **315**

Teece, David. See Monteverde & Teece (1982). **154n**

Telser, Lester (1966) "Cutthroat Competition and the Long Purse" *Journal of Law and Economics.* October 1966. 9: 259–77. **288**

Thisse, Jacques. See d'Aspremont et al. (1979). **272, 280**

Tirole, Jean (1986) "Hierarchies and Bureaucracies: On the Role of Collusion in Organizations" *Journal of Law, Economics, and Organization.* Fall 1986. 2, 2: 181–214. **153**

Tirole, Jean (1988) *The Theory of Industrial Organization.* Cambridge, Mass: MIT Press, 1988. **5, 129n, 306n**

Tirole, Jean. See Caillaud et al. (1988), **175, 176n, 177n**; Freixas et al. (1985), **177n**; Fudenberg & Tirole (1983), **237, 242n, 243n**; Fudenberg & Tirole (1986a), **5, 281, 306n**; Fudenberg & Tirole (1986b), **80, 81n**; Fudenberg & Tirole (1988), **141, 157**; Laffont & Tirole (1986), **155n, 176n**; and Maskin & Tirole (1987), **75**

Titman, Sheridan. See Hirshleifer & Titman (unpub). **128n**

Tucker, Albert (unpub) "A Two-Person Dilemma" Stanford University mimeo. May 1950. Reprinted in Straffin (1980). **13, 38n**

Tucker, Albert. See Dresher et al. (1957), Kuhn & Tucker (1950, 1953), and Luce & Tucker (1959).

Tukey, J. (1949) "A Problem in Strategy" *Econometrica.* July 1949 (supplement). 17: 73 (abstract). **81n**

Tullock, Gordon (1967) "The Welfare Costs of Tariffs, Monopolies, and Theft" *Western Economic Journal.* June 1967. 5, 3: 224–32. **298**

Umbeck, John. See Staten & Umbeck (1982, 1986). **154n, 221n**

Van Damme, Eric (1983) *Refinements of the Nash Equilibrium Concept.* Berlin: Springer-Verlag, 1983. **5**

Van Damme, Eric (1987) *Stability and Perfection of Nash Equilibrium.* Berlin: Springer-Verlag, 1987. **128n**

Varian, Hal (1984) *Microeconomic Analysis,* second edition. New York: W. W. Norton, 1984. **7, 66n, 174, 281n, 309**

Vickrey, William (1961) "Counterspeculation, Auctions, and Competitive Sealed Tenders" *Journal of Finance.* March 1961. 16, 1: 8–37. **173, 179n, 250, 256, 257n**

Vishny, Robert. See Shleifer & Vishny (1986). **128n, 303**

von Neumann, John (1928) "Zur Theorie der Gesellschaftsspiele" *Mathematische Annalen.* 1928. 100: 295–320. Translated by Sonya Bargmann as "On the Theory of Games of Strategy" in Luce & Tucker (1959), 13–42. **104n**

von Neumann, John & Oskar Morgenstern (1944) *The Theory of Games and Economic Behavior.* New York: Wiley, 1944. **13, 17n, 53, 82n**

Waldman, Michael (1987) "Noncooperative Entry Deterrence, Uncertainty, and the Free Rider Problem" *Review of Economic Studies.* April 1987. 54(2), 178: 301–10. **173**

Waldman, Michael. See Haltiwanger & Waldman (unpub). **281n**

Weber, Robert. See Milgrom & Weber (1982). **250, 255, 256n, 257n, 258n**

Weiner, E. (1984) *The Oxford Guide to the English Language.* Oxford: Oxford University Press, 1984. **18**

Weiss, Andrew. See Guasch & Weiss (1980), **202n**; Stiglitz & Weiss (1981), **197, 199**; and Stiglitz & Weiss (unpub), **221n**

Weitzman, Martin (1974) "Prices vs. Quantities" *Review of Economic Studies.* 1974. 41, 4: 477–91. **281n**

Werlang, Sergio. See Tan & Werlang (unpub). **66n**

Weston, J. Fred. See Copeland & Weston (1983). **90, 156n, 218**

Whinston, Michael. See Bernheim et al. (1987), **101n**; Bernheim & Whinston (1986), **157n**; and Bernheim & Whinston (1987), **101n**

White, E.B. See Strunk & White (1959). **18n**

Wicksteed, Philip (1885) *The Common Sense of Political Economy.* New York: Kelley, 1950. **16**

Williams, J. (1966) *The Compleat Strategyst: Being a Primer on the Theory of Games of Strategy.* New York: McGraw-Hill, 1966. **5**

Williamson, Oliver (1975) *Markets and Hierarchies: Analysis and Antitrust Implications: A Study in the Economics of Internal Organization.* New York: Free Press, 1975. **153**

Willig, Robert. See Schmalensee & Willig (forth). **5, 176n, 306n**

Wilson, Charles (1980) "The Nature of Equilibrium in Markets with Adverse Selection" *Bell Journal of Economics.* Spring 1980. 11, 1: 108–30. **188, 193, 221n**

Wilson, Robert (1971) "Computing Equilibria of n-Person Games" *SIAM Journal of Applied Mathematics.* July 1971. 21, 1: 80–87. **124**

Wilson, Robert (1979) "Auctions of Shares" *Quarterly Journal of Economics.* November 1979. 93: 675–89. **257n**

Wilson, Robert (unpub) Stanford University 311b course notes. **119**

Wilson, Robert. See Kreps & Wilson (1982a), **285, 286, 306, 306n**; Kreps & Wilson (1982b), **14, 109, 110, 127**; and Kreps et al. (1982), **14, 118, 127**

Winter, Ralph. See Mathewson & Winter (1985). **157n**

Wolfe, Philip. See Dresher et al. (1957).

Wolfson, M. (1985) "Empirical Evidence of Incentive Problems and their Mitigation in Oil and Tax Shelter Programs" In *Principals and Agents: The Structure of Business*, John Pratt & Richard Zeckhauser, eds. Boston: Harvard Business School Press, 1985. 101–25. **154n**

Wolinsky, Asher. See Binmore et al. (1986). **242n**

Wydick, Richard (1978) "Plain English for Lawyers" *California Law Review.* 1978. 66: 727–64. **18n**

Yellen Janet. See Akerlof & Yellen (1986). **177n**

Zahavi, Amotz (1975) "Mate Selection: A Selection for a Handicap" *Journal of Theoretical Biology.* September 1975. 53, 1: 205–14. **221n**

Zamir, S. See Mertens & Zamir (1985). **66n**

Zenger, Todd. See Rasmusen & Zenger (unpub). **179n**

Zermelo, E. von (1913) "Über eine Anwendung der Mengenlehre auf die Theorie des Schachspiels" *Proceedings, Fifth International Congress of Mathematicians.* 1913. 2: 501–4.

Zimmerman, Jerold. See Gaver & Zimmerman (1977). **178n**

Subject Index

This index includes subject entries only. For names of authors see the References and Citation Index. For mathematical terms see the Mathematical Appendix.

The first page number (or series of page numbers) listed after an entry shows the most important reference. The letters "n" and "d" stand for "notes" and "definitions/explanations." Many terms are listed only under the headings "Games" and "Equilibrium concepts."

Action, 22d
Action combination, 23d
Action set, 22d
Adverse selection, 181–203, 133–6d, 222n
  screening, 215–16
Advertising
  fat-cat, 282n
  signalling, 221–2n
Affiliated variable, 258nd
Agent, 134d
  gender, 154n
Anonymity, 231d
Assessment, 109d
Asymmetric information, 53d
  existence, 127
  subcategories, 133–6
Asymmetric information existence theorem,
    127d
Auctions, 245–58
  principal–agent, 157n
Auditing, 168d
  rewarding truth, 178n
Axelrod tournament, 119–20d, 128n

Backward induction, 88d
Bad news, 156nd
  auctions, 252
Bargaining, 227–43
  legal signalling, 221n
Battle of the Sexes, 34–5d, 39n
  mixed strategies, 80–1n

Bayesian games, 59d
Bayes's Rule, 56–8d
Behavior strategy, 80nd
Beliefs, 57, 207, 208–10, 238
  Predatory Pricing, 307–8n
  prior, 57d, 66n
Bertrand, 263–8d, 40n, 125, 126, 281n
Best reply, 28d
Best response, 28d
  Cournot, 77
  reaction function, 78
Bid–ask spread, 197–8, 202n
Bimatrix game, 39nd
Biology
  Boxed Pigs, 32–4, 40n
  evolutionary equilibrium, 121–3
  peacocks, 221
  perfectness, 100–1n
  references, 128n
Boiling in Oil, 148–50d, 169
Boundary, 36d, 41
Bourgeois strategy, 123d
Branch, 45d
Brouwer fixed point theorem, 315d
Budget balancing, 172d, 175
  risk aversion, 179n

Capacity
  Bertrand, 264–7
  excess, 177n, 290–4

Capacity (*cont.*)
  two-stage game, 281n
Cardinal utility, 38n
  mixed strategies, 81n
Certainty, 51–2d
Chainstore paradox, 88d, 94, 101n
Cheap talk, 76d
Chicken, 73–4d, 81n
  innovation, 295
Closed loop, 100nd
Coalition, 242nd
Coarse, 50d
  auctions, 255
Coase conjecture, 283nd
Coinsurance, 146d
Collusion
  auctions, 251, 257n
  price vs. quantity, 281n
  trigger strategies, 103n
Colonel Blotto game, 81nd
Common knowledge, 50–1d, 66d, 116–18
Common-value auction, 251–6, 246d
Communication, 37, 75–6
Comparative statics, 124
Competition constraint, 141–2, 155n, 161
  bargaining, 227
  Product Quality, 99
Complete information, 55d, 51
Complete robustness, 115d
Completely mixed strategy, 70d
Conditional likelihood, 66nd
Congestion, 281nd
Conjectural variation, 262–3d, 281n
Continuous time, 81n
Continuum of players, 81n
  bargaining, 237
  equilibrium concepts, 109–10
  mixed strategies, 72
Contraction mapping theorem, 315d
Cooperative game theory, 29d, 229, 242n
Coordination, 35–7, 38–41n
  teams, 179n
Correlated strategies, 75d
  Bourgeois strategy, 123
  Prisoner's Dilemma, 39n
Correlated-value auction, 246d
Cournot, 76–9, 126, 259–62
  differentiated products, 268–9
  Entry for Buyout, 292
  merger, 280n
  two-stage capacity game, 281n
Cream skimming, 192d
Credit markets

credit rationing, 196
Czarist debt, 94
Diamond model, 288–90

Decision time, 47d
Deductible, 146d
Depreciation, 200n
Differential games, 81nd
Dimensionality condition, 92–3d
  alternative condition, 104n
Discontinuities, 172–3
  takeovers, 302
Discontinuity existence theorem, 126d, 129n
  Edgeworth paradox, 266
  Hotelling Location Game, 273
Discoordination game, 37, 40n
Discount factor, 89d
Discount rate, 89d
Discounting, 89–91d
  bargaining, 232–4
  Folk Theorem, 92
  infinite payoffs, 102n
Dominant strategy, 28d, 38n
  repeated Prisoner's Dilemma, 89
Dominant strategy mechanism, 174d
Dominated strategy, 28d
Duels, 81nd
Duopoly, 263–72, 76–80, 259–61
Durable monopoly, 276–80d, 283n
Dutch auction, 249–50d
  second-price, 257n
Dynamic consistency, 100nd

Edgeworth paradox, 266d
Efficiency wages, 166–7d, 169, 177n
  Ford, 196
Efficient rationing, 281nd
End node, 45d
End point, 45d
English auction, 247d, 250
  reserve price, 257n
Epsilon truthfulness, 161d
Epsilon-equilibrium, 101nd
Equilibrium, 26d
Equilibrium concepts, 27d
  Bayesian, 59d
  biology, 121–3
  coalition-proof Nash, 101nd
  conjectural variation, 262–3d, 281n
  correlated strategy, 75d
  Cournot–Nash, 78d
  credible neologisms, 128n
  divine, 128n

dominant strategy, 28d, 39n
  Prisoner's Dilemma, 89, 95
  repeated Prisoner's Dilemma, 39
epsilon-equilibrium, 101nd
evolutionarily stable strategy, 121d
evolutionary equilibrium, 121–3d
intuitive, 128n
  Education I, 208
  PhD Admissions, 115d
iterated dominant strategy, 31d, 39n
  Prisoner's Dilemma, 95
list, 38n
maximin, 103nd
minimax, 103nd
Nash, 33d, 40n
  continua, 78n, 179n, 228, 240
  ESS, 121–2
  repeated Prisoner's Dilemma, 89, 104n
  weak, 33–4d, 177n
perfect Bayesian, 110d
  examples, 206–7, 209, 238–9
  existence, 127
perfect sequential, 128n
proper, 128n
rationalizable, 40nd
reactive, 194d, 202n, 217
redefining game tree, 87, 193, 208–9
refinements, 128n
sequential, 110d
  existence, 127
stability, 128nd
Stackelberg, 79d, 82
  conjectural variation, 263
  perfectness, 101n
subgame perfect, 85–8d
  Stackelberg, 101n
trembling hand perfect, 109d
  existence, 127
  subgame perfectness, 85
Wilson, 194d, 216–17
Equilibrium outcome, 27d
Equilibrium path, 83d
  bargaining, 240–1
  Predatory Pricing, 287
Equilibrium point, 40nd
Equilibrium strategy, 26d, 33
Evolutionarily stable strategy, 121d
Evolutionary equilibrium, 121–3d
Ex ante efficiency, 157nd
Ex post efficiency, 157nd
Executive compensation, 167–9, 177–9n
Existence, 123–7, 27, 129n
  adverse selection, 193

Cournot, 259, 280n
Edgeworth paradox, 266
Hotelling Location Game, 272, 282n
Hotelling Pricing Game, 269
screening, 215
Extensive form, 46d
Extrinsic uncertainty, 75d

Fan's Lemma, 315d
Fat-cat effect, 282nd
Feedback, 100nd
Fine partition, 50d
  auctions, 255
First-best, 147d
First-mover advantage, 35d
  bargaining, 228
  Entry for Buyout, 307n
First order condition approach, 141d, 155n
First-price open-cry auction, 247d
First-price sealed-bid auction, 247–50d
Fixed point theorems, 315–16
Focal point, 36–7
  bargaining, 228, 240
  class ranking, 176n
Folk Theorem, 92–4d, 103–4n
  incomplete information, 119–20
Forcing contract, 138d, 164
Ford, 196
Forgiving, 120d
Franchises, 157n
Free rider
  examples, 172–3
  takeovers, 301–2
Fringe, 82nd
Full insurance, 143d
Fully revealing, 160d

Game tree, 46d
Games
  Alternating Offers, 232–6d, 242n
  Bargaining with Incomplete Information,
    237–41d
  Battle of the Bismarck Sea, 30–2d, 39n
  Battle of the Sexes, 34–5d, 40n
    mixed strategies, 80–1n
  Boxed Pigs, 32–3d, 40n
  Bridge, 80n
  Broadway Game, 148–51d
    MLRP, 155–6n
  Chicken, 73–4d, 81n
    innovation, 295
  Colonel Blotto, 81nd
  Cournot Game, 76–9d, 259–62

Games (*cont.*)
  Customer Switching Costs, 274–5d
  Durable Monopoly, 276–80d
  Education
    I, 206–8d
    II, 208–9d
    III, 209–10d
    IV, 212d
    V, 213–15d
    VI, 215–16d
  Entry Deterrence
    I, 86–7d
    II, 107–12d
    III, 112–13d, 108
    IV, 116d, 108
    V, 116–18d
  Entry for Buyout, 290–4d
  Follow the Leader
    I, 44–6d, 83–5
    II, 52–3d
    III, 54–60
  Free Rider Problem in Takeovers, 301–2d
  Greenmail to Attract White Knights, 303–6d
  Hawk–Dove Game, 122–3d
  Highway Game, 40nd
  Hotelling Location Game, 272–3d
    reaction function, 125
  Hotelling Pricing Game, 269–72d
  Imitation with Bertrand Pricing, 294d
  Imitation with Profits in the Product Market,
    295d
  Insurance Game
    I, 143–4d
    II, 145–7d
    III, 189–95d, 215–16
  Lemons
    I, 183–4d
    II, 184–6d
    III, 186–7d
    III', 201nd
    IV, 187–8d
  Matching Pennies, 40nd
  Nash Puzzle, 33–4d
  one-sided Prisoner's Dilemma, 94–6d, 39n,
    104n
  Opec Model
    I, 22–5d
    II, 25–6d
  Patent Race for a New Market, 296–8d
  Patent Race for an Old Market, 298–300d
  PhD Admissions, 113–16d
  Png Settlement Game, 60–5d
  Predatory Pricing, 286–8d, 306n

  Prisoner's Dilemma, 28–9d, 35, 38–9n, 94–6
    minimax strategy, 104n
    tournament, 168
    two-tier tender offer, 302
  Product Quality, 96–9d, 105n
  Production Game
    I, 137d
    II, 137–8d
    III, 138–42d, 154n
    IV, 159–60d
    V, 181–2d
  Pure Coordination, 35d, 47
  repeated Prisoner's Dilemma, 88–92d
    ESS, 123
  Salesman Game, 163–6d
  Splitting a Pie, 227–9d
  two-sided Prisoner's Dilemma, 94–6d, 104n
  Underpricing, 218–20d
  War of Attrition, 74–5d, 81n
    Lemons I, 184
  Welfare Game, 69–72d, 124
    minimax strategy, 104n
Games of timing, 81nd
Gang of Four, 118–19n
  Predatory Pricing, 285–8
Glicksberg Theorem, 125d
Good news, 156nd
  auctions, 252
Gratitude, 104nd
Greenmail, 302–6
Gresham's Law, 199nd
Grim strategy, 91d, 93
Groves mechanism, 173–5d, 179n
  auctions, 249

Harsanyi transformation, 54–7d, 66n
Hawk–Dove Game, 122–3d
Hidden actions, 133–58d
Hidden information, 159–79, 133–6d, 154n
Hidden knowledge, 154nd
Hold-up potential, 153d
Hotelling models, 269–73

Imperfect information, 51d
Implement, 173d
Incentive compatibility, 140–1d, 161
  Product Quality, 98
  signalling, 207, 214
Incomplete information, 53–4d
Incomplete information folk theorem, 119d
Independence of irrelevant alternatives, 230–1d
Individual rationality, 140d, 155n
Infinite horizon, 96d, 102n

Inflation, 100n
Information partition, 50d, 117
   Broadway Game, 151
   Greenmail to Attract White Knights, 303
Information set, 48d, 23
Informed player, 134d
Innovation, 294–300, 307n
   hold-up potential, 153
Insurance, 142–7, 156–7n, 189–96, 202n
   law, 152
Intensity rationing, 265d
   efficiency 281n
Interim efficiency, 157n
Intuitive criterion, 128n
   Education I, 208
   PhD Admissions, 115d
Invariance, 229d
Invasion, 121d
Inverse intensity rationing, 265d
Invertibility, 158nd
Iterated dominant strategy, 31d, 39n

Kakutani fixed point theorem, 315d
   Glicksberg Theorem, 125

Law
   accidents, 152, 176n
   boundaries, 36
   contractibility, 140, 145
   durable monopoly, 283n
   insurance litigation, 157n
   plea bargaining, 221n
   settlement, 60–5
   testing, 178n
Lexicographic preferences, 161d
Likelihood, 58d, 66n
Linear contract, 139d, 155n
Linear distribution function condition, 155–6n
Location models, 269–73, 282n

Madman theory, 101nd
Marginal likelihood, 58d, 66n
Markov strategy, 102nd, 80n, 275
Matching Pennies, 40nd
Matrix game, 39nd
Maximin strategy, 103nd
Mean preserving spread, 156n
Mechanism, 173–5d, 176n
Mediation, 37
Mergers, 280n
Minimax criterion, 103nd
Minimax payoff, 92–3, 119
Minimax strategy, 104nd

Minimax theorem, 104nd
Mixed extension, 70d
Mixed strategy, 69–76d, 80–1n
   acceptance assumption, 228, 233, 238
   difference from random action, 157n
   Hawk–Dove, 122
   perfect information, 66n
   solving for distribution, 296
Monitoring, 168–9d, 178n
Monopoly slack, 167–8
Monotone Likelihood Ratio Property, 155nd
Moral hazard, hidden actions, 133–58d
Moral hazard, hidden information, 133–6d,
   159–79
Most-favored-nation contract, 283nd
Move, 22d

Nash bargaining solution, 229–31d, 242n
   Entry for Buyout, 292
Nash equilibria, 33d, 40n
   continua, 179n, 228, 240
   ESS, 121–2
   repeated Prisoner's Dilemma, 89, 104n
   weak, 33–4d, 177n
Nash existence theorem, 124d, 129n
Nature, 22d, 66n
Niceness, 120d
No-fat modelling, 14–16d
   dangers, 261
Node, 45d
Noncontractibility, 140d, 145, 198–9
Noncooperative game theory, 29d
Non-player, 22d
Nonpooling constraint, 161d
Nonresponders, 281nd
Nonseparating constraint, 161d
Normal form, 43–4d, 65n

Odd number theorem, 124d
Oligopoly, 261–2, 37, 82n, 281n
   trigger strategies, 103n, 173
One-shot game, 88d
One-sided asymmetry, 242nd
Open-cry auction, 247d
Open exit auction, 247d
Open loop, 100nd
Opportunism, 153d
Order of play, 23d
Ordinal utility, 38–9n, 81n
Ostracism, 198, 102–3n
Out-of-equilibrium behavior, 63
   bargaining, 238–40
   PBE, 110

Out-of-equilibrium behavior (*cont.*)
  signalling, 206–7, 210
  subgame perfectness, 85
Outcome, 25d
Outcome matrix, 43–4d, 65n
Overlapping generations model, 274
Overtaking criterion, 102nd

Pareto dominance, 40nd, 34–5, 77
  bargaining, 230
Participation constraint, 140d, 161
  boiling in oil, 148–9
  individual rationality, 155n
  signalling, 206–7, 213
  tournament, 168
Partition, information, 50d, 117
  auctions, 255
  Broadway Game, 151
Passive conjectures, 112d
  bargaining, 238
  Education I, 207
  PhD Admissions, 114
Patent race, 295–300d, 307n
  auctions, 257n
  discontinuity, 173
Path, 46d
Payoff, 24d
  existence, 124
Perfect Bayesian equilibrium, 110d
  examples, 207, 209, 238–9
  existence, 127
Perfect information, 51d, 66n
Perfect information existence theorem, 126d
Perfectness, 85d, 100n, 109–10, 210
  Stackelberg, 101n
Player, 22d
Poker, 54, 66n
Pooling, 160d
  almost pooling, 209
  existence, 191
  PhD Admissions, 113
Posterior belief, 58d, 66n
Predatory Pricing, 286–8
  Chainstore Paradox, 88–9
  Entry Deterrence I, 85–8
Predecessor node, 45d
Price discrimination, 179n
  most-favored-nation, 283n
Price dispersion, 195
Principal, 134d, 154n
Principal–agent model, 134–6d
  empirical work, 154n
Prior belief, 57d, 66n

Prisoner's Dilemma, 28–9d, 35, 38–9n, 94–6
  Axelrod tournament, 119–20
  minimax strategy, 104n
  one-sided, 96
  repeated, 88–92d, 123
  tournament, 168
  two-sided, 94–6d, 104n
  two-tier tender offer, 302
Private information, 53d
  auctions, 255
Private-value auctions, 246d
Product differentiation, 267–73
Proper subgame, 85d
Proportional rationing, 265d
Provokability, 120d
Pure strategy, 69d

Quality, 96–9
  advertising, 221n
  durability, 279
  efficiency wages, 105n
  one-sided Prisoner's Dilemma, 95
  price as signal, 188–9
Quasi-concavity existence theorem, 125d
QWERTY, 40n

Rao–Blackwell Theorem, 155nd
Rat race, 199n
Ratchet effect, 176–7nd
Rational expectations, 40n
Rationing, 264–6, 281n
Reaction function, 78d
Reaction function existence theorem, 125d, 129n
Reactive equilibrium, 194–5, 202n, 217
Real time, 47d
Realization, 22d
  auctions, 250
  Predatory Pricing, 288
Refinement of equilibrium, *see* Equilibrium
      concepts
Refinement of information, 50d
  Broadway Game, 151
  monitoring, 168
Rentseeking, 298
Reputation, 94–6, 169
  auctions, 254
  empirical work, 104–5n
  signalling, 210
Reservation utility, 137d, 140, 155n
  efficiency wages, 167–8
Reserve price, 257nd
Residual claimant, 150
  teams, 172
Responders, 281nd

Revelation principle, 161–2d, 164–5, 176n
Revenue equivalence theorem, 250d
Risk aversion, 53, 151, 157n, 201n, 257n
   adverse selection, 189
   budget balancing, 179n
   carrot and stick, 178n
   common-value auctions, 254
   indifference curves, 144
   maximin strategies, 104n
   private-value auctions, 251
   signalling, 217
Rules of the game, 22d

Screening, 212–17, 133–6d
   bargaining, 238
Second-best, 147d
   Salesman Game, 165
Second-price sealed-bid auction, 249d, 250, 254
Self-selection constraint, 161d, 167
Selling the store, 150d, 157n, 169
   teams, 171
Separable utility
   across time, 102n, 276
   effort and money, 155n, 156n, 157n
Separating, 160
   bargaining, 239–40
   PhD Admissions, 113
   signalling, 208
Sequential equilibrium, 110d
   existence, 127
Shapley value, 242nd
Share auction, 257nd
Shifting support scheme, 149–51d, 169
Side-payments, 29d
Signalling, 218–20
Singleton, 50d
Sleeping patent, 298–300
Solution concept, 27d
Spanning condition, 156n
Stability, 82, 128n
   Cournot, 78
   evolutionary equilibrium, 121, 123, 128n
Stackelberg, 79d, 82
   conjectural variation, 263
   perfectness, 101n
Standards, 35, 40n, 178n
Starting node, 45d
State of the world, 55d, 176n
State-space diagram, 142–7d, 191–3
Stochastic dominance, 156nd
Strategic complements, 281nd
Strategic substitutes, 281nd
Strategically equivalent, 249–50d

Strategy, 24d
Strategy combination, 24d
Strategy set, 24d, 102n, 124
Strategy space, 24d, 102n, 124
Subgame, 85d
   infinite, 100n
   supergame, 104n
Subgame perfectness, 85–8d
   Stackelberg, 101n
Successor node, 45d
Sufficient statistic condition, 155nd, 149
Sunspot models, 75d
Supergame, 104nd
Switching costs, 274–5
Symmetric equilibrium theorem, 127d
Symmetric information, 53d, 51
Synergism, 281nd

Tarski fixed point theorem, 315d
Teams, 170–2d, 179n
Testing, 168d, 178n, 198, 202n
   signalling, 211
Three-step procedure, 141d, 149
Threshold contract, 139d
   teams, 171, 179n
Time consistency, 100nd
Time line, 47d
   Predatory Pricing, 287
Timing, games of, 81nd
Tit-for-tat, 91–2d, 120
   Eye-for-an-Eye, 26
   Markov strategy, 102n
Tournament, 167–8d, 177–9n
   auctions, 256
   patent race, 295
Tremble, 85d, 209
Trembling hand perfectness, 109d
   existence, 127
   subgame perfectness, 85
Trigger strategies, 103nd
Truthtelling constraint, 162d
Two-by-two game, 28d, 39n
Two-sided asymmetry, 242nd
   Gang of Four, 119
Type, 55d, 200n, 202n, 306n
   changing, 170

Uncertainty, 51–2d
Uninformed player, 134d
Uniqueness, 27, 129n, 177n
Unravelling, 162d, 176n
   adverse selection, 185
   auctions, 255

Utility function, 155–6n, 158n
  expected, 52–3, 66n
  location model, 282n
  ordinal vs. cardinal, 38

Von Neumann–Morgenstern utility functions,
  52–3d, 66n

War of Attrition, 74–5d, 81n
  Lemons I, 184

Weakly dominant strategy, 31d
Wilson equilibrium, 194d, 216–17
Winner's Curse, 252–5d

Yardstick competition, 167d

Zero-profit constraint, 142d
Zero-sum game, 32d, 39–40n